DEFECT-ORIENTED TESTING FOR NANO-METRIC CMOS VLSI CIRCUITS

2nd Edition

FRONTIERS IN ELECTRONIC TESTING

Consulting Editor
Vishwani D. Agrawal

Books in the series:

Digital Timing Measurements – From Scopes and Probes to Timing and Jitter
Maichen, W., Vol. 33
ISBN 0-387-32418-0
Fault-Tolerance Techniques for SRAM-based FPGAs
Kastensmidt, F.L., Carro, L. (et al.), Vol. 32
ISBN 0-387-31068-1
Data Mining and Diagnosing IC Fails
Huisman, L.M., Vol. 31
ISBN 0-387-24993-1
Fault Diagnosis of Analog Integrated Circuits
Kabisatpathy, P., Barua, A. (et al.), Vol. 30
ISBN 0-387-25742-X
Introduction to Advanced System-on-Chip Test Design and Optimi...
Larsson, E., Vol. 29
ISBN: 1-4020-3207-2
Embedded Processor-Based Self-Test
Gizopoulos, D. (et al.), Vol. 28
ISBN: 1-4020-2785-0
Advances in Electronic Testing
Gizopoulos, D. (et al.), Vol. 27
ISBN: 0-387-29408-2
Testing Static Random Access Memories
Hamdioui, S., Vol. 26
ISBN: 1-4020-7752-1
Verification by Error Modeling
Redecka, K. and Zilic, Vol. 25
ISBN: 1-4020-7652-5
Elements of STIL: Principles and Applications of IEEE Std. 1450
Maston, G., Taylor, T. (et al.), Vol. 24
ISBN: 1-4020-7637-1
Fault injection Techniques and Tools for Embedded systems Reliability…
Benso, A., Prinetto, P. (Eds.), Vol. 23
ISBN: 1-4020-7589-8
Power-Constrained Testing of VLSI Circuits
Nicolici, N., Al-Hashimi, B.M., Vol. 22B
ISBN: 1-4020-7235-X
High Performance Memory Memory Testing
Adams, R. Dean, Vol. 22A
ISBN: 1-4020-7255-4
SOC (System-on-a-Chip) Testing for Plug and Play Test Automation
Chakrabarty, K. (Ed.), Vol. 21
ISBN: 1-4020-7205-8
Test Resource Partitioning for System-on-a-Chip
Chakrabarty, K., Iyengar & Chandra (et al.), Vol. 20
ISBN: 1-4020-7119-1
A Designers' Guide to Built-in Self-Test
Stroud, C., Vol. 19
ISBN: 1-4020-7050-0
Boundary-Scan Interconnect Diagnosis
de Sousa, J., Cheung, P.Y.K., Vol. 18
ISBN: 0-7923-7314-6

DEFECT-ORIENTED TESTING FOR NANO-METRIC CMOS VLSI CIRCUITS

2nd Edition

by

Manoj Sachdev
University of Waterloo
Ontario, Canada

and

José Pineda de Gyvez
Philips Research Laboratories, and
Eindhoven University of Technology
Eindhoven, The Netherlands

A C.I.P. Catalogue record for this book is available from the Library of Congress.

ISBN-10 0-387-46546-4 (HB)
ISBN-13 978-0-387-46546-3 (HB)
ISBN-10 0-387-46547-2 (e-book)
ISBN-13 978-0-387-46547-0 (e-book)

Published by Springer,
P.O. Box 17, 3300 AA Dordrecht, The Netherlands.

www.springer.com

Printed on acid-free paper

Dedication

"To Santosh and Baldev Sachdev;
Savitri and Dharm Bir Sawhney"

Manoj Sachdev

"To my teachers Jochen Jess
and Edgar Sánchez-Sinencio
for their invaluable knowledge"

José Pineda de Gyvez

Contents

Preface

Defect-oriented testing methods have come a long way from a mere interesting academic exercise to a hard industrial reality. Many factors have contributed to its industrial acceptance. Traditional approaches of testing modern integrated circuits have been found to be inadequate in terms of quality and economics of test. In a globally competitive semiconductor market place, overall product quality and economics have become very important objectives. In addition, electronic systems are becoming increasingly complex and demand components of the highest possible quality. Testing in general and defect-oriented testing in particular help in realizing these objectives.

For contemporary System on Chip (SoC) VLSI circuits, testing is an activity associated with every level of integration. However, special emphasis is placed for wafer-level test, and final test. Wafer-level test consists primarily of dc or slow-speed tests with current/voltage checks per pin under most operating conditions and with test limits properly adjusted. Basic digital tests are applied and in some cases low-frequency tests to ensure analog/RF functionality are exercised as well. Final test consists of checking device functionality by exercising RF tests and by applying a comprehensive suite of digital test methods such as I_{DDQ}, delay fault testing, stuck-at testing, low-voltage testing, etc. This partitioning choice is actually application dependent.

The relevance of defect-oriented testing in nanometer regime is more than ever. Higher packing density, ever larger systems on chip configurations, increased process complexity and process spread are making designs sensitive to subtle manufacturing defects. Tests professionals are expected to face numerous challenges in their quest to improve quality, reliability and yield of contemporary integrated circuits. Some of these challenges are mentioned below, and described through the book.

For economic reasons, test simplification is needed for SoC VLSI circuits. It is not unusual that an analog test engineer spends 20% of his/her efforts on software development, 30% on hardware test debugging and 50% on tester RF measurements. For a digital test engineer this workload is reversed, e.g. the digital test engineer spends most of his/her time at devising the appropriate test methodologies, designing the DfT, and generating test patterns. His/her post-silicon tasks are primarily concerned with product debugging and ensuring low test escapes. This can translate into several months of test development depending upon the maturity of the device and fabrication process. Test strategies may also be driven by a time to market window. Under this scenario wafer test is geared to improve yield only and most of the attention is devoted on final test.

Until now RF functionality has been provided by individual ICs such as mixers, PLLs, Multiple Output PLL, transceivers. Often, functionality and specifications are tested with "laboratory" or test-bench-like methods. Future ICs, in either silicon or Multi Chip Module (MCM) integration form, will force us to deliver more integrated functions with new tests challenges. Since these RF IPs will be embedded in the SoC, it will be difficult to access all RF ports and as such current RF test practices will not longer be applicable or will need to be revised. It is also evident that RF test times need to be reduced to acceptable limits within the digital-testing time domains through incorporation of DfT, BIST and silicon debug techniques. In addition, the RF testing will need to shrink the gap between customer needs in terms of PPM and testing methods.

Due to the device and voltage scaling scenarios for present and future nanometer CMOS technologies, it is inevitable that the attention will shift to testing parametric defects. As we know, the nano-metric regime brings new technological problems that did not exist before or that were not relevant in the past. Elevated leakage current, and signal integrity issues in interconnects are examples of new problems in modern technologies. Similarly, there are design and test challenges that are on the horizon. For example, transistor gate leakage, V_t mismatch, excessive substrate noise, etc. These issues, if left unattended, have the potential to erode yield, quality and reliability of integrated circuits. To deal with them, there is a need for (i) debugging, (ii) diagnosis, (iii) system-oriented testing, and (iv) "technology-oriented" test methods. Traditional stuck-at testing, will have difficulties catching many of these new "process-related defects" and as such comprehensive nano-metric test methodologies are imperative.

Without loss of generality, any comprehensive test program has the following challenges:

- Design for test.
- Need to deliver known good die (KGD).
- Need to guarantee low test escapes.
- Need to achieve very low cost testing.
- Need to diagnose failures

In this second edition, we have made an attempt to provide the reader with current trends in the field of defect-oriented testing. The target audience of this book consists of design and test professionals. However, this book may also be used as a reference book for graduate level courses on VLSI testing, or on VLSI quality and reliability. Our motivation to write the second edition comes from two diverse sources. Firstly, the field of defect-oriented testing is more than two decades old. However, the information on the subject is fragmented and distributed in various conferences and journal papers. Secondly, there is a wide disparity among various companies as well as academic institutions on the level of knowledge on this subject. A vast majority of research is carried out by a few companies and academic institutions. Therefore, it is intended that this book will help in spreading the knowledge of the subject.

Manoj Sachdev and José Pineda de Gyvez
July 2006

Foreword

Almost ten years would have passed since the publication of the original book by the time this second edition reaches the hands of a reader. Let's ask what has changed in the last decade. We find that chips have become systems. Besides, nanotechnology has crept into them. Testing, traditionally considered at the gate level, has advanced in two opposite directions. Upward, on one hand, to the system level and downward, on the other hand, to the nano-device level. A test engineer now needs to reach out in both directions. This second edition brings the updates to allow us to stretch upward as well as downward.

I can summarize the main differences from the first edition as follows:

1. A new chapter on *functional and parametric defect models* is added.
2. Enhancements to the chapter on *fault models* include inductive fault analysis for nano-metric technologies, radiation induced faults, and defects causing delay faults.
3. The chapter on testing of *RAMs* is extensively updated. New material on strategies of design for testability and test algorithms for weak cells in embedded SRAMs and address decoder faults has been added.
4. The chapter on analog testing is thoroughly revised. Notably, new test techniques, such as a power supply ramp testing method, have been added.
5. A new chapter on yield engineering is added.

This edition will be useful to those who work or plan to work in the area of VLSI testing, namely, practicing engineers and students. I thank the authors for their timely effort. I must, however, remind them that technology is never static. The changes in the next decade may be even more rapid than in the past. I hope the authors, Manoj Sachdev and Jose Pineda de Gyvez, will continue this work.

Vishwani D. Agrawal
vagrawal@eng.auburn.edu
August 2006

Foreword for the First Edition

We have made great strides in designing complex VLSI circuits. A laborious design verification process ensures their functional correctness. If no defects occur in manufacturing then testing will not be required. However, the world is not so perfect. We must test to obtain a perfect product.

An exact repetition of the verification process during manufacture is too expensive and even impossible. So, we test for a selected set of modeled faults. There is no unified modeling procedure for the variety of VLSI chips we make. Stuck-at model applies only to some types of digital circuits. Besides, there are problems, such as, (a) some stuck-at faults cannot occur in the given VLSI technology and (b) some actual manufacturing defects have no stuck-at representation. Numerous known problems with the present-day test procedures point to a defect-oriented testing. This simply means that we use the knowledge about the manufacturing process to derive tests. Such tests provide the greatest improvement in the product quality for the minimum cost of testing.

Dr. Sachdev has done original work on defect-oriented testing. He takes experimental defect data and applies the inductive fault analysis to obtain specific faults for which tests should be derived. His work is done in an industrial setting and has been put to practice at Philips Semiconductors and elsewhere. The material in this book is collected from his PhD dissertation, research papers and company reports.

A strength of this book is its breadth. Types of designs considered include analog and digital circuits, programmable logic arrays, and memories. Having a fault model does not automatically provide a test. Sometimes, design for testability hardware is necessary. Many design for testability ideas, supported by experimental evidence, are included.

In addition to using the functional and other conventional tests, Dr. Sachdev takes full advantage of the defect-isolating characteristics of non-functional tests. Imagine taking a multiple-choice examination. All of us can remember making a guess some time and succeeding. Suppose, I connect you to a lie detector while you checked those choices. The lie detector may tell me to fail you even on some correct answers. Also, given the new procedure, we can design special tests. Current measurements similarly bring out the internal conflicts whose effects may not be visible by conventional logic tests. Such tests, though non-functional, improve the defect coverage. Current measurement is an important subject discussed in this book.

Non-functional tests are not without their pitfalls. Not much is accomplished if one who is going to be an electrical engineer passes or fails an examination in history. Clearly, there is need for matching the test with the function. In electronic circuits a non-functional test, designed to isolate a real defect, can reject a circuit with some other functionally acceptable defect. This phenomenon, known as yield loss due to non-functional tests, impacts costs similar to the design for testability overhead. In both cases, the costs are associated with quality improvement. A central theme in this book is to minimize such costs and it wonderfully succeeds in putting the economics of test and manufacture into practice.

Vishwani D. Agrawal
Bell Labs
Murray Hill, New Jersey
va@research.bell-labs.com
September 1997

Acknowledgements

During the various phases of the work we had the opportunity to exchange ideas and learn from experts in the field. We would like to acknowledge the following people of Philips Research. Guido Gronthoud for his continuous support and encouragement on various topics such as analog testing, and process-aware testing. Rodger Schuttert for providing us with the data for the yield engineering section. Paul Volf and Rosa Rodríguez Montañez for their challenging minds and insight during the development of the work on resistive vias. Maurice Lousberg for his always open mind and interest in our work. Rene Segers for his support on various topics. Stefan Eichenberger and Bram Kruseman for their vast experience in defect-oriented testing. The young and challenging "analog and RF testing" minds of Amir Zjajo, Estella Silva and Shaji Krishnan deserve also being mentioned. We also like to acknowledge the support of Kees Veelenturf and Paul Simon, of Philips Semiconductors, for our early research on DfM.

The book would not be in this form if several of graduate students and post doctoral fellow in the electrical and computer engineering department at University of Waterloo had not participated tirelessly. In particular, contributions of Derek Wright, S. M. Jahinuzzaman, Oleg Semenov are worth mentioning. Authors would also like to thank Chuck Hawkins of University of New Mexico for reviewing some of the chapters of the book. Finally, the management support of Ad ten Berg at Philips Research is gracefully recognized.

Manoj Sachdev, and José Pineda de Gyvez

Chapter 1

INTRODUCTION

This chapter introduces some key test issues, namely test complexity, quality, reliability, and economics faced by the semiconductor industry. These issues form a basis for subsequent chapters.

1. EVOLUTION OF CMOS TECHNOLOGY

The microelectronics industry has been growing at an astounding pace in the last two decades, primarily due to the integration capability of complementary metal oxide semiconductor (CMOS) manufacturing processes. Ever increasing clock speeds of micro-processors and bigger, cheaper dynamic random access memories (DRAMs) are enabling applications that were unthinkable just a few years ago. A foray of CMOS technology in numerous application domains such as telecommunications, computing, and consumer applications continues at the cost of other manufacturing technologies such as bipolar, GaAs, etc. This trend is likely to continue for some time as we move forward in the 21st century [22].

The concept of a metal oxide semiconductor (MOS) transistor was independently described by Lilienfeld, and Heil, respectively in 1930s [27,20]. However, it could not be manufactured owing to poor Si-SiO_2 interface. The bipolar junction transistor (BJT) was invented at Bell laboratories in 1947 [6,45]. It took several years to exploit the transistor's true potential with the invention of integrated circuits (ICs) in the late 1950s by Jack Kilby [25]. Modern ICs owe their root to Frank Wanlass. He invented the concept of CMOS logic in 1963 [52] and called it nanowatt logic [53]. However, CMOS technology did not gain popularity until the late 1970s. Since then, CMOS has been the technology of choice for a vast majority of applications owing to its relatively simple, inexpensive manufacturing process, integration capability, and extremely small power consumption compared to other integrated circuit technologies.

The recent surge in information technology related industries is largely enabled by our abilities to design and manufacture complex ICs. The semiconductor industry is unique in having sustained such a spectacular growth over a significantly long period. As a result, industry has provided electronic products at substantially lower cost per function with higher performance year after year.

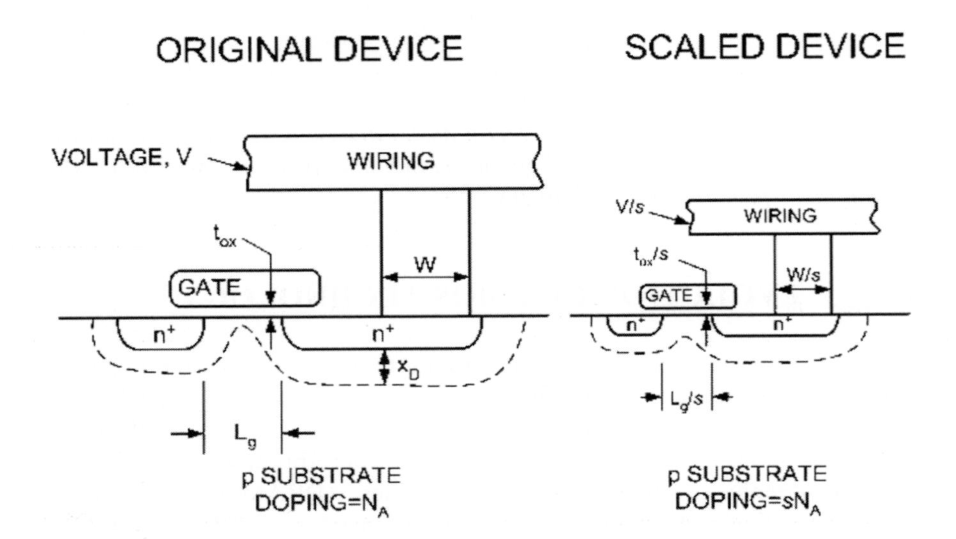

Figure 1-1. MOS transistor scaling.

For several technology generations, the shrinking of metal oxide semiconductor (MOS) transistors has been governed by the concept of scaling [33,10,5,13]. Figure 1-1 depicts the concept. All dimensions of a MOS transistor are scaled by a factor s ($s > 1$) to produce a smaller transistor while preserving its behavior. If all the dimensions and voltages are reduced by a factor s, and doping densities are increased by s, the electric field inside the device remains as before. This type of scaling is known as *constant electric field scaling* (CFS). Since the electric field remains constant, this type of scaling does not result in device damage due to excessive electric field. As evident from the second column of Table 1-1, scaling results in higher relative gate density (s^2), lower gate delay ($1/s$), and reduced power dissipation ($1/s^2$).

Table 1-1. Scaling concepts for MOS transistor.

Parameter	Relation	Constant Electric Field Scaling	Constant Voltage Scaling	General Selective Scaling
W, L, t_{ox}		$1/s$	$1/s$	$1/s$
V_{DD}, V_t		$1/s$	1	$1/g$
Area	WL	$1/s^2$	$1/s^2$	$1/s^2$
C_{ox}	$1/t_{ox}$	s	s	s
C_{gate}	WL C_{ox}	1/s	$1/s$	$1/s$
I_{sat}	$C_{ox}WV$	$1/s$	1	$1/g$
Gate delay	VC_{ox}/I_{sat}	$1/s$	$1/s$	$1/s$
Power dissipation	$I_{sat}V$	$1/s^2$	1	$1/g^2$
Power Density	Power/Area	1	s^2	s^2/g^2

Constant electric field scaling is not always possible. Very often, power supply voltage is determined due to system considerations, or to keep newer devices compatible with existing parts. Therefore, earlier devices (until 0.8 μm) followed the *constant voltage scaling* (CVS) path. However, it was subsequently abandoned in favor of CFS owing to higher electric fields inside the device and its implications on long-term device reliability. Deep sub-micron devices often follow *general selective scaling* (GSS) where the device dimensions and voltages are scaled by different factors. Several intrinsic voltages inside a MOS transistor such as built-in junction potential are material parameters, while others such as threshold voltage (V_t) cannot be scaled by the same factor. Therefore, voltage should be scaled less aggressively by a factor g (where $g > 1$). GSS offers the performance benefits of CFS or CVS while its power dissipation is in between CFS and CVS.

It appears that the industry is now in a deep sub-micron regime, and a number of technical challenges threaten the continuation of what is known as Moore's Law [33]. The difficulty of design and manufacture has increased to a point where exploitation of its full potential seems to be unrealistic. For example, the above mentioned scaling scenarios assume insignificant leakage current increase with scaling. However, this component is significantly large in sub-0.18 μm technologies. Increased leakage current consumption in modern ICs is causing long term reliability concerns. Elevated leakages result in increased power dissipation which, in turn causes higher junction temperature. Recently, Semenov et al. [46] estimated a 1.45 times increased in junction temperature per technology generation under nominal operational conditions. Higher junction temperature is one of the major contributors to poor device reliability.

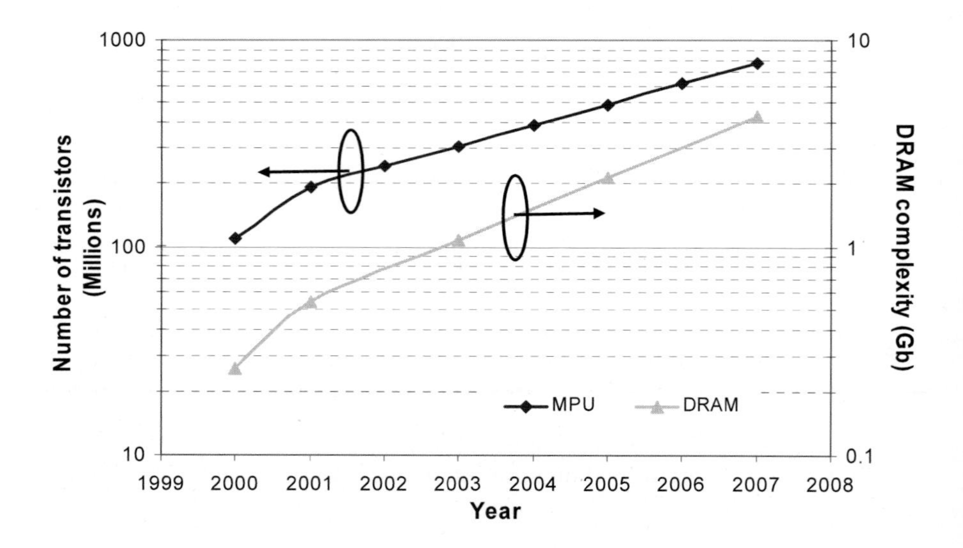

Figure 1-2. Giga scale integration [22].

Similarly, nominal functionality of scaled transistors is extremely susceptible to natural manufacturing process spreads. Varying impurity densities, gate oxide thickness, and junction depths may cause transistors parameters such as V_t to shift resulting in abnormal delays and leakages. Finally, as more transistors are crammed per unit area, tiny defects and imperfections created during the manufacturing process can cause failures.

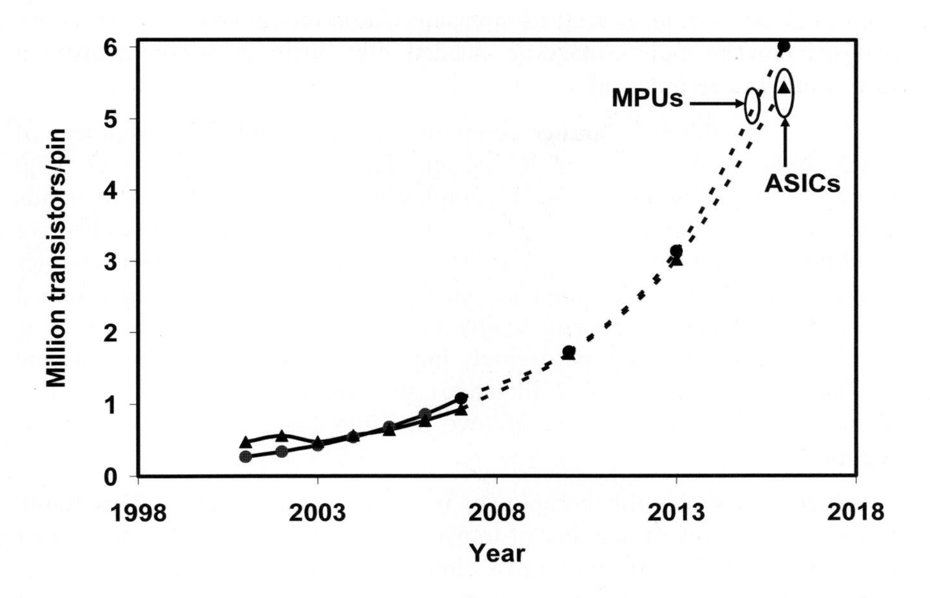

Figure 1-3. Number of transistors per IO pin for microprocessors and ASICs [22].

2. THE TEST COMPLEXITY

Imperfections in the manufacturing process necessitate testing of the manufactured ICs. The fundamental objective of the testing is to distinguish between good and faulty ICs. This objective can be achieved in several ways. Earlier, when ICs were relatively less complex, this objective was achieved by functional testing. Functional tests are closely associated with the IC function. Therefore, these tests are comparatively simple and straightforward. A 4-bit binary counter can be exhaustively tested by $2^4 = 16$ test vectors. However, as the complexity of the fabricated ICs increased, it was soon discovered that the application of a functional test is rather expensive on test resources and is inefficient in catching the manufacturing process imperfections (or defects as they are popularly known). For example, a digital IC with 32 inputs has only a modest design complexity by today's very large scale integration (VLSI) standards, but will require 2^{32} = 4,294,967,296 test vectors for exhaustive functional testing. If these are applied at the rate of 10^6 vectors per second, it will take 71.58 minutes to test a single IC. The test becomes even longer if the IC contains sequential logic. Moreover, exhaustive testing may not be enough to detect defective parts if the faulty behavior becomes sequential. In this case, newer test methods such as delay testing as well as stressing conditions including temperature and low power supply voltage are needed. Obviously, it is too expensive a test solution to be practical.

The test problem is further compounded by the rapid development of CAD tools in the areas of IC design and manufacturing, which help engineers to design and fabricate complex ICs. For example, recent trends towards silicon reuse (core based systems on chip (SoC) design styles) are resulting in shrinking design cycles. However, test and testability Computer Aided Design (CAD) tools are lagging. The need for simulation tools for test and testability analysis became visible only when testing was recognized as a bottleneck in achieving increasingly important quality, reliability and time to market goals. Figure 1-2 illustrates this complexity vividly. As it is abundantly clear from the graph, we are into the giga-scale integration regime.

Figure 1-3 shows this complexity from the test perspective. This figure illustrates the growing number of transistors per IO pin for microprocessors and application specific integrated circuits (ASICs). The number of input and output pads or pins has not been able to keep up with increased integration. This packaging limitation puts severe additional constraints on the testing of complex ICs. For example, the number of transistors on a chip continues to double every 1.5-2.0 years. However, the number of package pin/balls grows at an annual rate of approximately 11% [22]. Typically, larger, bigger ICs require an increased number of pads and pins to allow data flow to and from the IC. Additionally, more pads and pins are required to provide adequate power and noise immunity. The issue of power delivery and power supply noise is critical in high performance circuits. Approximately two-thirds of all pads are dedicated to power and ground so as to deliver excess of 100 W of power to hungry transistors. In high performance ASICs the situation is better and only approximately half of the total number of pads is for power and ground.

Irrespective of application domains, the number of transistors per signal pad is growing rapidly, and Figure 1-3 illustrates its projections. Figure 1-4 depicts the growing transistor density of ICs. Effectively, the depth of logic that is to be accessed from primary pins increases for each successive generation of chips. In other words, controllability and observability objectives become much more difficult to achieve for modern ICs from outside the chip. As a result, test vector sequences are becoming longer and are adding to the test cost. At the same time, the cost of general-purpose automatic test equipment (ATE) is also increasing significantly. A state of the art ATE can now cost a few million dollars. The expensive ATE and the longer test vector sequences push the test costs to unacceptable levels.

The test complexity can also be segregated in terms of (i) quantitative issues, and (ii) qualitative issues. Tens of millions of transistors on a chip must be tested in a reasonable and economically viable test time. The built-in self test (BIST) has become a de-facto standard testing of embedded memories, while significant progress has been made on BIST for logic and analog circuits.

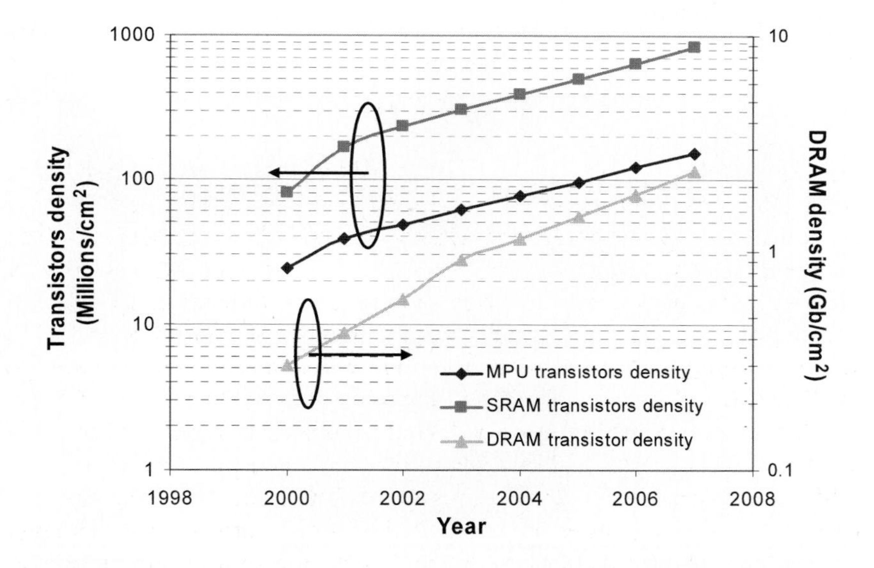

Figure 1-4. Growing transistor density [22].

The above-mentioned scenario matches with the evolution of the semiconductor memory market. Each successive DRAM generation has grown in complexity by a factor of four and the access time has decreased by a factor 0.8 for each new generation. Therefore, testing time is increased by 3.2 times for each new generation. This results in a tremendous increase in testing cost which prevents the cost per bit from coming down despite increased integration. Small feature size and huge chip size result in an enormous critical area [12] for defects. Since RAMs must be mass produced, their test strategies are under severe pressure to ensure the quality of the tested devices while maintaining the economics of the production. In other words, testing of RAMs in an efficient, reliable, and cost effective manner is becoming an increasingly challenging task [29].

For example, a study of DRAMs identified the test cost, along with process complexity, die size, and equipment costs as a major component in future DRAM chip costs [23]. The test cost of a 64 Mbit DRAM was projected to be 240 times that of a 1 Mbit DRAM. For a 64 Mbit DRAM, the test cost to total product cost ratio was expected to be 39%. If the conventional test methods are used, test costs will grow at a rapid rate. The SRAM test cost was also expected to follow a similar trend. Moreover, RAMs are the densest and one of the biggest chips ever to be tested. DRAM chip size has grown by 40 to 50% while the cell size has decreased by 60 to 65% for each successive generation. The chip size of 64 Mbit DRAM is in the range of 200 mm^2.

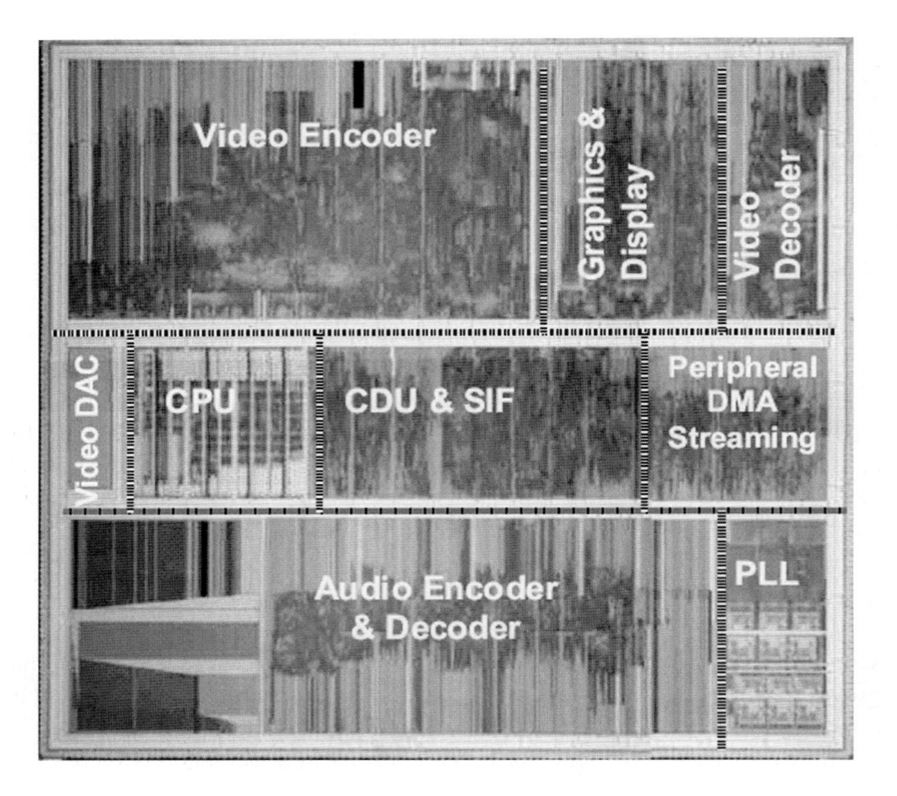

Figure 1-5. A typical syetm on chip [17].

This IC test cost explosion is not limited to RAMs. There has been a dramatic increase in SoC designs, which can include digital, analog, RF, mixed-signal, and memory all on the same die, as is illustrated in the die shown in Figure 1-5 [18]. It is a single chip MPEG-2 decoder for use in DVD players that illustrates how both digital (CPU), mixed-signal (video DAC), and analog (PLL) circuits now reside on the same die. Since each of these circuit types requires different tester capabilities, testers must now be able to test different kinds of functionality. Also, there are an increasing number of chips per wafer, which necessitates either testers with more channels for testing, or fewer channels with more touchdowns (the number of times the probes of the tester have to move to a new location). The end result is a dramatic increase in testing time and costs, to the point where in some cases the cost of testing dominates the overall cost of manufacturing [51]. For example, the test development time for complex single chip television ICs manufactured by Philips is reported to be many man years! Such developments have caused a surge of interest in the economics of test [3]. A number of studies have been reported on test economics [1,8,57]. Dislis et al. [8] demonstrated that economic analysis can be a powerful aid in the selection of an optimal set of design for test (DfT) strategies, and in the organization of production test processes.

3. QUALITY AND RELIABILITY AWARENESS

Ever since the invention of the transistor in late 1940s, the semiconductor industry has grown into diverse applications areas. These range from entertainment electronics to space applications. Computers and telecommunication are other notable applications. Irrespective of the application areas, the quality and reliability demands for semiconductor devices have significantly increased [15,16]. This requirement is not difficult to understand.

It is a well known rule of thumb that if it costs one dollars to test a defective component at the chip level, it will cost ten dollars at the board level and hundred dollars at the system level to test, diagnose and replace the same defective component. Therefore, economically it makes a lot of sense to build a system with high quality components. As a well known example, a version of the Pentium processor was released with an undetected error in the floating-point unit. This design flaw was discovered only after the processor had been integrated into systems and sold to consumers as desktop computers. The replacement and lost inventory charges cost Intel Corporation $475M. Better design verification and testing could have detected this error very early in the design phase for a fraction of the cost [38]. ICs for the automotive branch are just another example of the need for quality and reliability with preferably zero ppm levels. DfT strategies have an important role to play in reducing high costs associated with testing and debugging at the sub-system and/or system level. Researchers have shown that such strategies improve quality and decrease test costs by an order of magnitude [36].

Pulat and Streb [34] put numbers into the escalating cost of building products with quality and reliability. Imagine a component with 1% test escapes. It will cause a shipment of 10,000 defective parts per million items produced. If 30 such components are required to make a product, each with 1% test escape, the overall product yield would be only 74%. Hence, modest failure rates at the component level may result in a significant likelihood of failure at the board or system level. The increasing system complexities require still better quality from IC suppliers so as to make economic, quality systems.

On the other hand, market economics forced what were known as purely digital ICs to incorporate embedded memories as well as analog blocks so as to offer cheaper and more reliable SoC solutions. As mentioned in the previous section, these SoCs have many different functional blocks all on the same substrate, which makes circuits such as RAMs, analog blocks more susceptible to a variety of manufacturing process defects. Higher degree of integration, though far reaching in terms of market penetration, caused anxiety amongst design, process, and test professionals.

As systems became more complex, their upkeep, maintenance, and repair became more costly. Often specialists are required for such functions. Therefore, reliable system operation over its lifetime became another absolute requirement. These developments led to slogans like Design for Quality and Design for Reliability. The terms quality and reliability are often misunderstood. Here, for the sake of clarity, we must distinguish between the terms quality and reliability.

According to Hnatek [15]; the words *"reliability" and "quality" are often used interchangeably as though they were identical facets of a product's merit; however, they are different. Quality pertains to the population of faulty devices among the good ones as they arrive at the user's plant. Or, in another view, quality is related to the population of faulty devices that escape detection at the supplier's plant... Reliability is the probability that an IC will perform in accordance with expectations for a predetermined period of time in a given environment. Thus reliability is quality on a time scale, so to speak, and testing (screening) compresses the time scale.*

4. BUILDING QUALITY AND RELIABILITY

Design, manufacturing, and test form three major activities in the development of an IC. It is futile to believe that overall quality of any IC can be achieved considering only design, manufacturing, or test alone. In other words, robust design, controlled manufacturing process, and effective test strategy together result in a quality product.

The role of design and manufacturing in building IC quality and reliability has been investigated in depth and is the focus of further investigations [50]. From the manufacturing standpoint, fabrication process and device technologies in the deep sub-micron region (90-32 nm) are approaching practical limits, and therefore concurrent achievements in high performance, high packing density, and high reliability are expected to become increasingly difficult. Besides, quality and reliability issues for VLSI (with as many as 10^9 transistors on a chip) are becoming more stringent due to required escape rates of less than 100 parts per million (PPM) and required failure rates of less than 10 failures in time (FIT) [9,44]. One device failure in 10^9 device-operating hours is termed as one FIT. Furthermore, due to the large initial investment required by the fabrication process complexity, it has recently become a matter of considerable debate whether such an investment is profitable. Similarly, contribution of design to improve quality and reliability of ICs has been outstanding, and is beyond the objectives of this book.

The often-stated objective of testing is to ensure the quality of the designed systems. Testing is the last check-post before the product is shipped to its destination. In other words, it is the last opportunity to prevent the faulty product from being shipped. Pulat and Streb [34] stressed the need for component (IC) testing in total quality management (TQM). In a large study spreading over three years and encompassing 71 million commercial grade ICs, Hnatek [15] reported differences in quality seen by IC suppliers and users. One of the foremost conclusions of the study was that IC suppliers often do not do enough testing. *How thorough must the functional testing of digital ICs be to guarantee adequate quality? Is fault grading necessary?* If yes, how high must the single-stuck fault coverage be for a given quality? These were the objectives of a study conducted by Agrawal et al. [2]. They described a model-based technique for evaluating the fault coverage requirement for a given field escape rate (PPM). In their subsequent paper [47], the authors showed that the fault simulation results with tester data can also predict the yield and fault coverage requirements for a given PPM for an IC. It was shown that for 1000 PPM, about 99% fault coverage will be needed. Similar results were obtained by McCluskey and Buelow [31]. The result of their theoretical analysis as well as experimental evidence indicated that logic production test fault coverage of greater than 99% is necessary for manufacturing and selling high quality ICs.

At the same time, it was discovered that classical voltage based test methods for digital CMOS ICs are grossly inadequate in ensuring the desired quality and reliability levels [11,35]. Many commonly occurring defects like gate oxide defects often are not detected by logic tests [11,42]. Therefore, such escaped defects are quality and reliability hazards. This increased quality awareness brought in new test techniques like quiescent current measurements (QCM), or I_{DDQ} as it is popularly known, in the test flow for digital CMOS ICs [35,4,19,30]. Arguably, I_{DDQ} is the most effective test method in catching manufacturing process defects. Perry [35] reported that with the implementation of I_{DDQ} testing on ICs, the system failure rate dropped by a factor of six. Gayle [17] reported that with implementation of I_{DDQ} testing the defect rate had fallen down from a high 23,000 parts per million to a more acceptable 200 parts per million. Similarly, Wiscombe [54] reported improvement in quality levels.

In the late 1990s, several researchers [41,55,56] identified that increasing MOSFET off-currents (I_{OFF}), together with a higher degree of integration is going to erode the benefits of I_{DDQ} testing. Contemporary MOSFETs are scaled using the concept of general selective scaling, as depicted in Table 1-1. Hence, V_{DD} and V_t are scaled down proportionately with scaling of the MOSFET dimensions. An 80-100 mV reduction in the V_t of a MOSFET increases its I_{OFF} by a factor of 10. In recent years with each successive technology generation, the V_t was reduced by 100-200 mV. As a consequence, the I_{OFF} was increased between 10-100 times for a given transistor width with scaling. As the total chip leakage current approaches the mA range, the defect-free and defective I_{DDQ} distributions begin to overlap, hence reducing its effectiveness.

Despite the decreasing effectiveness of traditional I_{DDQ} measurements, researchers continue to devise new current-based test methods that are effective in deep sub-micron technologies [43]. Some of these methods, namely ΔI_{DDQ} and I_{CCQ}, exploit differential measurement to cancel the increasing common-mode leakage current [32, 26]. Maxwell et al. argued that both approaches are based on some threshold of current differences and therefore, they suffer from the effects of process variation. Setting a threshold based on either maximum allowable current or the difference between currents will be difficult because of large vector-to-vector, or die-to-die, variations. Maxwell et al. suggested plotting I_{DDQ} in ascending order as a function of test vectors, and characterizing it [28]. Some researchers also explored the feasibility of transient current measurements and characterized its effectiveness compared to I_{DDQ} [42].

Analog test complexities are different from that of digital circuits. The application of digital DfT schemes has been largely unsuccessful in the analog domain [40]. As a result, a vast majority of analog circuits are tested by verifying the functionality (specifications) of the device. Since different specifications are tested in different manners, it makes analog functional testing costly and time consuming. Moreover, often extra hardware is needed to test various specifications. Limited functional verification does not ensure that the circuit is defect-free and escaped defects pose quality and reliability problems. Defect-oriented testing provides a structured analog test methodology which improves the quality, reliability, and economics of tested devices.

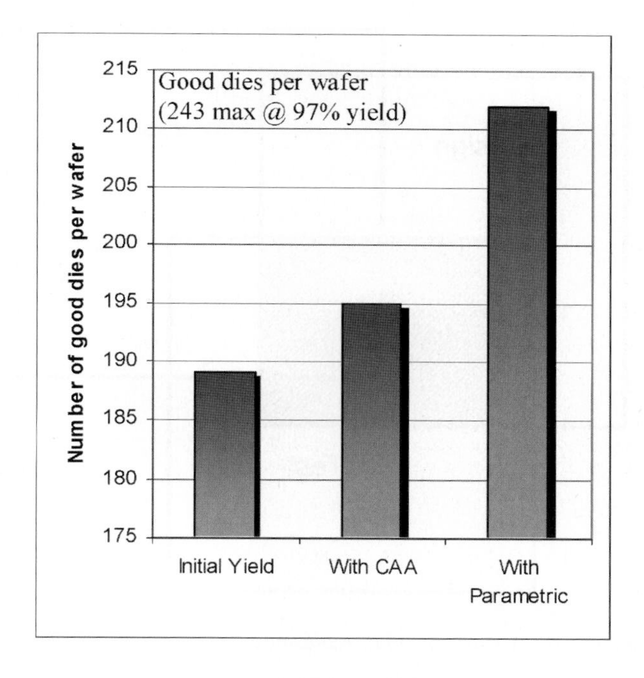

Figure 1-6. DfY techniques improve yield [54].

A recent trend towards a closer relationship between design, manufacturing, and test is called design for yield (DfY). A design and its layout are implemented to be insensitive to the most common manufacturing defects for a given process. Similarly, the testing strategies are devised to catch the likely defects. Critical area analyses (CAA) of layouts help to find areas where faults are likely to occur. Parametric analysis helps designers to estimate the impact of process variation on the performance of analog circuits. A study reported by Rencher illustrates the benefits of DfY strategies. A summary of results are depicted in Figure 1-6. The yield of a product is improved significantly with the help of critical area and parametric analysis tools [37].

Testing is required to improve the quality and reliability of manufactured ICs. As devices become smaller, integration becomes higher, and economics dictate even better quality, hence testing has moved from being an afterthought of designers to a forefront issue of IC design and manufacture. High-level functional testing with a limited number of test vectors has evolved into to full defect modeling, BIST, I_{DDQ}, DfT, DfY, etc. Improved testing is increasingly critical in the race to extend Moore's Law.

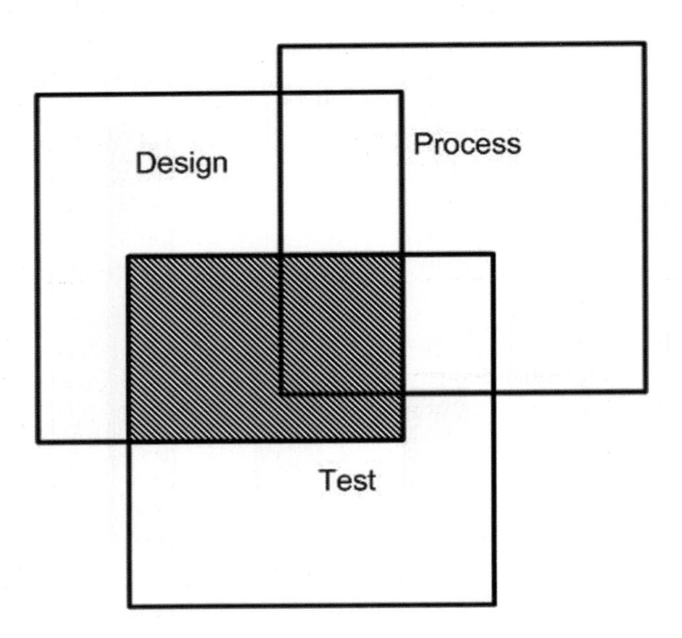

Figure 1-7. Major steps in IC realization and the focus of the book.

In modern semiconductor facilities, "closing the loop" between design and fabrication is needed to accelerate yield maturity. While in the past process control monitors (PCM) and yield engineering monitors (YEM) were primarily used for yield ramp up, the complexity of modern technologies is such that the use of this kind of monitors is not longer sufficient. Namely, results of testing actual chips are used to identify process weaknesses, and test data is used to guide in-situ failure analysis in pinpointing the exact location of the problems. Testing plays, thus, a critical role in "closing the loop" between design and manufacturing [21].

5. OBJECTIVES OF THIS BOOK

Design, fabrication process and test constitute three major steps in the realization of an IC. In an idealized environment these three steps should perfectly match. For example, an ideal design realized in an ideal fabrication process environment will have 100% yield. Therefore, test in an ideal environment is redundant and not required. The real world is far from an idealized one where all these steps have a certain amount of uncertainty associated with them. Figure 1-7 symbolically illustrates the non-idealized IC realization process with the three major steps having partial overlap with each other. The partial overlap signifies an imperfect relationship amongst the steps. In other words, only a subset of fabricated ICs is defect-free and only a subset of defective ICs is caught by the test. As a result, a set of design, test, and process professionals have to make a conscious effort to strive for a near optimum relationship for better product quality and economics. For example, the test should cover all the likely defects in the design, or the design should work within the constraints of the process, or the test should incorporate the process defect information for optimum utilization of resources.

In this broad spectrum, this book focuses on the darkened area of Figure 1-7. The primary objective of the book is to make readers aware of process defects and their impact on test and quality. The target audience of this book is practicing VLSI design and test professionals. The motivation of the book comes from the fact that costs of IC testing have risen to absurd levels and are expected to rise further for SoCs. According to experts, design and test professionals have to focus on defects rather than higher level fault models to reduce the test cost while improving the quality and reliability of products. It is a daunting task given the complexity of modern ICs. Furthermore, shrinking technology makes circuits increasingly prone to defects. Millions of dollars are spent in state of the art manufacturing facilities to reduce particulate defect count (the primary cause of yield loss), defect monitoring, yield improvement, etc. Therefore, in such a scenario, knowledge of what can go wrong in a process should be advantageous to design and test professionals. This awareness can lead to robust design practices such that the probabilities of many types of defects are reduced, or alternatively their detection is simplified. Similarly, test solutions for dominant types of defects may be generated to rationalize test costs.

There are a number of defect types that may occur in a circuit and often different circuit types have to co-exist on the same die. Depending upon the circuit type (dynamic, static, digital, RAM, PLAs, or analog) defects influence the operation differently. Hence, such circuits should be addressed separately and optimum test solutions for each circuit type should be evolved. For example, certain classes of defects are not detected by logic testing, however, are readily detected by I_{DDQ}. A good DfT scheme is the one that works within the constraints of a given circuit type. A few of these schemes are suggested in subsequent chapters and may be used to create test modes such that the defect coverage of the test is enhanced or very few tests are needed for defect detection.

Why this book? The field of defect-oriented testing is nearly two decades old. The information on defect-oriented testing is fragmented and distributed in various conference and journal papers. There is hardly any book providing a cohesive treatment this field deserves. In this book an attempt is made to bridge this gap and provide an overview of this field. Our focus in this book is to study the impact of defects on various circuit types and draw conclusions on defect detection strategies. This book does not pretend to include all the work done in this area. However, an effort is made to include the most practically relevant information in this area and present it in a readable format. The book is written keeping practical VLSI aspects in mind. The DfT strategies described in the book are realizable in CMOS technology and many have actually been implemented at Philips Semiconductors and elsewhere.

The relevance of defect-oriented testing in nano-metric regime is more than ever. Higher packing density, ever larger systems on chip configurations, increased process complexity and process spread are making designs sensitive to subtle manufacturing defects. Traditional approaches of testing are inadequate, and practicing engineers have to focus on defects in their quest to improve quality, reliability and yield of contemporary integrated circuits.

6. BOOK ORGANIZATION

A wealth of knowledge is available on manufacturing defects. *Chapter 2* provides an overview of defects. Defects are segregated into several categories. The chapter addresses the modeling issues of defects and their circuit impact is described. Particular attention is devoted to process variability in nano-metric geometries and its impact on circuit performance. The second half of the chapter is devoted to the fundamental concepts of the inductive fault analysis (IFA) techniques for realistic fault model development. The relationship between process deformations and IC failures is illustrated.

In *Chapter 3*, a review of digital fault models is provided. These fault models are classified according to the level of abstraction. The relative merits and shortcomings of these methods are also reviewed. The differences between functional and structural testing are brought out and the impracticality of functional testing for complex VLSIs is highlighted. Attention is paid on delay fault models, and temporary faults (e.g., soft errors) which are becoming prominent in nano-metric technologies.

"How do different models fare in real life?" is the focus of *Chapter 4*. This chapter provides a summary of some of the important studies conducted on defect-oriented testing in the last two decades. The earlier work on defects in simple NMOS and CMOS logic circuits is studied. Early studies on the effectiveness of the stuck-at (SA) fault model in detecting defects in CMOS circuits are discussed and their conclusions are summarized. Work on Maly et. al., on the effectiveness of IFA is highlighted and the pioneering work on gate oxide defects and its impact on IC quality and reliability by Hawkins and Soden is presented. Often such defects are not detected by voltage testing and I_{DDQ} measurements are needed to detect them. Subsequently, the studies on Boolean and I_{DDQ} testing are described and important conclusions are noted. Enhanced leakage current and delay effects of realistic defects in CMOS circuits are illustrated. Finally, how IFA is being used in nano-metric technologies described. It is worth noting that researchers have analyzed Pentium microprocessor using IFA tools.

Random access memories (RAMs) are integral parts of modern ICs as well as systems. Proliferation of microprocessor, DSP, and micro-controller based systems require a large amount of embedded and dedicated RAMs. As far as their testing is concerned, RAMs suffer from quantitative issues of digital testing as well as qualitative issues of analog testing. In *Chapter 5*, we address the application of defect-oriented test method to RAMs. The application of this method results in efficient algorithms whose effectiveness is demonstrated with silicon test data. Particular attention is paid on address decoder defects and stability faults in SRAMs. The latter is becoming a growing concern with technology scaling. Transistors in SRAM cells are susceptible to process variations owing to their small geometries. Traditional test approaches are unlikely to detect such parametric failures. In this chapter, causes of poor SRAM stability due to process and manufacturing defects and circuit techniques to test them are described.

In *Chapter 6*, the defect-oriented test methodology is applied to find non-specification based analog test methods. Owing to the non-binary nature of their operation, analog circuits are influenced by process defects in a different manner than digital circuits. In this chapter, we demonstrate with the help of real CMOS circuits that simple test stimuli, like DC, transient, and AC can detect most of the modeled process defects. This test methodology is structured and simpler, and therefore results in substantial test cost reduction. Furthermore, we tackle the issue of analog fault grading. The quality of the test, and hence the tested device, depends heavily on the defect (fault) coverage of the test vectors. Therefore, it is of vital importance to quantify the fault coverage. We demonstrate how the IFA technique can be exploited to fault grade given (conventional) test vectors. Once, the relative fault coverage of different blocks is known for given test vectors, an appropriate DfT scheme can be applied to the areas where fault coverage of existing test methods is relatively poor.

Chapter 7 discusses issues related to manufacturing yield. Manufacturing of integrated circuit is extremely expensive venture where manufacturing yield plays a crucial role. Manufacturing defects and its knowledge is important to yield ramp up and yield improvement.

Finally, in *Chapter 8* conclusions on defect-oriented testing are given. Its advantages and limitations are outlined. Some potential research directions are recommended.

References

1. M. Abadir, and A.P. Ambler, Economics of Electronic Design, Manufacture and Test, Boston: Kluwer Academic Publishers, 1994.
2. V.D. Agrawal, S.C. Seth, and P. Agrawal, “Fault Coverage Requirement in Production Testing of LSI Circuits”, IEEE Journal of Solid State Circuits, vol. SC-17, no.1, pp. 57–61, February 1982.
3. A.P. Ambler, M. Abadir, and S. Sastry, Economics of Design and Test for Electronic Circuits and Systems, New York: Ellis Horwood, 1992.
4. K. Baker, ‘‘QTAG: A Standard for Test Fixture Based I_{DDQ}/I_{SSQ} Monitors,’’ Proceedings of the IEEE International Test Conference, pp. 194–202, 1992.
5. G. Baccarani, M.R. Wordeman, and R.H. Dennard, “Generalized Scaling Theory and its Application to a ¼ micrometer MOSFET design,” IEEE Transactions on Electron Devices, vol. ED-31, pp. 452–462, Apr. 1984.
6. J. Bardeen, and W. Brattain, “The Transistor, a Semiconductor Triode,” Phys. Rev., vol. 74, pp. 230, Jul. 1948.
7. S.D. Brown, Field-Programmable Devices: Technology, Applications, Tools, 2nd Edition, Los Gatos: Stan Baker Associates, 1995.

8. C. Dislis, J.H. Dick, I.D. Dear, and A.P. Ambler, Test Economics and Design For Testability, New York: Ellis Horward, 1995.

9. D.L. Crook, ''Evolution of VLSI Reliability Engineering,'' Proceedings of the IEEE International Reliability Physics Symposium, pp. 2–11, 1990.

10. R.H. Dennard, F.H. Gaensslen, H.N. Yu, V.L. Rideout, E. Bassous, and A.R. Leblanc, "Design of ion-implanted MOSFETs with very small physical dimensions," IEEE Journal of Solid State Circuits, vol. SC-9, pp. 256–268, Oct. 1974.

11. F.J. Ferguson, and J.P. Shen, ''Extraction and Simulation of Realistic CMOS Faults using Inductive Fault Analysis,'' Proceedings of the IEEE International Test Conference, pp. 475–484, 1988.

12. A.V. Ferris-Prabhu, ''Computation of the critical area in semiconductor yield theory,'' Proceedings of the European Conference on Electronic Design Automation, pp.171–173, 1984.

13. D.J. Frank, R.H. Dennard, E. Nowak, P.M. Solomon, Y. Taur, and H.P. Wong, "Device Scaling Limits of Si MOSFET and Their Application Dependencies," Proceedings of the IEEE, vol. 89, no. 3, Mar. 2001.

14. C. Hawkins, and J. Soden, ''Reliability and Electrical Properties of Gate Oxide Shorts in CMOS ICs,'' Proceedings of the IEEE International Test Conference, pp. 443–451, 1986.

15. E.R. Hnatek, ''IC Quality – Where Are We?'' Proceedings of the IEEE International Test Conference, pp. 430–445, 1987.

16. E.R. Hnatek, Integrated Circuits Quality and Reliability – 2nd Edition, New York: Marcel Dekker, Inc., 1995.

17. R. Gayle, ''The Cost of Quality: Reducing ASIC Defects with I_{DDQ}, At-Speed Testing and Increased Fault Coverage,'' Proceedings of the IEEE International Test Conference, pp. 285–292, 1993.

18. J. Geerlings, E. Desmicht, and H. de Perthuis, "A single-chip MPEG2 CODEC for DVD+RW", IEEE International Solid-State Circuits Conference, vol. 1, pp. 40–41, 2003.

19. R. Gulati, and C. Hawkins, I_{DDQ} Testing of VLSI Circuits, Boston: Kluwer Academic Publishers, 1993.

20. O. Heil, "Improvements in or relating to electrical amplifiers and other control arrangements and devices", British Patent no. 439,457, 1935.

21. C. Hora, R. Segers, S. Eichenberger, M. Lousberg, "An effective Diagnosis Method to Support Yield Improvement," Int. Test Conference, pp. 260–269, Oct. 2002

22. International Technology Roadmap for Semiconductors [Online], Available: http://public.itrs.net.

23. M. Inoue, T. Yamada, and A. Fujiwara, ''A New Testing Acceleration Chip for Low-Cost Memory Test,'' IEEE Design and Test of Computers, vol. 10, pp. 15–19, Mar. 1993.

24. J. Khare, and W. Maly, From Contamination to Defects, Faults and Yield Loss, Boston: Kluwer Academic Publishers, 1996.

25. J. Kilby, "Semiconductor Device-and-Lead Structure," U.S. Patent no. 2,981,877, 1959.

26. J. P. M. van Lammeren, "I_{CCQ}: A Test Method for Analogue VLSI Based on Current Monitoring," IEEE International Workshop on I_{DDQ} Testing, pp. 24–28, 1997.

27. J. E. Lilenfeld, "Method and apparatus for controlling electric currents", U.S. Patent no. 1,745,175, 1926.

28. P. Maxwell, P. O'Neill, R. Aitken, R. Dudley, N. Jaarsma, M. Quach, D. Wiseman, "Current ratios: a self-scaling technique for production I_{DDQ} testing," Proceedings of the IEEE International Test Conference, pp. 1148–1156, 2000.

29. P. Mazumder, and K. Chakraborty, Testing and Testable Design of High-Density Random-Access Memories, Boston: Kluwer Academic Publishers, 1996.

30. S. D. McEuen, ''I_{DDQ} Benefits,'' Proceedings of the IEEE VLSI Test Symposium, pp. 285–290, 1991.

31. E.J. McCluskey, and F. Buelow, ''IC Quality and Test Transparency,'' Proceedings of the IEEE International Test Conference, pp. 295–301, 1988.

32. A. C. Miller, "I_{DDQ} Testing in Deep Submicron Integrated Circuits," Proceedings of the IEEE International Test Conference, pp. 724–729, 1999.

33. G. Moore, "Cramming More Components into Integrated Circuits," Electronics, vol. 38, no. 8, Apr. 1965.

34. B. Mustafa Pulat, and L. M. Streb, ''Position of Component Testing in Total Quality Management (TQM),'' Proceedings of the IEEE International Test Conference, pp. 362–366, 1992.

35. R. Perry, ''I_{DDQ} testing in CMOS digital ASICs,'' Journal of Electronic Testing: Theory and Applications, vol. 3, pp. 317–325, Nov. 1992.

36. J. Rajski, "DFT technology for low cost IC manufacture test", Electronic Engineering Design, pp. 21, Jun. 2002.

37. M. Rencher, "Yields can be improved via design techniques", EE Times special publication: "What's Yield Got To Do With IC Design?," 2003.

38. J. Roberts, and S. Burke, "Intel to take $475M charge to cover costs in Pentium recall", Computer Reseller News, iss. 614, pp. 135, Jan. 1995.

39. M. Sachdev, ''A Defect Oriented Testability Methodology for Analog Circuits,'' Journal of Electronic Testing: Theory and Applications, vol. 6, no. 3, pp. 265–276, Jun. 1995.

40. M. Sachdev, ''Reducing the CMOS RAM Test Complexity with I_{DDQ} and Voltage Testing,'' Journal of Electronic Testing: Theory and Applications, vol. 6, no. 2, pp. 191–202, Apr. 1995.
41. M. Sachdev, "Deep Sub-micron I_{DDQ} Testing: Issues and Solutions," Proceedings of the European Design and Test Conference, pp. 271–278, 1997.
42. M. Sachdev, P. Janssen, and V. Zieren, "Defect detection with transient current testing and its potential for deep sub-micron CMOS ICs," Proceedings of the IEEE International Test Conference, pp. 204–213, 1998.
43. M. Sachdev, "Current-Based Testing for Deep-Submicron VLSI", IEEE Design and Test of Computers, pp. 77–84, Mar. 2001.
44. A. Schafft, D. A. Baglee, and P. E. Kennedy, ''Building-in Reliability: Making it Work,'' Proceedings of the IEEE International Reliability Physics Symposium, pp. 1–7, 1991.
45. W. Schockley, "The Theory of pn-junction in Semiconductors and pn-junction Transistors," BSTJ, vol. 28, p. 435, 1949.
46. O. Semenov, A. Vassighi, M. Sachdev, A. Keshavarzi, and C. F. Hawkins, "Effect of CMOS technology scaling on thermal management during burn-in", IEEE Transactions on Semiconductor Manufacturing, vol. 16, iss. 4, pp. 686–695, Nov. 2003.
47. S.C. Seth, and V.D. Agrawal, "Characterizing the LSI Yield Equation from Wafer Test Data," IEEE Transactions on Computer-Aided Design, vol. CAD-3, no. 2, pp. 123–126, Apr. 1984.
48. J.M. Soden, C.F. Hawkins, R.K. Gulati, and W. Mao, ''I_{DDQ} Testing: A Review,'' Journal of Electronic Testing: Theory and Applications, vol. 3, pp. 291–303, Nov. 1992.
49. M. Syrzycki, ''Modeling of Spot Defects in MOS Transistors,'' Proceedings of the IEEE International Test Conference, pp. 148–157, 1987.
50. E. Takeda et al., ''VLSI Reliability Challenges: From Device Physics to Wafer Scale Systems,'' Proceedings of the IEEE, vol. 81, no.5, pp. 653–674, May 1993.
51. "Test and Test Equipment", 2003 International Technology Roadmap for Semiconductors [Online], Available: http://public.itrs.net/Files/2003ITRS/Test2003.pdf.
52. F. Wanlass, "Low Stand-By Power Complementary Field Effect Circuit," U.S. Patent no. 3,356,858, 1963.
53. F. Wanlass, and C. Sah, "Nanowatt Logic Using Field-Effect Metal-Oxide Semiconductor Triodes," ISSCC Digest of Technical Papers, pp. 32–33, February 1963.
54. P. Wiscombe, ''A Comparison of Stuck-At Fault Coverage and I_{DDQ} Testing on Defect Levels,'' Proceedings of the IEEE International Test Conference, pp. 293–299, 1993.
55. T. W. Williams, R. H. Dennard, and R. Kapur, "Iddq Analysis: Sensitivity Analysis of Scaling," Proceedings of the IEEE International Test Conference, pp. 786–792, 1996.

56. T. W. Williams, R. Kapur, and M. R. Mercer, "Iddq Testing for High Performance CMOS – The Next Ten Years," Proceedings of the European Design and Test Conference, pp. 578–583, 1996.
57. P. Varma, A.P. Ambler, and K. Baker, "An Analysis of the Economics of Self-Test," Proceedings of the IEEE International Test Conference, pp. 20–30, 1984.

Chapter 2

FUNCTIONAL AND PARAMETRIC DEFECT MODELS

Defects are undesired features in the silicon layer structure of an IC. They can take form of missing or extra pieces of material, as well as of random and systematic shifts in the outcome of the semiconductor process. This chapter addresses the modeling of such defects and their impact on the behavior of the circuit. Particular attention is given to parametric and catastrophic faults. Although unusual, delay and leakage are considered examples of parametric faults. This stems from the importance that process variability has on deep submicron technologies and the role that low-power circuits have in today's consumer electronic applications. With the very high transistor densities accomplished in modern technologies, a quick estimation of the IC robustness against spot defect is needed. Many of the yield losses arise from the interconnect which is susceptible to shorts and opens. This chapter also builds from fundamentals of layout critical areas to the estimation of the layout's defect sensitivity of a complex IC to evaluate catastrophic faults.

1. BRIEF CLASSIFICATION OF DEFECTS

Defects can be classified as local or global. The latter classification concerns disturbances that affect complete regions of a wafer, while the local class concerns random regional disturbances within an IC. Under the global class one encounters defects like over(under) etching, mask misalignments, non-uniformity of critical dimensions, shifting of dopants, etc. The local class deals primarily with spot defects. Spot defects are local disturbances of the silicon layer structure caused by dust, process variations, etc. The general assumption is that spot defects are in essence random phenomena occurring with certain stochastic frequency and size, and a certain stochastic spatial distribution [39, 40, 41]. Not all defects are due to lithographic processing steps. Some defects arise from process variability such as incomplete step coverage. Therefore the way in which individual process steps are executed is of critical importance to the outcome of the IC. Table 2-1 shows a brief classification of spot defects. A sample of defect types is illustrated in Figure 2-1 as well.

The impact that a defect has on circuit behavior is rendered as a fault. Faults in turn can be classified as catastrophic, or parametric. A fault is catastrophic when the functional behavior of the IC is incorrect. On the other hand, parametric faults are those faults for which the IC is functional but it fails to meet its specifications, e.g. timing, power budget, leakage, etc.

Table 2-1. Brief classification of defects.

	Description
Type 1	Round or long shaped bubble
Type 2	Big particles
Type 3	Flakes
Type 4	Shaped defects
Type 5	Long defects
Type 6	Resist residues
Type 7	Fallen landing pad
Type 8	Shallow irregular dips
Type 9	Big porous particles
Type 10	Irregular shapes
Type 11	Very small particles on top of metal

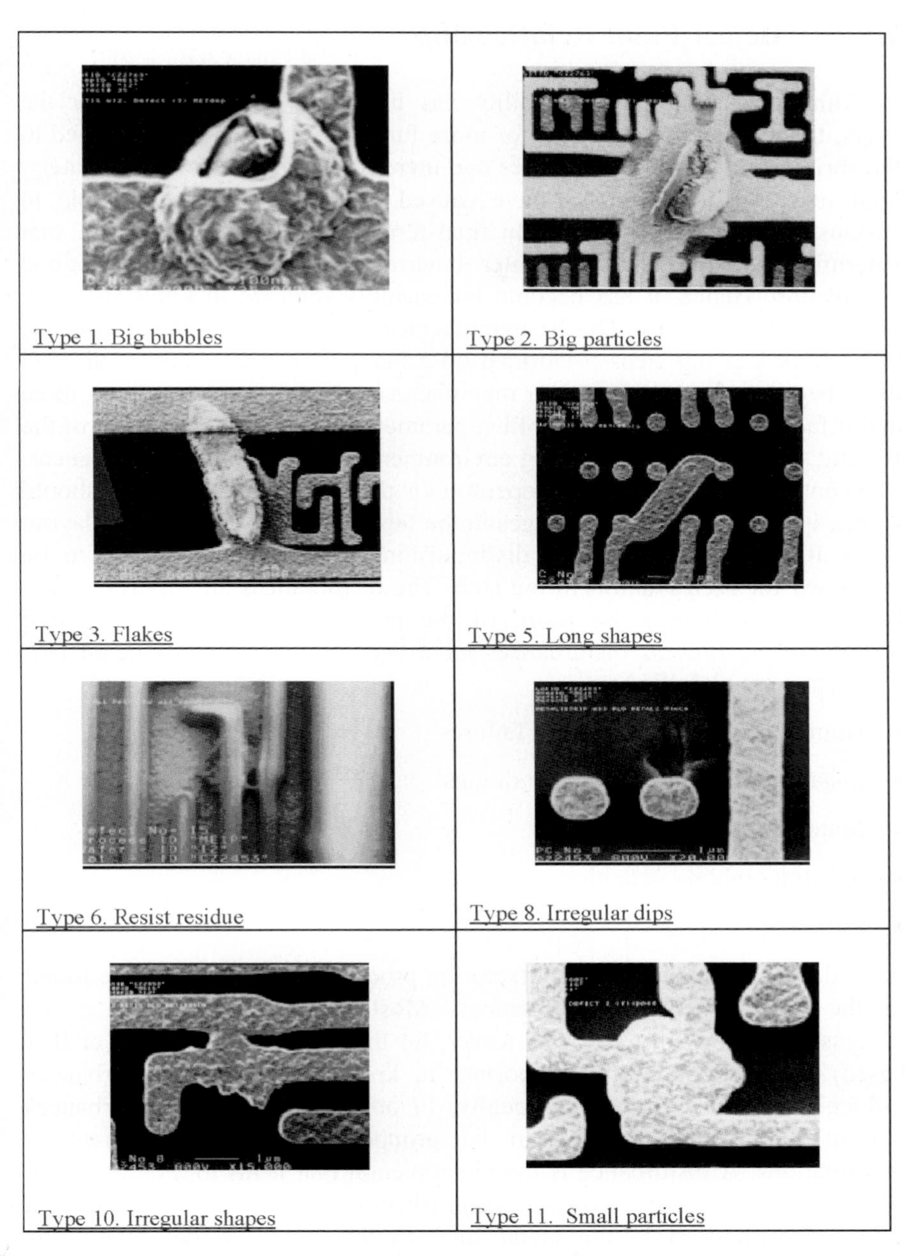

Figure 2-1. Various spot defects.

1.1 Defect-Fault Relationship

Although the process capability has improved dramatically over the years, the ever increasing quest for more functions on a single IC has led to the shrinkage of device geometries and increase in chip area. Unfortunately, both of these developments have caused ICs to become susceptible to various yield loss mechanisms. In final terms, it is the yield of an IC that determines whether or not greater integration is an economically good proposition. Hence, it has become increasingly relevant to know different yield loss mechanisms. The IC manufacturing process involves a sequence of basic processing steps performed on a batch of wafers. Maly et al. [21] described that the outcome of a manufacturing operation depends on three major factors: the process controlling parameters or control, the layout of the IC, and some randomly changing environmental factors, called disturbances. The control of a manufacturing operation is the set of parameters that should be manipulated for desired changes in the fabricated IC structure. The layout of an IC is the set of masks distinguishing IC areas which need to be processed for each manufacturing step. The disturbances are environmental factors that influence the result of the manufacturing operation. These manufacturing process disturbances have been studied in great detail [17, 21,42] and are classified as:

- Human errors and equipment failures
- Instabilities in the process conditions
- Material instabilities
- Substrate inhomogeneities
- Lithography spots

A detailed treatment of manufacturing process disturbances can be found in the above mentioned references. Most disturbances influence the processed topology of the IC. However, for the purpose of realistic (or IFA based) fault modeling it is important to know that not all disturbances influence the IC performance equally. In other words, these disturbances deform the IC and hence can be grouped according to classes of deformations. A disturbance is the phenomenon that leads to a deformation in an IC. For example, a contamination (disturbance) on the wafer causes a break (deformation) in the metal line. In this case the deformation is geometrical in nature. Similarly, a poor temperature control (disturbance) during the growth of gate oxide may results in lower threshold voltage (electrical deformation). In general, all process disturbances can be classified into geometrical and electrical deformations [21].

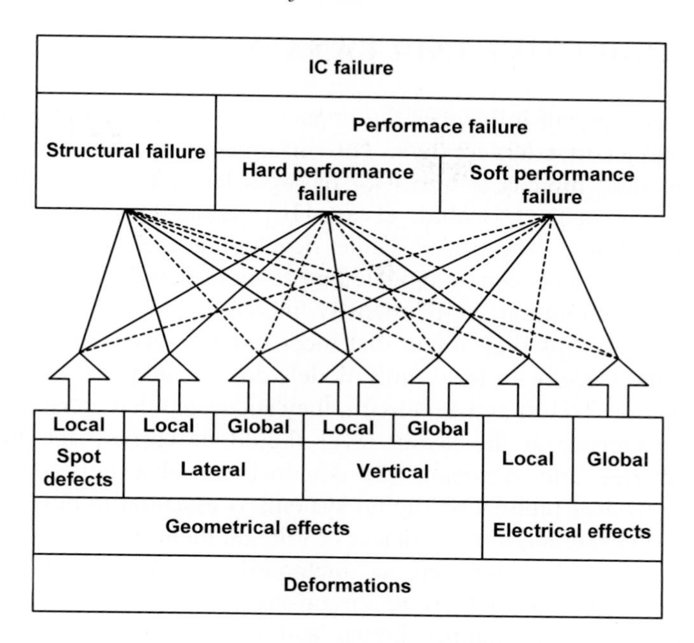

Figure 2-2. IC manufacturing process deformations and their relationship with IC faults [21].

Figure 2-2 shows this classification and its relationship with different IC failures. The lower half of the figure shows the classification of physical phenomena that cause yield loss and the upper half of the figure shows the fault classification (structural and performance faults). Geometrical and electrical deformations have local as well as global influences on IC function and/or performance. A global influence occurs when a particular parameter, say the transistor threshold, is affected over the complete wafer. The term local is used when the influence on the parameter is limited to a region smaller than a wafer. Often these local deformations are called defects like break and short in conductors. In addition, spot defects that are primarily lithographic in nature form a part of geometrical deformations. In principle, each class of physical deformation is capable of causing a variety of faults. However, some are more likely (solid lines) than others. For example, all global effects are more likely to cause soft performance failures. Similarly, spot defects are more likely to cause structural or hard performance failures. Since, the impact of global deformations affects the complete or a large part of a wafer, they are quickly detected by test structures designed for them. Furthermore, in a well controlled fabrication environment such problems are kept under control. Therefore, for IFA based fault modeling and testing, only the local deformations or defects are taken into account.

2. INDUCTIVE FAULT ANALYSIS

The circuit layout influences the impact of a defect, and thus the faulty circuit behavior to a large extent. This information is often ignored while developing fault models at transistor or logic level. In one of the earliest papers on the subject, Galiay et al. [16] pointed out that the layout information should be utilized while developing the fault model. They argued that all failures can be modeled by stuck-at-0 and stuck-at-1, will not be a sound assumption as IC density increases. In a study over 4-bit microprocessor chips they found that most failures were due to shorts and opens. Furthermore, gate level fault models do not adequately represent the faulty behavior. To test a given circuit, they suggested that an analysis of the failure mechanisms at the layout level should be carried out. In another study, Banerjee and Abraham [2] demonstrated that understanding the effects of physical failures on digital systems is essential to design tests for them and to design circuitry to detect or tolerate them. In yet another study, Maly, et al. [34] proposed a methodology of mapping physical manufacturing process defects to circuit-level faulty behavior caused by these defects. In this manner layout and technology specific faults are generated and ranked according to their likelihood. One conclusion of their analysis is that manufacturing process defects can give rise to a much broader range of faults than can be modeled using single line stuck-at fault model. Similarly, other studies point out that a lot of information can be extracted from the layout [10, 24, 7].

These early publications [10, 24, 7, 33] on the subject formed the basis of what is known today as the inductive fault analysis (IFA). The IFA differentiates itself from the conventional fault modeling approaches of assuming faulty behaviors on interconnects or logic gates. It derives the circuit or logic-level fault model starting from particular physical defects. In other words, a higher- level fault model is formulated by examining defects at lower level or defects are induced by simulation of the defect creation process. Hence the word "inductive" which means the higher-level fault information is induced from lower level defects. Often IFA is referred to as realistic fault analysis and fault modeling based on IFA is referred to as realistic fault modeling. The term realistic signifies that each fault has a physical basis, i.e., a defect. A test approach based on IFA is also referred to as defect-oriented testing.

In order to fully exploit the potential of IFA, it is important to understand the relationship between manufacturing defects and IC faults. For example, many defects that influence the IC performance are tested before the functional testing. Therefore, they are not included in IFA based test generation. Similarly, manufacturing defects in an IC and their impact on performance is strongly influenced by the IC design and layout. IFA can also be exploited to find out areas in the design that are difficult to test. Using this information the design robustness and yield can be improved. In the following sub-sections, an overview of defect-fault relationship and IC design and layout related sensitivity will be given.

2.1 IC Design and Layout Related Defect Sensitivity

The design of a modern IC is a complex task. Often designs tend to exploit the very maximum of what a manufacturing process can offer. Issues like time to market, time to profitability, and reduced product life cycle further complicate the decisions in the design. Higher operational frequency, higher complexity, smaller area, and lower power consumption usually are the design objectives. Often these objectives are in conflict with each other and are rarely in agreement with what is known as robust design practices. Typically, it takes a lot of product engineering and a couple of design iterations to stabilize the design for reasonably good yield. In spite of all this effort, it has been observed that yields of certain designs, for a given chip area, are lower compared to those of others. Design related sensitivity to yield can be divided into two major classes that are further divided into subclasses.

2.2 Defect Sensitive Design

The operational frequency of a digital IC is determined by its critical path. A critical path is a data path in an IC with the largest delay. For a correct operation of the chip, the critical path delay should be less than the clock period. However, the actual delay of the critical path is governed by the process. The design margin between the critical path and the clock frequency should be reasonable, otherwise parametric process variations or spot defects may result into timing (parametric) failures. Furthermore, how much physical area does the critical path have on the chip is also an important issue because higher area will increase the probability of a defect landing onto the critical path. An otherwise innocuous defect in the critical path may increase its parasitic capacitance and/or resistance resulting in a timing failure. Similarly, the logic implementation also has an influence over the timing related sensitivity of the design. For example, the timing critical design aspects are much higher in the dynamic logic implementation compared to the static logic implementation. Furthermore, in the dynamic logic implementations, often logic levels are defined under the dynamic conditions. Many defects influence the dynamic behavior and hence cause failures. On the other hand, impact of such defects on the static logic is not so severe, but they cause increased delay. If this delay is not in the critical path, it may not lead to a failure. Moreover, the type of synchronous logic implementation, random logic, on board memories and PLA, all have yield related repercussions because they have different inherent sensitivities to defects.

2.3 Basic Concepts of IFA

IFA is a systematic approach for determining what faults are likely to occur in a VLSI circuit. This approach takes into account the technology of the implementation, the circuit topology and the defect statistics of the fabrication plant. Shen et al. [34] formalized concepts of IFA as follows: IFA is a systematic procedure to predict all the faults that are likely to occur in a MOS integrated circuit or sub-circuit. The three major steps of the IFA procedure are; (1) generation of physical defects using statistical data from fabrication process, (2) extraction of circuit level faults caused by these defects, and (3) classification of fault types and ranking of faults based on their likelihood of occurrence.

The major steps of the IFA are shown in Figure 2-3. The circuit layout and the manufacturing defect statistics form inputs to the analysis. The IFA methodology takes into account only local deformations or defects. The size and the probability of defects is defined by the defect density distribution (DDD). For an IC manufacturing fab, DDDs are normally available for each layer and defect types. DDDs define how large the probability of a defect in a certain layer is with respect to other layers, and how the probability of defects in a layer depends on the size of the defect. For a typical double metal single poly CMOS process, these defects include:

- Extra and missing material defects in conducting and semiconducting layers
- Presence of extra contacts and vias
- Absence of contacts and vias
- Thin and thick oxide pinholes
- Junction leakage pinholes

The defects are sprinkled onto the layout in a random manner. For this purpose, a CAD tool like VLASIC (Vlsi LAyout Simulation for Integrated Circuits) or DEFAM (DEfect to FAult Mapper) is utilized. The defects are modeled as absence or presence of material on the layout. Only a small subset of all defects cause a change in the circuit connectivity. For example, a short is created between two nodes or an open defect causes a break in the connectivity. These defects are extracted and their impact on circuit behavior is modeled at an appropriate level of abstraction for fault simulation. Subsequently, the abstracted defects (i.e., faults) are simulated with given test stimuli. The fault simulation information is exploited for providing DfT solutions, building fault tolerance into the circuit, etc. Finally, simulation results are verified by the silicon data.

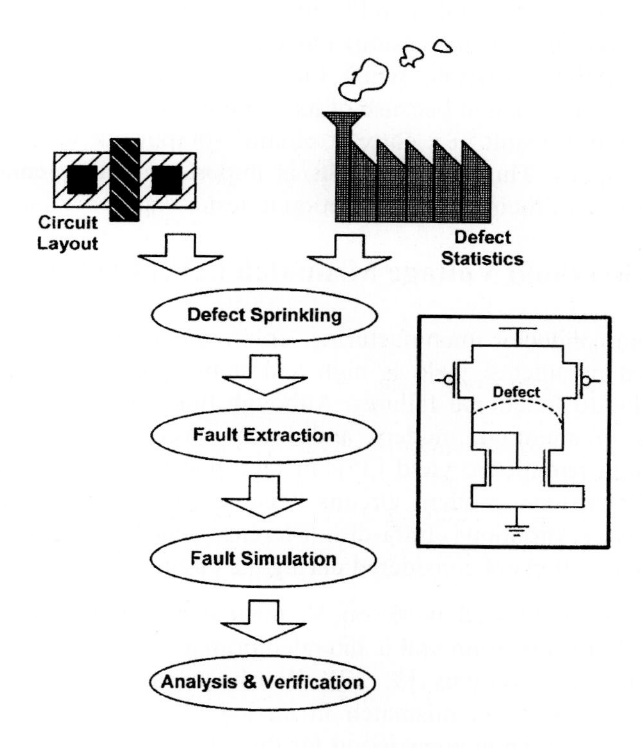

Figure 2-3. A graphical representation of IFA.

The IFA technique takes into account only a subset of process defects, namely local defects. The global defects or deviations cause widespread failures or near failures. Since, the impact of global defects is present over a relatively large area, these are detected rather quickly. Moreover, there are special test structures to test for these erroneous conditions. Hence, such defects are detected early in the manufacturing process before the functional/structural testing. Furthermore, in a well controlled and monitored process, major process errors causing global defects are relatively easily detected and solved.

3. PARAMETRIC DEFECT AND FAULT MODELS

As was previously mentioned, modern deep submicron circuits are affected not only by spot defects but also by process variations [21]. Process variations can have both global and local effects on an IC. The interconnect resistance increasing (or decreasing) radial gradient across the wafer is an example of global variation, while threshold voltage mismatch can be considered a local variation because of its random nature. Faults arising from process variability result in a correfct circuit's output but without meeting performance specs. This type of faults is important to the semiconductor industry because of their direct association to test escapes.

3.1 Threshold Voltage Mismatch (ΔV_t) Fault Modeling

In a semiconductor manufacturing environment it is conventionally assumed that parametric yield is high and stable and that the main yield losses are due to functional failures. Although functional yield remains the main focus of attention, modern and future circuits may not have the presumed high parametric yield [35]. In fact, due to the use of submicron transistor dimensions, modern circuits become quite sensitive to intra-die (process) device variations. Intra-die differences, such as random local V_t fluctuations are often not considered during the circuit design process.

Most of the published work on V_t mismatch consists of device and technology characterization and is intended primarily for analog and mixed-signal circuits and systems [9, 27]. Worth stressing is that while the detrimental effects of V_t mismatch in analog circuits are well described [26,28], little is known in comparison for digital circuits. Previous works on digital-design V_t-variability focus primarily on global effects [3, 4, 19], i.e. they consider a common augmenting or decreasing value of the nominal V_t. This means that fundamental device limitations such as Vt differences (mismatch) among transistors are not accounted for [6, 14, 18, 20, 25]. A statistical approach for intra-die V_t variability for critical-path delay analysis was performed in [8] and a study of the impact of process variability on leakage current levels was done in [11].

As in deep-submicron technologies the power supply voltage decreases, the effect of the variation of some transistor properties become more important. In this section, the effect of the V_t mismatch on a 0.18um CMOS technology is presented. The reason why threshold voltage was chosen as a parameter for study is because this parameter captures in essence many of the variations that occur during processing, e.g. variations in dopant concentration, gate oxide thickness, gate length critical dimension, etc. The impact on both static and dynamic behavior is evaluated in terms of quiescent current consumption and timing effects, respectively.

3.2 Sources of Threshold Voltage Variability

Essentially, it is possible to identify two sources of V_t variations. One of them due to *global* manufacturing variations and the other due to *local* random fluctuations. In long channel transistors V_t variations are mainly due to *global* variations on dopant diffusions, ion implantation, gate dielectric thickness, etc. For short channel transistors, on the other hand, V_t variability is predominantly determined by the transistor's geometries, in particular L_{eff}. This can be considered a global variation as well.

Let us now consider a pair of transistors. Experimental results have shown that there can be random V_t (mismatch) differences between two closely placed "identical" transistors in the order of about 100μV to 10mV. This is a *local* effect that becomes extremely important in deep submicron transistors as the mismatch magnitude is generally observed to be inversely proportional to the square root of the transistor's area. This local effect is random and is due among other things to a statistical distribution of dopant atoms per unit area. A simple rule of thumb to estimate threshold voltage mismatch is to assume a variation of 1mVμm of square root of active area per nanometer of gate oxide thickness. For a technology with 4nm of oxide thickness and minimum channel length and width of 0.18μm this corresponds to a threshold voltage standard deviation of approximately 22mV. Experimental measurements of threshold voltage due to (global) inter-die variations show that it can be considered as normally distributed with a mean μ_G and standard deviation σ_G. Threshold voltage mismatch observations based on transistor pairs can be described as well with a normal distribution with mean μ_Δ and standard deviation σ_Δ, see Figure 2-4.

If we assume that global and intra-die variations are independent, it follows then that the total V_t variation of a single transistor can be calculated as [27]

$$(\sigma V_t)^2 = (\sigma_G)^2 + 0.5(\sigma_\Delta)^2 \qquad (2.1)$$

with

$$\sigma_{\Delta} = \frac{A_{VT}}{\sqrt{WL}} \tag{2.2}$$

where A_{VT} is a technology conversion constant (in mVμm), and WL denotes the product of the transistor's active area. The 0.5 factor in (2.1) arises because only one transistor from the pair is considered at a time. The remaining analyses are based on a 0.18μm CMOS process that yields an AV_t value of 7mVμm for the NMOS transistors provided they do not operate in the subthreshold regime. Thus, for NMOS transistors with minimum dimensions, e.g. W/L = 0.28μm/0.18μm, the estimated statistical V_t mismatch is σ_{Δ}31.18mV. Worth pointing out is that σ_G is of the order of 30mV. It follows then that threshold voltage mismatch is very important for deep submicron technologies.

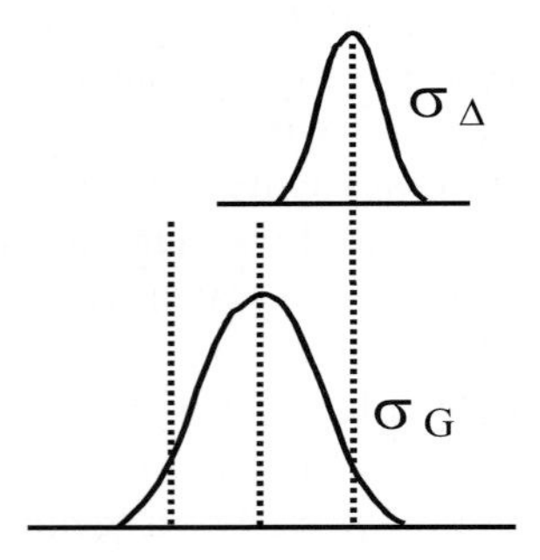

Figure 2-4. Global (systematic) and local (random) variability of the threshold voltage. The resulting V_t is obtained after the addition of both contributions.

3.3 Leakage Current due to V_t Mismatch

Let us now investigate the role of threshold voltage mismatch on the off-state current of a typical digital CMOS cell. Assume without loss of generality a cell with two inputs. Let us further assume a local random variation in intrinsic V_t's, e.g. one arising from fluctuation of random dopant diffusion, dopant clustering, or interface states, etc. Let us assume V_{t1} and V_{t2} as the threshold voltages of two transistors. Since $V_{t1} \neq V_{t2}$ their V_t mismatch and average values can be expressed as in (2.3) and (2.4) respectively.

$$\Delta V_{T21} = V_{T2} - V_{T1} \tag{2.3}$$

$$V_{To} = \frac{V_{T1} + V_{T2}}{2} \tag{2.4}$$

After some algebraic manipulation it is possible to express individual V_t's as a function of their mismatch and average values as indicated in (5) and (6)

$$V_{T1} = V_{To} - \frac{\Delta V_{T21}}{2} \tag{2.5}$$

$$V_{T2} = V_{To} + \frac{\Delta V_{T21}}{2} \tag{2.6}$$

Let us consider two stacked transistors with V_1 and V_2 being their gate voltages and consider the following cases: 1) $V_1 = 0, V_2 = 0$; 2) $V_1 = V_{dd}, V_2 = 0$; and 3) $V_1 = 0, V_2 = V_{dd}$. Figure 2-5 shows simulated results of the off-state current as a function of V_t mismatch obtained through PSTAR which is an internal SPICE-like simulator that uses Philips' MOS-9 compact transistor model. Transistor sizes used in the simulation were W/L = 1.4μm/0.18μm.

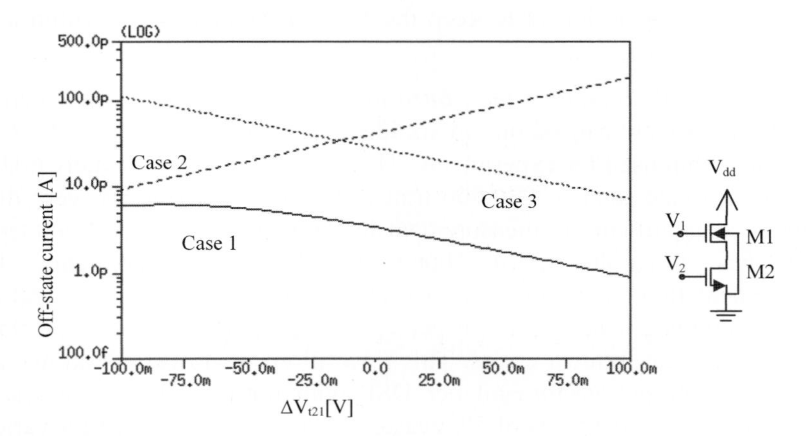

Figure 2-5. Leakage current as a function of V_t mismatch.

Figure 2-5 shows the known behavior of many digital cells, i.e. there is an obvious dependence of the leakage current on the cell's inputs. For nominal conditions, e.g. $\Delta V_t = 0$, the input state that shuts off all the transistors of a digital cell is the one that renders the lowest leakage current; for other input states the off.state current level is simply higher. This plot shows the dependence of the leakage current on the V_t mismatch between the cell's transistors. The increase in current for increasing ΔV_{t21} (and starting from a negative mismatch) is because we assumed that V_{t2} and V_{t1} have positive and negative mismatches as in eqs. (2.5, 2.6), respectively. A reverse behavior would have been observed if we had assumed the opposite for V_{t2} and V_{t1}. In the presence of mismatch, i.e. $\Delta V_t \neq 0$, the cell's leakage current has a significant spread of approximately 1.5 orders of magnitude. This situation is not uncommon in modern digital ICs since a voltage mismatch of $|\Delta V_t|$ = 90mV amounts to an expected $3\sigma_\Delta$ tolerance window from the fabrication process and this is just considering only intra-die variations. Needless to say, the larger this current is the larger the chip's power consumption will be.

Statistical simulations were carried out to verify the impact of other parameters on the off-state current based on the same circuit for W/L = 1.4μm/0.18μm. These simulations involve parameter variations of up to 4σ of V_t, L, β and some sheet resistances. Results are shown in Figure 2-6. Since the scatter plots follow the trends of Figure 2-6 one can conclude that V_t variability is indeed a dominant factor. The histograms show the distribution of off-state currents; worth noticing is a common current spread of about 40% with respect to the mean value. These large spreads make it more difficult for designers to keep the IC performance target within a 3σ tolerance window.

Experimental results on intra-die variation. Leakage current measurements were carried out on six identical DSP cores named (A, B, … F) within a chip used for experiments. The DSPs have approximately 60,000 gates for a round total of 240,000 transistors. The experiments were done using the standard current measurement features of the Agilent 93000 tester which has a resolution of 100nA for 100uA measurements. The measurement flow is simple: For each DSP apply 75 input vectors and log the corresponding leakage current per vector; repeat the procedure 50 times to minimize measurement errors. This flow allows us to extract an average leakage current per vector and per DSP with minimized instrumentation error. Two independent sets of 50 measurements were performed for various dies on the wafer. Figure 2-7a shows the normalized leakage current, sorted in ascending order, per vector and per DSP. The normalization is with respect to the average leakage current per DSP.

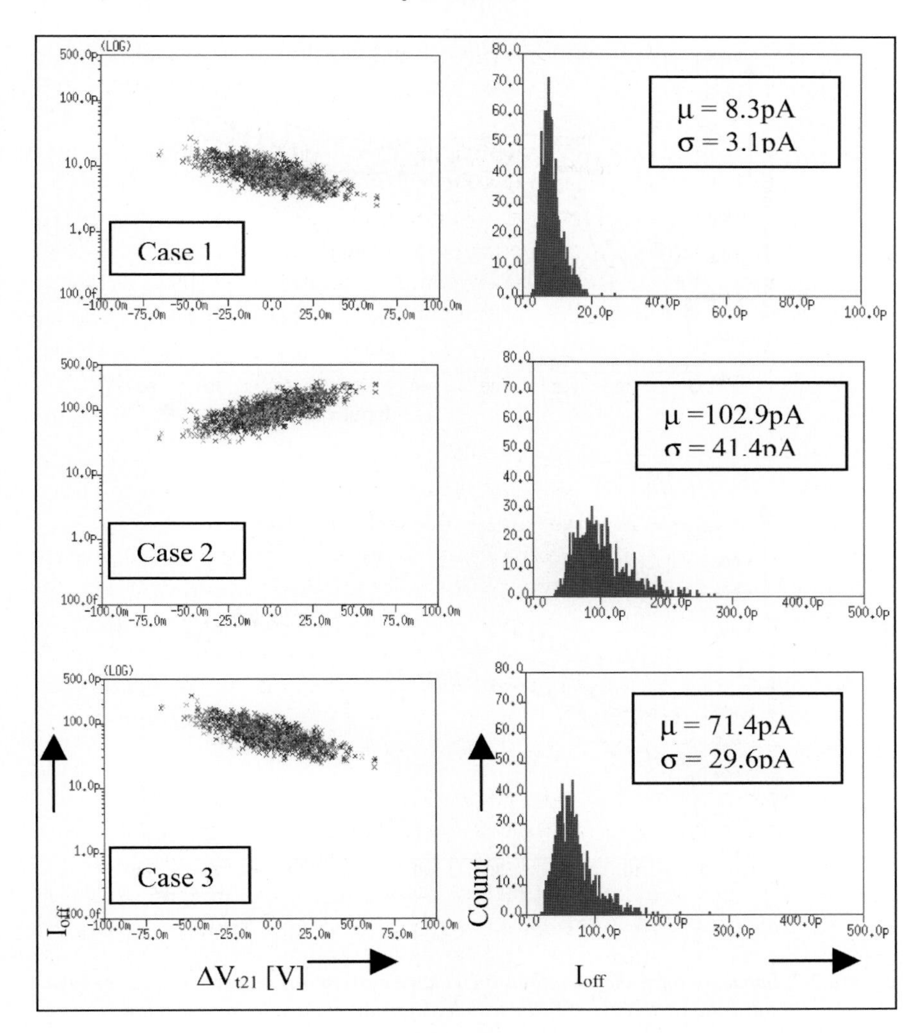

Figure 2-6. Monte Carlo simulations to estimate intra-die off-state current. Scatter plots present points for off-state current vs. V_t mismatch. Histograms show the corresponding distribution of off-state currents.

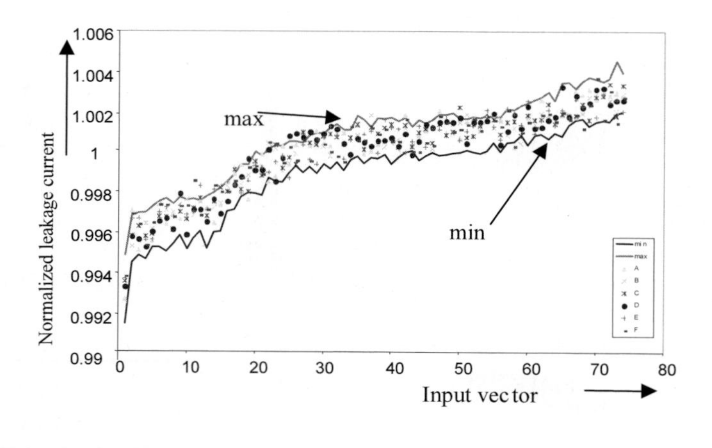

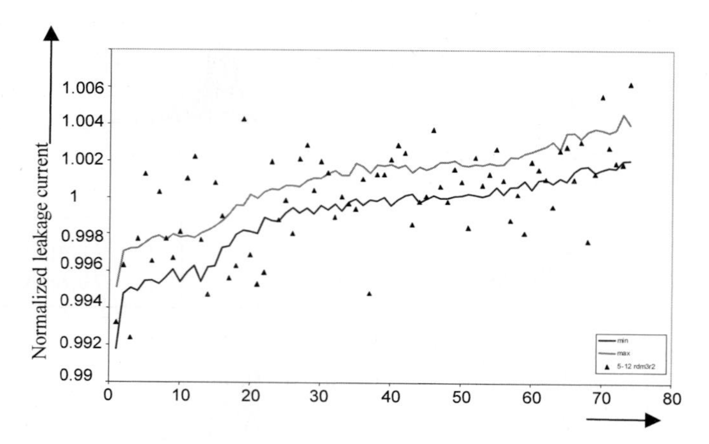

Figure 2-7. Impact of intra-die variations on leakage current. (a) expected correct behavior (b) intra-die variations of DSPs with leakage out of the tolerance window.

The *min* and *max* boundaries were obtained from measuring four fault-free distinct dies. From this figure one can observe the state dependence of the leakage current, e.g. a different current for each input vector, and a "tolerance window" with the intra-die leakage variations of the DSPs. The average width of this window is 0.18% of the normalized average current. Figure 2-7b depicts a die for which, possibly, an excessive V_t mismatch is present in various cells giving origin to higher leakage currents. This excess can be attributed to a mismatch in the cells, such as the one hereby described, and to a mismatch from cell to cell [29]. The excess in current is small enough to discard the possibility of low-resistance shorts although not necessary the presence of very high resistance shorts, e.g. > $10^8\Omega$. Observe namely that the magnitude of the outlying points is comparable to the width of the tolerance band. One can also notice points above and below the tolerance band. This can be explained by recalling that the off state current of a mismatched cell increases or decreases depending upon the V_t mismatch shift and the input state. Let us carry out some simple calculations to infer the number of mismatched cells that would result in a current out of the tolerance window. Let us assume a normalized average leakage current per cell equal to 1/60000. Let us also assume that there is a group of cells that have a 3σ deviation from their mean value, e.g. a hundredth of the average value, and also that they have the same input state. Let us now consider an outlying point with a normalized current of 1.004. A simple arithmetic calculation results in 239 gates with mismatch. This can be interpreted as an equivalent of 239 cells whose effect did not average out with all other cells and that yielded a current deviation of 0.004. The calculated number of cells actually defies the law of probabilities in the sense that if a *strict* normal distribution of ΔV_t is considered along with a *strict* uniform distribution of input states, then the number of cells equal or in excess of a 3σ deviation is only 162. Nevertheless, the doubt to consider is the possibility of non strictly normal process parameter distributions, or a non-uniform switching state of the cells.

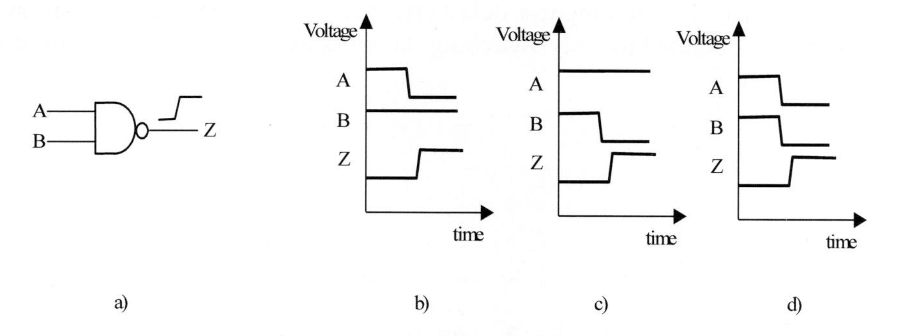

Figure 2-8. Input combinations to create a low-to-high transition at the output of a 2-input NAND gate.

3.4 Delay in Parallel-connected Networks

A 2-input NAND gate has been used as the vehicle of study. Three different input signal transitions have been considered as shown in Figure 2-8, namely, a high-to-low transition at the input *A*, a high-to-low transition at the input *B* and both transitions at the same time with no skew. The relative variation in the delay of the 2-input NAND gate for the same analysis is shown in Figure 2-9. The parallel network (PMOS-network in this case) plays a role in the low-to-high transition at the output of the gate. This plot depicts the difference between the expected (nominal) T_{PLHNOM} and the real delay value T_{PLH}.

To analyze this scenario let us consider a mismatch with a constant V_{tPB} and a varying V_{tPA} (see Figure 2-8). If only one input signal is being switched at a time, the change in delay ΔT_{pLH} depends only on the threshold voltage mismatch Δ_{VtP} of the PMOS transistor associated to the switched input. Assuming a fixed Vt_{PB}, the T_{pLH}-T_{pLHNOM} value for the *B* input transition case is a constant value (a horizontal line in Figure 2-9). Conversely, the T_{pLH}-T_{pLHNOM} behavior for a transition in the *A* input varies along the x axis; the delay increases if the P-MOSFET$_A$ transistor increases its V_{tPA} value, since it switches with a certain delay introduced by the increase in V_{tPA}.

The third situation that switches all (both) the inputs shows an intermediate variation since both transistors act during the transition of the gate. The absolute and/or relative change of the timing performance of a circuit binds the impact of the deviation from the nominal behavior. Based on the behavior illustrated in Figure 2-9 one can conclude that for a parallel-connected network of transistors and a given ΔV_{ti} for the *i-th* transistor (one of the transistors), the highest delay effect caused at the output transition of the gate is obtained by only switching the input associated to such transistor.

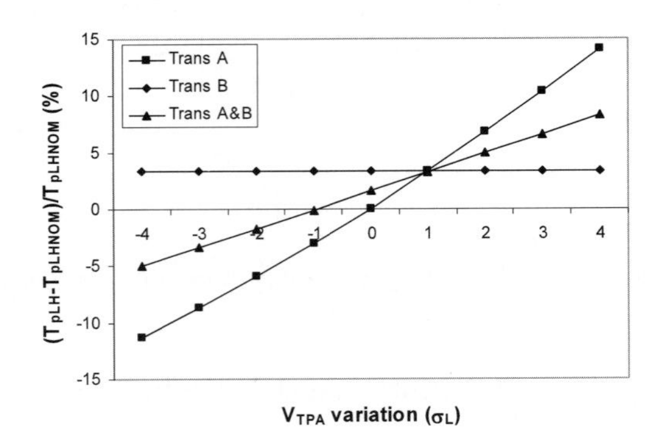

Figure 2-9. Delay for V_{tPB} constant and greater than the nominal value (Vt_{PBNOM}).

3.5 Delay Variation Model with ΔV_t for Parallel Transistor Networks

To predict the maximum delay variation caused in a CMOS gate by the V_t mismatch, the simple model shown in Figure 2-10 is used where K and α are constants of the technology and C_{out} models the output capacitance of the gate. The use of a Taylor expansion gives the following expression for the

$$\Delta T_{pLH} = \frac{K \cdot C_{out}}{W_{PMOS}} \left[\frac{\alpha}{\left(V_{DD} - V_{TNOM}\right)^{(\alpha+1)}} (\Delta V_T) + \frac{\alpha\ (\alpha+1)}{\left(V_{DD} - V_{TNOM}\right)^{(\alpha+2)}} (\Delta V_T)^2 \right]$$

delay variation of the gate.

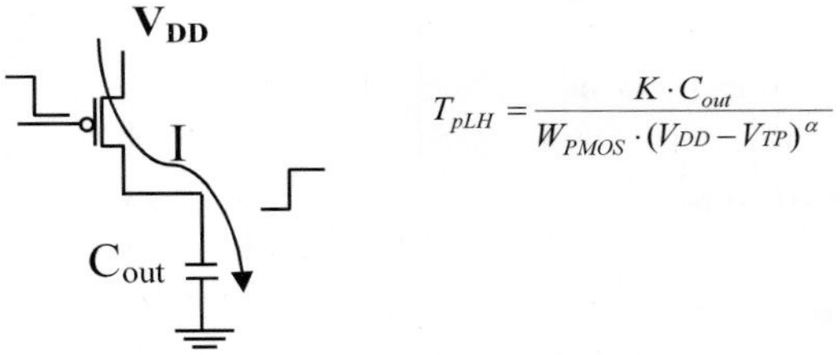

VDD

I

Cout

$$T_{pLH} = \frac{K \cdot C_{out}}{W_{PMOS} \cdot (V_{DD} - V_{TP})^{\alpha}}$$

Figure 2-10. Model for the delay variation due to the V_t mismatch.

A good agreement between the model and the simulation can be observed in the example illustrated in Figure 2-11 where the delay variation for several inverters with a number of different output loads (inverters) have been considered.

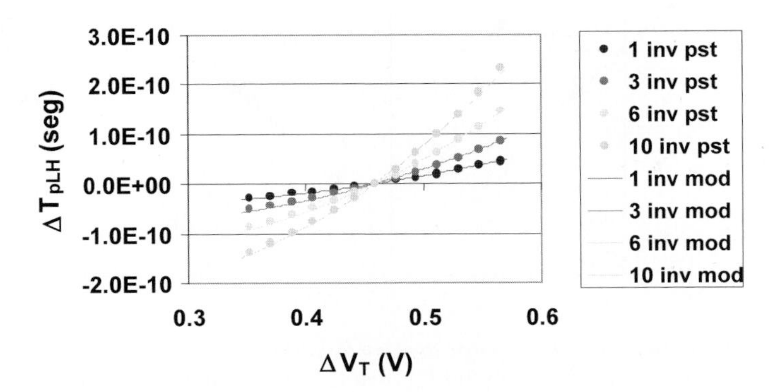

Figure 2-11. Transistor level simulations and the proposed model for the delay variation at the output of the NAND gate.

Table 2-2. High-to-low transition at the output of a 2-input NAND gate. Dependence on the initial capacitance state.

Case	Vector_1 (A, B) Z	Vector_2 A, B) Z	Vector_3 (A, B) Z	Vector_4 (A, B) Z	T_{pHL} (ps)
1	X	X	(0, 1), 1	(1, 1), 0	32.2
2	X	X	(1, 0), 1	(1, 1), 0	32,7
3	X	(1, 1), 0	(0, 0), 1	(1, 1), 0	40,1
4	(1, 1) 0	(1, 0), 1	(0, 0), 1	(1, 1), 0	42,9

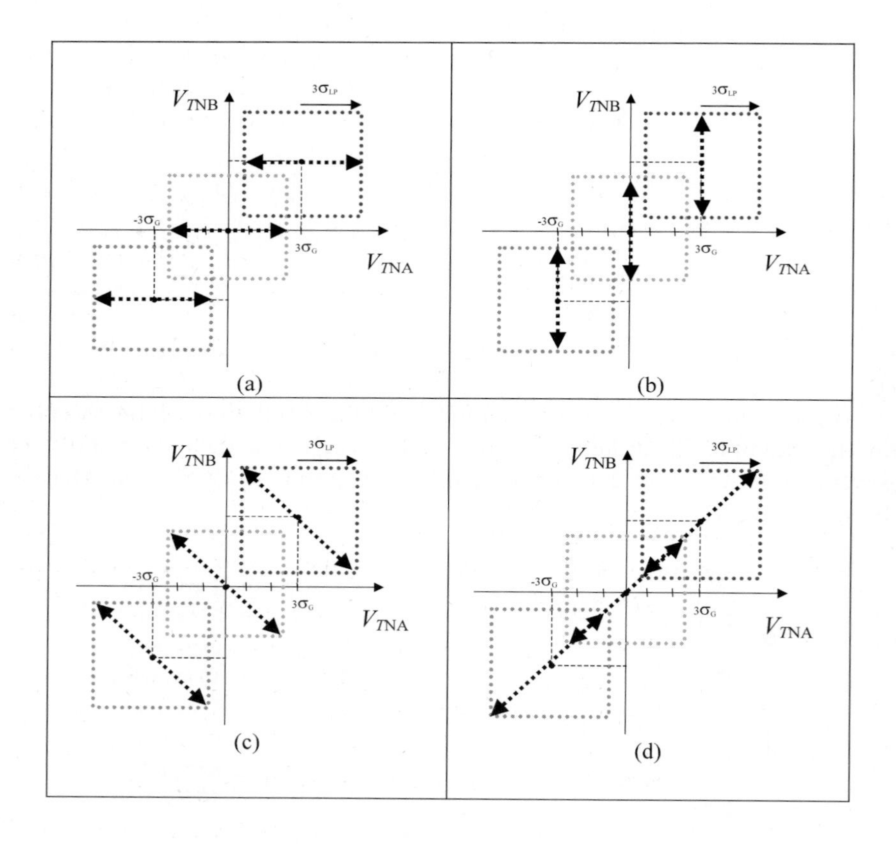

Figure 2-12. V_t mismatch analysis. (a) constant V_{tNB}, (b) constant Vt_{NA}, (c) V_{tNA} and V_{tNB} running in opposite directions, (d) V_{tNA} and V_{tNB} running in the same direction.

3.5.1 Delay in Series-connected Transistors

There are various ways of generating a high-to-low transition at the output of a 2-input NAND gate. One has to pay especial attention to the internal node capacitance between both NMOS transistors to account for node charging. For instance, assume that a target output transition is forced with a vector (A,B) = (1,1), then the internal node is discharged if the preceding vector was (A, B) = (1, 1); but if the preceding vector were (A, B) = (1, 0) the node is charged when the *A* and *B* transitions arrive.

A clear difference from the parallel-connected network case is the fact that the switching of more than one input at a time makes the response of the series-connected case slower. Indeed, in the case of parallel-connected transistors, the current flowing through each transistor is added to the rest helping to charge or discharge the output node. In the case of series-connected transistors, the same current flows through each one of the transistors.

If more than one transistor switches *on* at the same time, the output response takes longer than if only one does (and the rest are already *on*). In the former case, there is more than one transistor with an initial high resistance while in the latter only one transistor has an initial high resistance.

Let us assume that each Vector_*i* is applied at a previous time than Vector_*i+1*. A couple of delay times are shown in Table 2-2 to evaluate the influence of the initial state of the internal capacitance. Cases 1 and 2 correspond to a simple case when two input patterns are applied. Cases 3 and 4 show the effect when a sequence of input patterns is applied. The rightmost column shows the delay of the gate. Cases 3 and 4 are slower because two series-connected transistors are switching at the same time. Case 4 is even slower because the internal capacitance is initially (at the time when Vector_4 is applied) charged and, thus, has to be discharged.

There are four ways of analyzing the threshold voltage mismatch for the case studies explained above. They are shown in Figure 2-12. Each of the squares shows the space of V_t variation around a V_t centered at its nominal value or at either of its $\pm 3\sigma_G$ global values. Only V_t variations affecting the two NMOS transistors will be considered in this subsection.

In our analysis, four different ways of generating a high-to-low transition have been simulated for every sweep of points (V_{tNA}, V_{tNB}). They are the transition only at input *A*, the transition only at input B, the transition at both inputs but with initially discharged internal capacitance, and the transition at both inputs with charge in the internal capacitance. Let us first consider a case in which the threshold voltage of the $PMOS_B$ transistor (V_{tNB}) is kept constant although able to differ from the nominal value, see Figure 2-12a. Results for this case are shown in Figure 2-13.

The topmost set of points, labeled as "$3\sigma_G$", have an expression like $(V_{tNA}, Vt_{NB}) = (V_{tNA}, V_{tNB(NOM)} + 3\sigma_G)$, the second set labeled as "$0\sigma_G$" like $(V_{tNA}, V_{tNB}) = (V_{tNA}, V_{tNB(NOM)})$ and the third (labeled as "$-3\sigma_G$") as $(V_{tNA}, V_{tNB}) = (V_{tNA}, V_{tNB(NOM)} - 3\sigma_G)$. The threshold voltage of the $NMOS_A$ transistor is supposed to vary over the whole possible range, as illustrated in the same Figure. The delay variation at the output of the gate has been computed through transistor-level simulations for every point belonging to any of the three lines (arrows). The x-axis depicts the local mismatch, i.e. the difference between both threshold voltages, $V_{tNB}-V_{tNA}$.

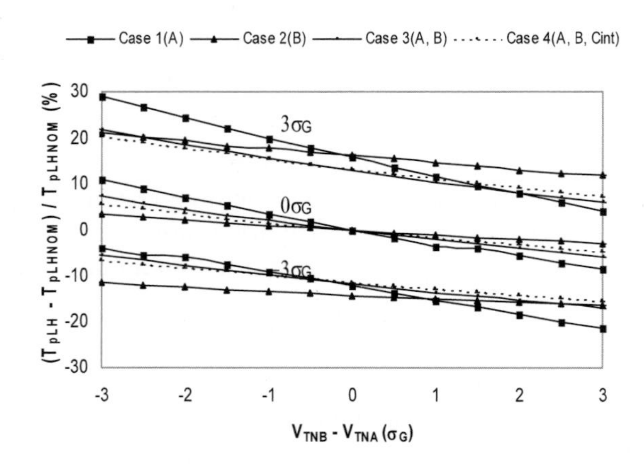

Figure 2-13. *Vt_{NA} variation considered for three different fixed Vt_{NB} values, namely, $Vt_{NB}=(Vt_{NB(NOM)}-3\sigma_G, Vt_{NB(NOM)}, Vt_{NB(NOM)}+3\sigma_G)$.*

For the V_t mismatch of Figure 2-12a, the difference in relative variations increases due to the fact that the nominal delay for cases 3 and 4 are longer than for cases 1 and 2. It is also to be underlined the fact that although V_{tNB} is constant for each one of the three sets of simulations, the delay for transition at *B* suffers from some variations (the smaller ones compared to the rest of cases). This is a difference compared to the parallel-connected structures. Case 1 suffers from the widest variations, as expected, since V_{tNA} varies significantly. For a mismatch as in Figure 2-12b we have that transistors A and B interchange the behavior and as such the delay variations are similar as for the ones of Figure 2-12a. For V_t sweeps like the ones of Figure 2-12c, the delay variation caused by switching both inputs at the same time (cases 3 and 4) is close to a constant value. In Figure 2-12d we have that the variation of threshold voltage is similar for both NMOS transistors. Indeed, the *local* variation is zero while the *global* variation goes from one end point to the other. For this combination of values there is not a big difference in the delay at the output of all four cases.

Experimental verification of delay mismatch. In this subsection we characterize the delay between two of the DSP's previously described. Here we analyze a batch of 25 wafers. In the e-sort test program all dsp-cores that pass the stuck-at test are further subjected to a delay fault test. The delay fault test uses special patterns that have two normal mode cycles between the scan-in and scan-out cycles. The minimum delay between these two edges can be measured by varying the active clock edge in the second normal mode cycle compared to the active clock edge in the first cycle. Excessive delays could be attributed to other defects such as weak opens[31]. Figure 2-14 shows the distribution of delay difference (mismatch) between two of these modules. The delays hereby shown, i.e. delay mismatches between -0.1ns and 0.1ns could be attributed to defects such as the ΔV_t defect described here.

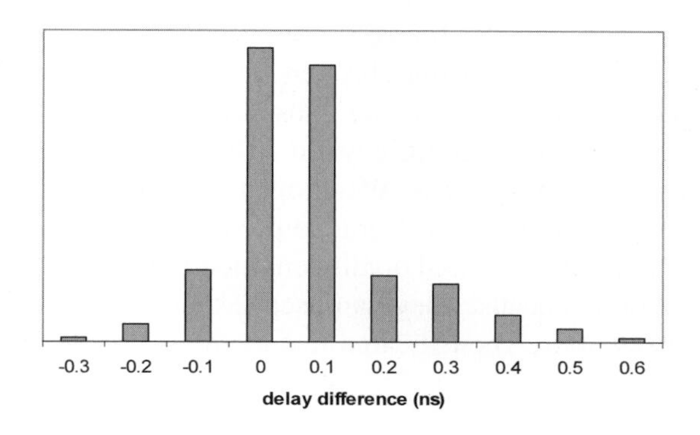

Figure 2-14. Delay distribution of identical cores.

3.6 Spot Defect Statistics: Resistive Opens

The detection of defects is a great concern in the semiconductor industry because of the need to eliminate malfunctioning circuits or candidates to be so in the near future. As shown in previous works based on Inductive Fault Analysis (IFA), extra material as well as lack of material deposition may cause most of the defects in present technologies. Depending on the isolating/connecting nature of the defectively deposited material, the electrical effect on the circuit may show an opposite behavior.

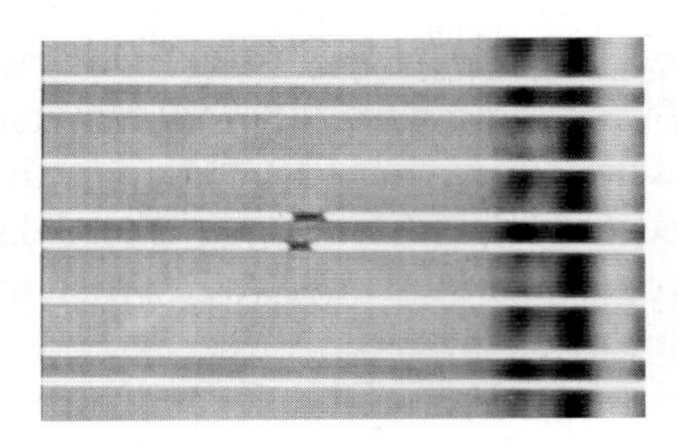

Figure 2-15. Open metal lines (a) Global top view of an open defect line. (b) Detailed cross-sectional view of the metal cavity and formation of the weak open defect due to the Ti barrier.

Traditionally, open defects have been considered as ideal non-connected nodes in a manufactured circuit (but connected in the original design). However, open defects can still connect the two end points of the net although in a weak way, e.g. by introducing a higher than expected (but finite) resistance between the linked points [9, 31]. Thus, open defects can manifest themselves as resistive broken lines or as resistive vias and contacts. In general, strong opens are caused by killer defects and have an immediate impact on the circuit's yield. These opens can be found by applying regular stuck-at patterns. Weak opens, on the other hand, allow the circuit to still work but exhibit a degraded performance in the form of signal delay [11]. From a reliability and quality engineering standpoint, weak opens are potential hazards because they can escape the boolean testing stage. To detect weak opens, more sophisticated test methods have to be applied, like delay fault testing.

Visual examples of open lines and vias are shown in the microphotographs of Figure 2-15a and Figure 2-15b. shows a metal line with a (weak) open defect. It basically depicts a missing metal section. The Titanium (Ti) barrier layer that remained in the cavity gave origin to a resistive open defect. Figure 2-16 illustrates the case of a strong open via as well as a resistive one. Figure 2-16a illustrates the open via defect. Basically, one can see that the via-hole was not etched deep enough to contact the next metal layer. Figure 2-16b shows the resistive via that came from making contact only at the two edges of the hole.

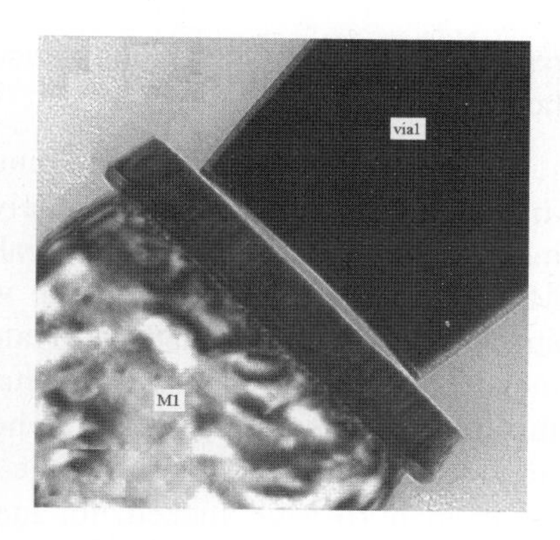

Figure 2-16. Open vias. (a) Strong open via. (b) Resistive via.

For the results presented in Table 2-3 a total of 7440 dies (x4 structures) were measured for each one of the six metal layers [31]. This corresponds to a total of 186 wafers from 12 lots. The detection of low resistive open defects was addressed as the main target of the work since these defects may cause a delay fault and may escape boolean based test techniques. The percentage of low resistive open defects was found using the method presented in the previous sections. In general, the average percentage of defective YEMs is less than 0.5%. Results are shown in Table 2-3. We consider that the open defect is a weak open if its resistance value is less than 10 MΩ.

Table 2-3. Resistance distribution of open line defects.

Resistance[Ω]	Metal 1	Metal 2	Metal 3	Metal 4	Metal 5	Metal 6
3k< R_M < 100k	0.4%	10.9%	12.0%	6.9%	0%	0%
100k < R_M < 1M	5.2%	3.5%	6.2%	4.3%	0%	0%
1M < R_M < 10M	9.2%	4.9%	6.2%	5.2%	0%	0%
Partial Total Weak Opens	14.8%	19.3%	24.5%	16.4%	0%	0%
10M < R_M < 100M	8.9%	5.7%	2.9%	9.5%	0%	0%
100M < R_M < 1G	7.0%	6.0%	1.9%	8.6%	0%	0%
R_M > 1G	69.8%	69.0%	70.7%	65.5%	100%	100%
Total	100%	100%	100%	100%	100%	100%

Worth pointing out is that dies that were consistently defective for all wafers were eliminated from the analysis. This set of repetitive defects on a particular die was not taken into account since the defect is expected not to be a random one.

An important difference is found between the incidence of open defects in the various metal layers. Namely, the percentage of defective structures made of metal M_5 and M_6 is noticeably lower than for metal M_1, M_2, M_3, and M_4 due to the difference in line width and thickness. The percentage of strong opens with resistances greater than 1GΩ is higher than 65% for all metal layers. Furthermore, for metal M_5 and M_6 all open defects belong to this range. These open defects behave as completely opened lines. An important percentage of open line resistances has been found with values lower than 10 MΩ. Indeed, for metal layers M_1 to M_4 levels, percentages between 15% and 25% have been detected with resistance belonging to this range.

A total of four lots with 25 wafers each were analyzed for open contacts, and a total of three lots were used for open vias. Empty fields are because of the small sample size. Since the number of weak open contacts per layer was extremely low, all data was summarized under the "contacts" row of Table 2-4. The general trend for open contacts is that the percentage of strong opens is higher than for metal open lines, more than 91%. The results for vias have a wide spread that depend on the type of via. This spread goes from 52% (via 5) up to 88% (via 2).

Table 2-4. Distribution of resistive open contacts and vias

Resistance[Ω]	Contacts	Via 1	Via 2	Via 3	Via 4	Via 5
$R_M < 10k$	0.8%	13.8%	1.45%	15.7%		7.3%
$10k < R_M < 100k$		6.9%	1.45%	2%	9.4%	6.00%
$100k < R_M < 1M$	0.8%	3.4%	4.35%	5.9%	3.1%	12.7%
$1M < R_M < 10M$	2.8%	17.2%	4.35%	11.8%	6.2%	10.0%
Partial weak opens	4.4%	41.4%	11.6%	35.3%	18.7%	36.0%
$10M < R_M < 100M$	4.4%			3.9%		12.0%
$R_M > 1G$	91.2%	58.6%	88.4%	60.8%	81.2%	52.0%
Total	100%	100%	100%	100%	100%	100%

As far as contacts concerns it is found that the incidence of weak opens among all opens is very small. For all types of contacts, i.e. contacts between metal M_1 and p+, n+, poly and LIL, this percentage is below 5%. This corresponds to approximately a probability of weak contact-failure of 10^{-9} considering that there are on average $2 \cdot 10^6$ contacts per structure. Statistics for all open vias are different depending upon the metal layers that are connected. For vias 1, 3 and 5, the percentage of weak opens ranges from

35% to 41%, while for vias 2 and 4 the percentage of weak opens decreases down to 19% and 12%.

As previously argued, delay faults are typically caused by weak opens. The techniques described in this section were applied to determine the probability of a weak metal open or a weak via open in one of the dsp-cores of the test chip [31]. Assuming that each weak-open in the back-end results in a delay fault, this hypothesis is compared with the actual measured percentage of delay faults. To achieve this, the following procedure was followed. For each lot and each metal layer, we computed the Probability of Failure (PoF): the probability that a piece of interconnect is hit by a weak open. This is possible, since our techniques give the number of weak opens (defined here as opens with a resistance below 10 MΩ) per lot, and we know the length of the meanders in the YEM structures. Because also the lengths of the interconnect in the dsp-cores per metal layer are known, we can compute the probability of a weak open for each metal layer, and thus for the entire dsp-core. The same is done for the via levels. Rodriguez, Volf and Pineda de Gyvez used the methods described in this section to compute for each lot and for each via level the probability of a weak open via (defined here as a via with a resistance between 10 kΩ and 10 MΩ). Normalizing for the lengths of the via-chains, and consequently multiplying by the number of vias in a dsp-core, gives us the probability that one via-level is hit by a weak via. Combining the probability of a weak via and the probability of a weak metal open results in the probability of a weak open in the back-end of the dsp-core.

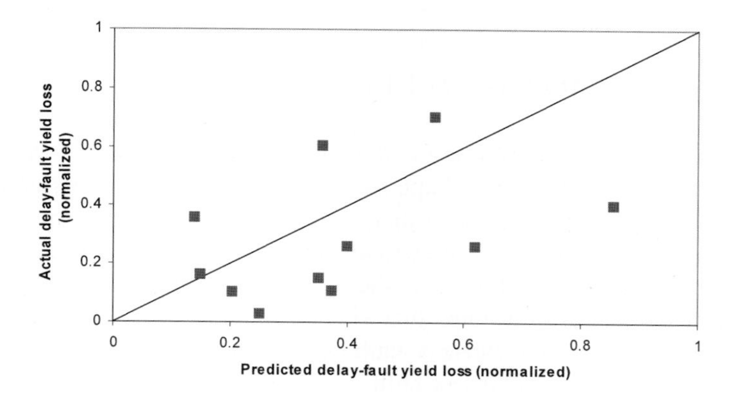

Figure 2-17. Correlation of predicted and actual normalized yield loss due to delay faults.

Rodriguez, Volf and Pineda de Gyvez applied the procedure described above on eleven different lots. Next, we counted, by means of the method described above, the number of delay faults on the dsp-cores of these lots. Each point in Figure 2-17 is a lot prediction (x-axis) against the actual measured percentage of delay faults (y-axis). Both axis have been normalized by the same factor, thus the order of the data before normalization is the same. If the model would describe the data perfectly, we should have obtained the straight line.

Clearly, the correlation in Figure 2-17 is rather weak. Its confidence interval is between –0.247 and 0.812. This is probably caused by the limited delay-fault model. On one hand, the prediction can be too low. Contact opens were not taken into account (due to the substantial workload), and also resistive bridges and numerous problems in the front-end that can cause delay faults. But on the other hand, this prediction can be too high. As mentioned before, weak opens in the via-chains could be caused by metal opens, giving place to a higher probability of failure. Furthermore, some weak opens will cause a delay in a non-critical path, and moreover the coverage of the delay-fault test might not be close to a 100%. Obvious improvements would include taking a considerably larger set of data and taking contacts into account. Furthermore, the presented methods give an accurate estimate of the resistance of the weak open, while we used a rough categorization here.

Despite all these shortcomings, it is clear that the prediction is in the correct order, and a trend can be observed: *more weak opens result in more delay faults*.

4. FUNCTIONAL DEFECT MODELS

Often enough defects reproduce the silicon layer structure of the IC yet they cause a deviation of the shape of such structure. The importance of a defect is determined by the effect that it has on the behavior of the IC. This effect is what we know as an electrical fault. It follows then that there are benevolent defects and of course the other kind, the catastrophic ones. Benevolent defects are the ones that change the shape of the silicon layer structure yet they do not cause a fault. Catastrophic defects, on the other hand, are the ones that do cause a fault.

Strictly speaking, spot defects are splotches of extra or missing material altering the ideal shape of their underlying pattern (see Section 1). Their presence can give origins to electrical faults among them the most common are shorts and opens. Figure 2-18 shows an example of a missing material defect affecting an ideal pattern structure. This is a "typical" defect without a well defined shape, e.g. it basically is a splotch. There could several approaches to model it. One could be a piecewise linear approximation as shown in Figure 2-18b. This model intends to capture the real shape and size of such defect. Figure 2-18c shows a model that is often used in the literature. Basically this model captures the size of the defect by tracing a circle with radius r measured from the splotch's center to the outermost protuberance of the shape. The circle can further be approximated by a square as shown in Figure 2-18d. From the above models the least desired is the piecewise linear model. Basically, it is very difficult to collect defect shape statistics in the manufacturing line. Furthermore, any software intended for yield prediction based on this model will incur in unnecessary computational resources. The circular model has been extensively used in the technical literature for explaining concepts such as critical area and deriving closed analytical expressions useful for yield prediction. However, from a computational standpoint, the square model is the easiest to deal with [30].

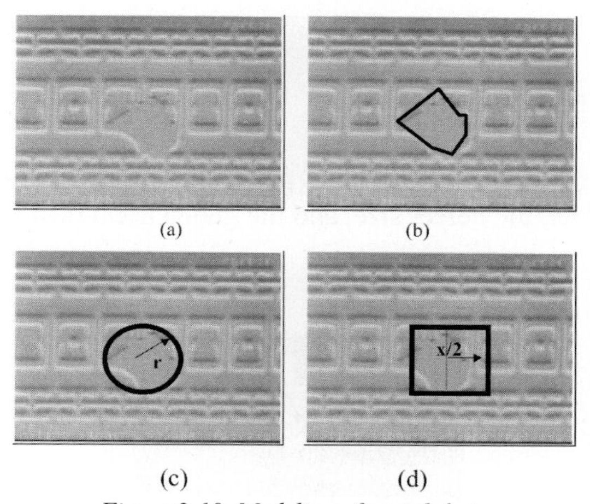

Figure 2-18. Modeling of spot defects.

Let us now investigate a probabilistic model to assess the chances that a defect becomes an electrical fault [22]. Without loss of generality let us consider a missing material defect. For this purpose, consider the pattern-defect setup shown in Figure 2-19. To make the example more realistic, let us also consider the impact of process variability on this pattern, in particular let us assume that there exists overetching during the fabrication process. Let g be a random variable representing the overetching process defect. Let the ideal layout corner points of the pattern be given as (x_1, y_1) and (x_2, y_2) for the left and right corners, respectively. Let us further assume that the width of the pattern is given as $c=x_2-x_1$. After fabrication and in the presence of overetching the actual silicon pattern is a shrunk version of the ideal layout pattern. Assume now a Euclidean coordinate system with origin on the lower left corner of the layout pattern. The new corner points of the silicon pattern with respect to its layout counterpart are now (x_1+g, y_1+g) and (x_2-g, y_2-g). Notice that we have validly assumed that overetching affects all sides of the pattern in a uniform manner. Let us now consider the presence of the missing material defect of size r and positioned at (x,y). Notice that because of the overetching effect the radius of the defect is extended to r+g; in other words there is even more missing material. The defect becomes catastrophic when its size entirely cuts the width c of the pattern. Taking into account the impact of overetching and assuming that $x_2 >> x_1$ one can calculate the probability of an open fault as follows [5]

$$P_{out} = prob \begin{Bmatrix} x-(r+g)-(x_1+g) \leq c \\ \wedge \\ x+r+g-(x_2-g) \geq c \end{Bmatrix} \tag{2.7}$$

$$= \int_{-\infty}^{\infty} f_g dg \int_{\frac{x_2-x_1-4g-2c}{2}}^{\infty} f_r dr \int_{x_2-2g-r-c}^{x_1+2g+r+c} f_x dx$$

where f_g, f_r, and f_x are the probability density functions of the overetching effect, the defect size, and the geometrical position of the defect, respectively.

Figure 2-19. Pattern and missing defect setup.

Typically, overetching statistics can be modelled using a Gaussian distribution while the distribution of defect sizes is modeled using an exponential distribution. These two distributions are "process dependent". That is to say, they are the result of the quality and clean of the fabrication line. The distribution of geometrical centers depends entirely on the specific layout and can be captured using the concept of critical areas.

4.1 Critical Areas

Whether a defect is catastrophic or not, depends on its size and its position in the IC. This very simple observation was kept in secret by IBM for many years until Maly and Stapper separately published essays on the sensitivity of photolithographic patterns to spot defects in 1983 [23, 37]. Defects are typically splotches with irregular shapes, however, for most cases, defects can be modeled as circles or squares. Using the square defect model, let us now analyze the effect of extra material defects [30]. Consider that we have two adjacent patterns of length L separated a distance s from each other. Let us assume that we have a defect of size x such that x > s. Then the critical area of this two-pattern layout is the area where the center of the defect has to be placed to create a bridge between both patterns, see Figure 2-20a. For this configuration the critical area is computed as

$$A(x) = (x + L)(x - s) \tag{2.8}$$

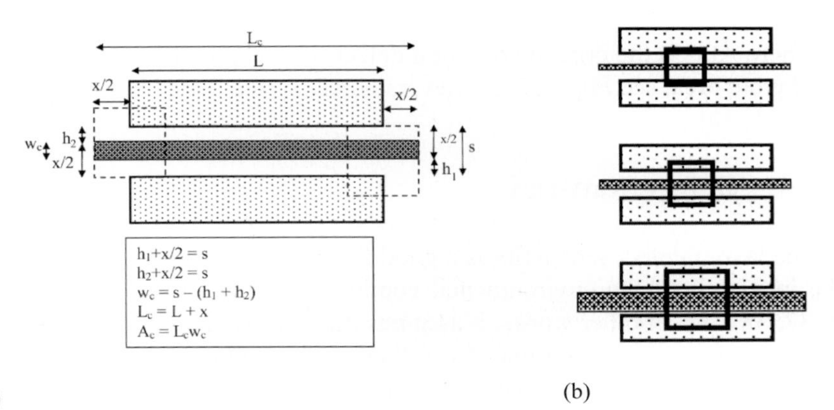

Figure 2-20. Critical area computation. (a) Setup and (b) Efect of defect size on critical area.

The dual of the previous analysis is the case of breaks. Consider a setup with one pattern of length L and width w. Assume a defect of size x such that x > w. Then the critical area for breaks for this pattern is the area where the center of the defect has to be placed to break the pattern. This critical area is computed as

$$A_B(x) = (L + w)(x - w) \quad (2.9)$$

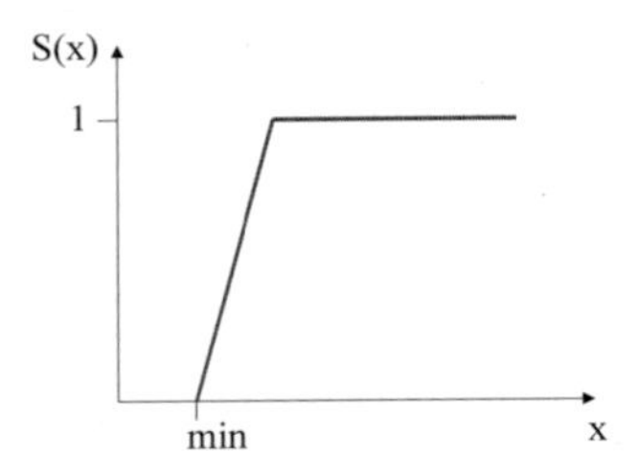

Figure 2-21. Layout sensitivity as a function of defect size.

Notice that the critical area increases with the size of the defect as it is illustrated in Figure 2-20b. The larger the defect size the more vulnerable the IC will be. To assert this statement let us define the layout defect sensitivity for a given defect size x as

$$S(x) = \frac{A(x)}{A_{layout}} \quad (2.10)$$

where A(x) is the critical area for a defect of size x, and A_{layout} is the total chip layout area, see Figure 2-21. Several CAD systems are built around this notion [1, 30].

4.2 Defect Statistics

The layout defect sensitivity is a good figure of merit however it does not take into account the environmental conditions of the silicon foundry [12, 13, 37, 38, 39]. In other words, it assumes that the probability of occurrence of every defect size is the same. Nevertheless, this figure of merit is good to estimate the robustness of a layout independent of where it is fabricated. In practice, and due to tight clean-room control, large defect sizes rarely occur. There is thus a distribution that characterizes the various defect sizes [13, 36, 37]. Probably, the most common one is the one that follows the $1/x^n$ law where x is the defect size and n is typically 3. This distribution takes the form [38].

$$D(x)=\begin{cases}\dfrac{2(n-1)x}{(n+1)x_o^2} & 0\le x\le x_o\\[2ex] \dfrac{2(n-1)x_o^{n-1}}{(n+1)x^n} & x_o\le x\le\infty\end{cases} \tag{2.11}$$

where x_o is the peak defect size. In other words, x_o represents the minimum lithographic feature size resolvable by the technology, e.g. the transistor's channel length. This distribution is illustrated in Figure 2-22. Notice that the distribution to the left of x_o is fictitious since it is not possible to electrically measure defects $x < x_o$. This distribution assumes that i) there are no random defects larger than some maximum size x_M, ii) the number of defects smaller than x_M increases monotonically up to some defect size x_o less than the minimum lithographic feature w, i.e. $x_o/w < 1$, at which it peaks, and iii) the number of defects smaller than x_o decreases monotonically to zero. The value of n is indication of how clean and mature is the manufacturing line.

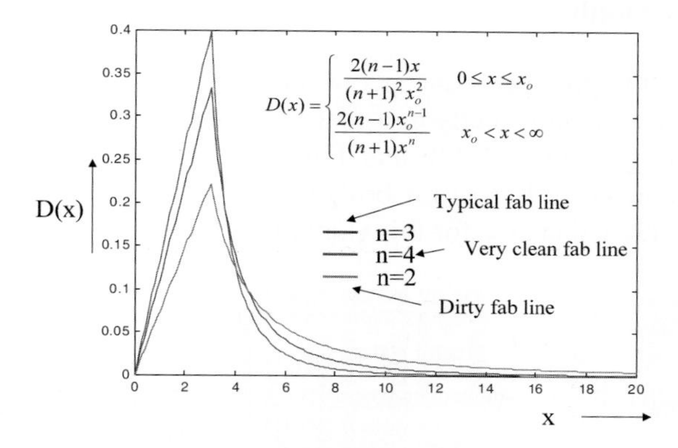

Figure 2-22. Defect size distributions.

With this distribution it is possible to create a probability of failure θ that takes into account the environmental condition of the manufacturing line and the sensitivity of the layout to be implemented. This probability of failure Θ is computed as follows [12].

$$\theta=\int_0^\infty D(x)S(x)dx \tag{2.12}$$

Figure 2-23 presents a graphical interpretation of the PoF. In addition to defect size statistics, another important parameter is the defect density. Defect density is defined as the number of defects per unit area.

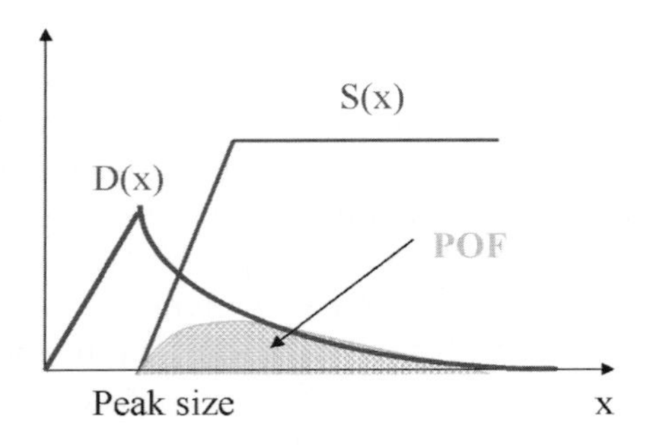

Figure 2-23. Graphical interpretation of PoF.

4.2.1 An Example

Let us apply the previous concepts to the layout of Figure 2-24a. This is the layout of four memory cells in a matrix arrangement implemented in a CMOS technology. Figure 2-24b and Figure 2-24c show snapshots of the critical area (depicted in black) for bridges and breaks, respectively, for the Metal 1 layer. The defect size for this example is 0.6μm.

Figure 2-24. Layout sensitivity for open and short defect of each metal layer.

Of interest is to compute the sensitivity of the layout. Figure 2-25 shows the sensitivity of each layer for open and bridging defect types. One can easily see that there are layers that are more sensitive to defects than others. Usually these layers are the ones used for the interconnect. Layers such as active-area or well definition are less prone to breaks or shorts. Let us now assume that this layout is implemented in a manufacturing environment that has a peak defect size of 1/6 of the minimum spacing. Figure 2-26 displays the probability of failure (PoF) for each defect mechanism in the layers of our layout.

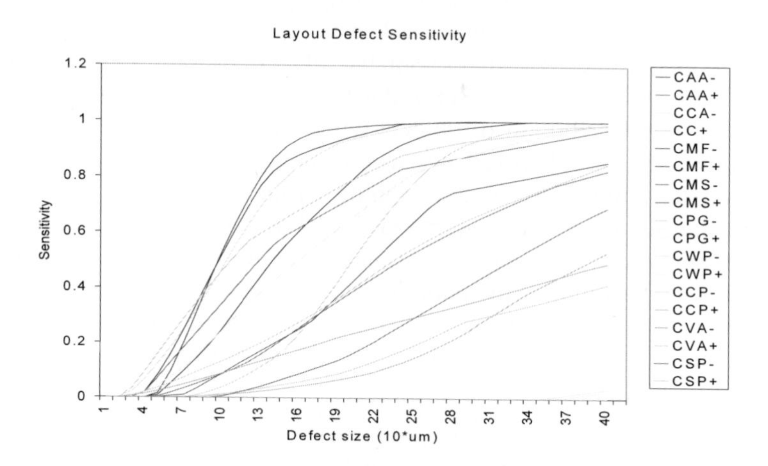

Figure 2-25. (a) Reference layout. (b) Critical areas for shorts in metal 1, (c) critical areas for opens in metal 1.

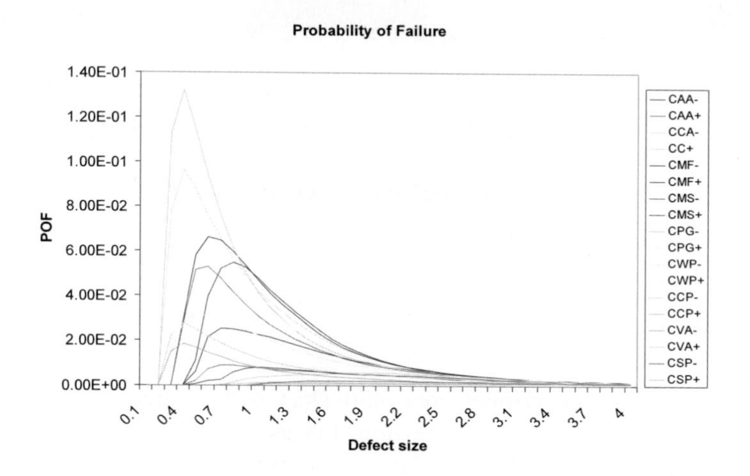

Figure 2-26. Probability of failure for each defect type.

4.3 Average Probability of Failure of Long Interconnects

Within the interconnect defects are usually caused by particle contamination and are divided into bridging defects, which join adjacent wires, and cuts, which result in broken wires. The probability of failure is therefore determined by the geometry of the routing channels and the distribution of defect sizes. Since the wire spacing and width are usually fixed, and the distribution of defects within a mature production facility is well known, the problem reduces to estimating individual wire lengths for cuts, and to estimating the overlapping distance that two wires share in neighboring sections of the routing grid for bridges. In the following we will look at analytical formulae to estimate the probability of failure of the interconnect. To simplify the complexity of the problem, and without loss of generality we are assuming that our interconnect consists of a finite number of parallel lines. This assumption holds valid for most large area ICs.

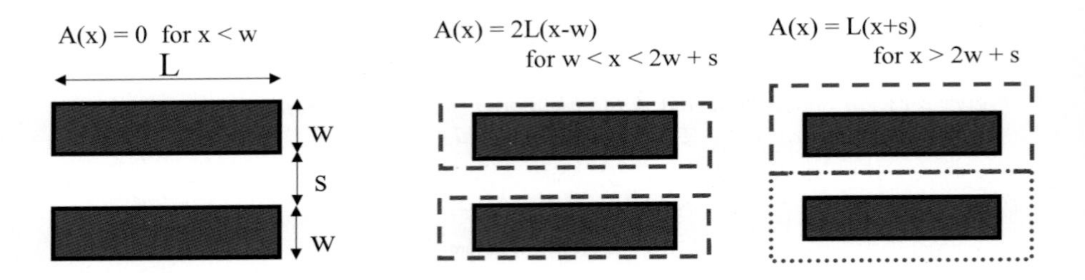

Figure 2-27. Setup to find the critical area of a pair of conductors.

Let us first estimate the expected number of faults in an IC as

$$\lambda = \int_0^\infty A(x)D(x)dx \tag{2.13}$$

where $A(x)$ is the critical area, and $D(x)$ is the defect size distribution extracted from the manufacturing line. Let us further model the defect size distribution according to the $1/x^3$ law as follows

$$D(x) = \frac{x_o^2 \overline{D}}{x^3} \tag{2.14}$$

where x_o is the peak defect size, D is the average defect density, and x is the defect size. Let us now consider a pair of conductors of width w, length L, and separation s between them as shown in Figure 2-27. Let us investigate the layout sensitivity as a function of defect size. One can observe that for defect sizes $w < x < 2w + s$ the critical area increases linearly as a function of the conductor's width. When the defect size is $2w+s$, the critical areas in between the patterns overlap each other and thus the critical area grows only from the non overlapping conductor edges. The corresponding layout sensitivity is shown in Figure 2-28.

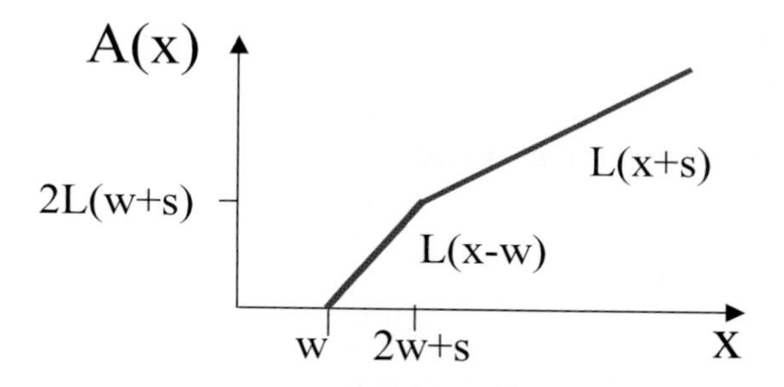

Figure 2-28. Critical area for two conductors.

Let us now investigate the impact of the manufacturing line on these conductors. Let us thus calculate the average probability of failure λ as follows

$$\lambda = \int_{w}^{2w+s} L(x-w)\frac{x_o^2\overline{D}}{x^3}dx + \int_{2w+s}^{\infty} L(x+s)\frac{x_o^2\overline{D}}{x^3}dx$$
$$= \frac{L(3w+2s)x_o^2\overline{D}}{2w(2w+s)} \quad (2.15)$$

Observe that the first integral basically finds the average probability of failure per pattern for both patterns while the second integral finds the average probability of failure of both patterns combined. This is an interesting result. Observe that if the defect density increases the chances of having a fault increase linearly. Observe also that if the peak defect size x_o increases the chances of having a fault increase quadratically and vice versa. Also notice that the average probability of failure is inversely proportional to the width of the conductors. One can extract the average critical area for both patterns from equation (9) to yield

$$\overline{A}_2 = \frac{L(3w+2s)x_o^2}{2w(2w+s)} \tag{2.16}$$

A similar analysis on a single conductor results in an average critical area of

$$\overline{A}_1 = \frac{Lx_o^2}{2w} \tag{2.17}$$

Combining both critical areas we have

$$\overline{A}_2 = \frac{3w+2s}{2w+s}\overline{A}_1 \tag{2.18}$$

Let us investigate now some properties of (12). When the separation s between conductors becomes very large we have that

$$\lim_{s\to\infty}\overline{A}_2 = 2\overline{A}_1 \tag{2.19}$$

which basically states that the critical area of both patterns is the sum of the individual areas. However, in typical interconnects with $w = s$ we have that

$$\overline{A}_2\Big|_{s=w} = \frac{5}{3}\overline{A}_1 \tag{2.20}$$

Notice that in this case the total critical area of both patterns is less than the sum of their individual components. In other words, a system of two closely placed patterns is less sensitive to open faults when compared to a system with two widely spread patterns.

4.4 Average Critical Area of N Conductors

Let us extend the previous results to a typical interconnect centric design consisting of N conductors as depicted in Figure 2-29. The critical area for various defect size ranges can be found as

$$\begin{aligned} &A(x) = 0 && 0 < x < w \\ &A(x) = NL(x-w) && w < x < 2w+s \\ &A(x) = L(x + (N-2)w + (N-1))s && x > 2w + s \end{aligned} \quad (2.21)$$

As in the case of two conductors, let us analyze the different the critical area for various defect ranges. As long as the defect size is less than $2w + s$ the critical areas do not overlap and the total critical area is the sum of the individual areas. For defect sizes beyond this threshold the critical areas begin to overlap. Therefore the total critical area is the area of the conductor and the space between conductors plus the critical area growing out of the topmost and bottom most patterns. The corresponding layout sensitivity is illustrated in Figure 2-30.

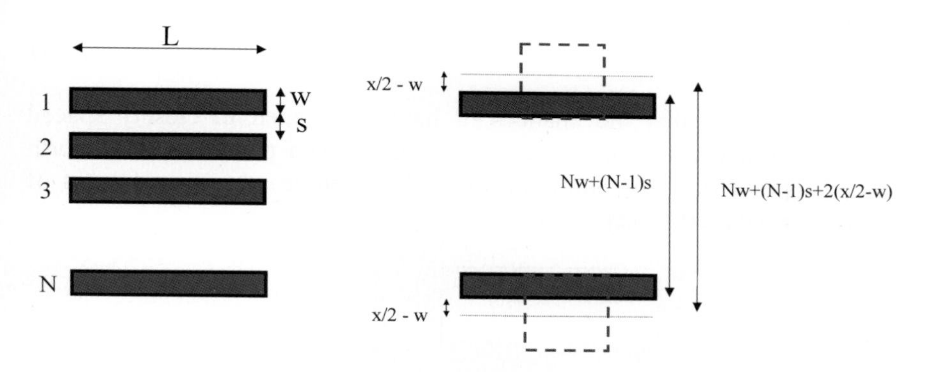

Figure 2-29. Setup to find the critical area of N conductors.

The average probability of failure of N conductors can be found as was done previously for two conductors. The result is

$$\lambda = \frac{L((N+1)w + Ns)x_o^2 \overline{D}}{2w(2w+s)} \quad (2.22)$$

Here we can see once again that the average probability of failure is directly proportional to the number of conductors, inversely proportional to their width, and that it has a quadratic dependence on the peak defect size. The corresponding average critical area can be calculated as

$$\overline{A}_N = \frac{L((N+1)w+Ns)x_o^2}{2w(2w+s)} \tag{2.23}$$

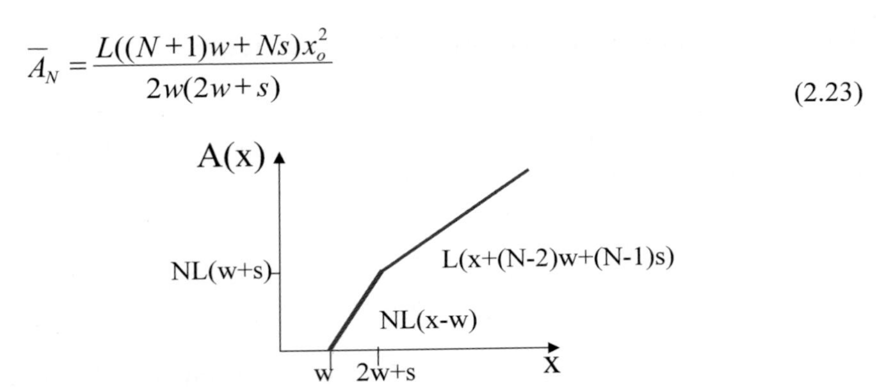

Figure 2-30. Critical area of N conductors.

By taking the limit when $s \rightarrow \infty$ it is easy to show that the total average critical area is the sum of the individual areas of each conductor. However, for typical interconnect in which $s \approx w$ and $N >> 1$, we have that

$$\overline{A}_N \Big|_{\substack{s=w \\ N>>1}} \approx \frac{2}{3} N\overline{A}_1 \tag{2.24}$$

which implies that the chances of having an open in closely spaced patterns is much less than if we had widely spaced patterns. Also, worth noticing is that $A_2 > A_N$ implying as well that there are more chances of breaking two patterns than N ones.

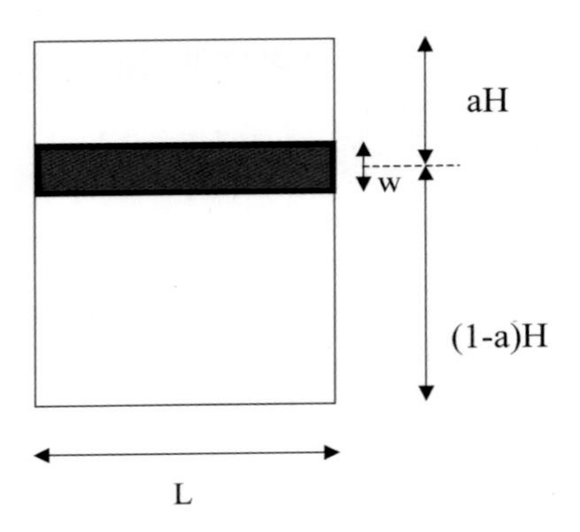

Figure 2-31. Setup to find the critical area of one conductor with chip boundary constraints.

4.4.1 Average Critical Areas with Bounded Chip Area

Let us investigate now how the chip boundaries impact the critical area. Observe that in fact the critical area cannot grow without limits since the impact of a defect is basically constrained to the chip boundaries. Let us consider the an interconnect layer of length L and height H. Let us further assume a conductor placed whose center is placed at some distance aH and $(1-a)H$ from the chip boundaries as depicted in Figure 2-31.

Consider a defect of size $w < x < 2aH+w$. Observe that as the defect size increases, the critical area increases vertically up and down until it saturates against the top chip boundary. For defect sizes beyond $x > 2aH+w$, the critical area will continue to increase only downwards until it also saturates with the bottom chip boundary. Huge defect sizes will give origin to a critical area equal to the chip area. This behavior is illustrated in the layout sensitivity of Figure 2-32.

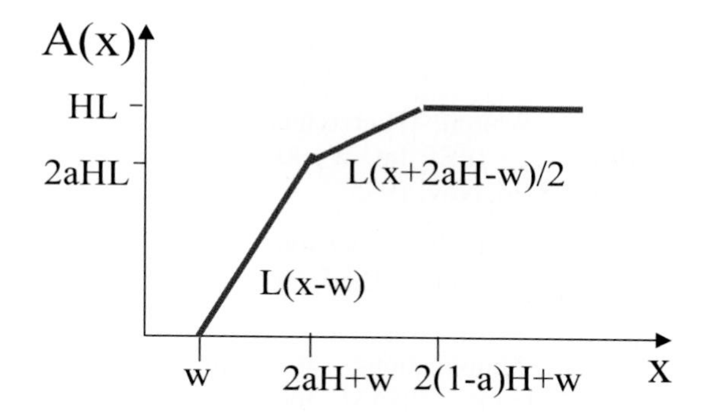

Figure 2-32. Critical area of one conductor with chip area constraints.

The average probability of failure taking into account defect statistics is

$$\overline{A}_{H'} = \frac{LH(w+4a(1-a)H)x_o^2}{2w(2w+2aH)(w+2(1-a)H)}$$
$$\lim_{H\to\infty} \overline{A}_{H'} = \frac{Lx_o^2}{2w} \tag{2.25}$$

Observe that for large H, e.g. no boundary conditions, this expression reduces to the average critical area of one conductor.

5. CONCLUSIONS

This chapter illustrated the importance of defect-oriented testing in common engineering practices. We started with and overview of "traditional sport defects" and their impact on circuit performance. It is a fact that in deep submicron technologies, parametric shifts as well as resistive defects are other sources of impairments on circuit behavior. As the transistor threshold voltage is a parameter that binds many electrical parameters, we chose it as an example for a new kind of defect that hurdles the performance of the circuit but not to the point of making it non functional.

Modern designs are interconnect centric. Upcoming CMOS technologies have seven to nine metal layers to enable the routing of such SoCs. That is why this chapter closes with fault prediction techniques that quickly enable the designer or test engineer to estimate the circuit's robustness against spot defects. Obviously, for very accurate predictions, sophisticated tools that extract the layout defect sensitivity and combine the results with fab. statistics are needed.

References

1. G.A. Allan and J.A. Walton, "Hierarchical critical area extraction with the EYE tools, " Proc. 1995 IEEE Int. Workshop Defect and Fault Tolerance in VLSI Systems, pp. 28-36, Nov. 1995.
2. P. Banerjee, and J. A. Abraham, "Characterization and Testing of Physical Failures in MOS Logic Circuits," *IEEE Design and Test of Computers*, vol. 1, pp. 76-86, August 1984.
3. D. Burnett et al., "Implications of fundamental threshold voltage variations for high-density SRAM and logic circuits," *Symp. on VLSI Tech.,* pp. 15-16, 1994.
4. D. Burnet and A-W Sun, "Statistical Threshold Voltage Variations and its impact on supply-voltage scaling," *SPIE*, vol. 2636, pp. 83-90.
5. I. Chen and A.J. Strojwas, "RYE: A realistic yield simulator for structural faults," Proc. International Test Conference 1987, Washington, DC, USA, 1-3 Sept. 1987, pp.31-42.
6. Z. Chen, L. Wei, M. Johnson, K. Roy, "Estimation of Standby Leakage Power in CMOS Circuits Considering Accurate Modeling of Transistor Stacks."
7. R. Dekker, F. Beenker, and L. Thijssen, "Fault modeling and Test Algorithm Development for Static Random Access Memories," *Proceedings of the IEEE International Test Conference*, pp. 343-352, 1988.

8. M. Eisele et al., "Intra-die device parameter variations and their impact on digital CMOS gates at low voltages", *IEDM Tech. Dig.*, pp. 67-70, 1995.

9. M. Niewczas, "Characterization of the Threshold Voltage Variation: A Test Chip and the Results," *IEEE Int. Conference on Microelectronics Test Structures*, vol. 10, pp. 169-172, 1997.

10. F. J. Ferguson, and J. P. Shen, "Extraction and Simulation of Realistic CMOS Faults using Inductive Fault Analysis," *Proceedings of the IEEE International Test Conference*, pp. 475-484, 1988.

11. A. Ferre, "On Estimating Leakage Power Consumption for Digital CMOS Circuits," Ph.D. Dissertation, Universitat Politecnica de Catalunya, Spain, Nov. 1999.

12. A.V. Ferris Prabhu, "Modeling the critical areas in yield forecasts," IEEE J. Solid State Circuits, vol. SC-20, no. 4, pp. 874-878, 1985.

13. A. V. Ferris-Prabhu, "Role of defect size distribution in yield modeling," IEEE-Transactions on Electron Devices, vol.ED-32, no.9, p.1727-36, Sept. 1985.

14. J. Figueras, "Possibilities and Limitations of I_{DDQ} Testing in Submicron CMOS," *Proc. IEEE Conference on Innovative Systems in Silicon,* pp. 174-185, 1997.

15. D. Gaitonde, and D. H. H. Walker, "Test Quality and Yield Analysis Using the DEFAM Defect to Fault Mapper," *Proceedings of the International Conference on Computer Aided Design*, pp. 202-205, 1993.

16. J. Galiay, Y. Crouzet, and M. Vergniault, "Physical Versus Logical Fault Models in MOS LSI Circuits: Impact on Their Testability," *IEEE Transaction on Computers*, vol. C-29, no. 6, pp. 527-531, June 1980.

17. S. K. Gandhi, "VLSI Fabrication Principles", *John Wiley and Sons*, 1983.

18. R.X. Gu and M.I. Elmasry, "Power Dissipation and Optimization of Deep Submicron CMOS Digital Circuis," *IEEE Journal of Solid State Circuits,* vol. 31, no. 5, pp. 707-713, May 1996.

19. T. Kuroda, T. Fujita, T. Nagamatu, S. Yoshioka, T. Sei and K. Matsuo, "A High Speed Low Power 0.3mu m CMOS gate array with Variable Threshold Voltage (Vt) Scheme," *Proc. Custom Integrated Circuits Conference,* pp. 53-56, 1996.

20. M.C. Johnson, D. Somasekhar and K. Roy, "Models and Algorithms for Bounds on Leakage in CMOS Circuits," *IEEE Trans. On Computer Aided Design of Integrated Circuits and Systems*, vol. 18, no. 6, pp. 714-725.

21. W. Maly, A. J. Strojwas, and S. W. Director, "VLSI Yield Prediction and Estimation: A Unified Framework," *IEEE Transactions on Computer Aided Design*, vol. CAD- 5, no. 1, pp. 114-130, January 1986.

22. W. Maly, "Modeling of lithography related losses for CAD of VLSI circuits" IEEE Trans. On Computer Aided Design, vol. CAD-4, no. 3, pp. 166-177, 1985.

23. W. Maly and J. Deszczka, "Yield estimation model for VLSI artwork evaluation, " Electron. Lett., vol 19, no. 6, pp. 226-227, 1983.

24. W. Maly, F. J. Ferguson, and J. P. Shen, "Systematic Characterization of Physical Defects for Fault Analysis of MOS IC Cells," *Proceedings of the International Test Conference*, 390-399, 1984.

25. P.C. Maxwell and J.R. Rearick, "Estimation of Defect-Free IDDQ in Submicron Circuits using Switch Level Simulation, "*Int. Test Conference,* pp. 80-84, 1997.

26. C. Michael and M. Ismail, "Statistical Modeling of Device Mismatch for Analog Integrated MOS Circuits," *IEEE Journal of Solid State Circuits,* vol. 27, no. 2, pp. 154-166, Feb. 1992.

27. M. J. M. Pelgrom et al., "Matching Properties of MOS transistors", *IEEE Journal of Solid-State Circuits,* vol. 24, pp. 1433-1440, 1989.

28. M. Pelgrom, H. Tuinhout, M. Vertregt, "Transistor matching in analog CMOS applications", *IEDM Tech. Dig.,* 1998.

29. J. Pineda and E. van de Wetering, "Average Leakage Current Estimation of Logic CMOS Circuits," *IEEE VLSI Test Symposium,* pp. 375-379, Apr. 2001.

30. J. Pineda de Gyvez, "IC defect sensitivity for footprint type spot defects," IEEE Trans. On Computer Aided Design, vol. 11, pp. 638-658, May 1992.

31. R. Rodriguez, P. Volf and J. Pineda de Gyvez, "Resistance Characterization of Weak Opens," *IEEE Design & Test Magazine,* pp. 18-26, Sept. 2002.

32. R. Rodriguez-Montanes and J. Pineda de Gyvez and R. Rodriguez-Montanes, "Threshold Voltage Mismatch (ΔV_T) Fault Modeling," *IEEE. VLSI Test Symposium,*pp. 145-150, Napa-Valley CA, April 2003.

33. D. R. Schertz, and G. Metze, "A New Representation for Faults in Combinational Digital Circuits," *IEEE Transactions on Computers*, vol. c-21, no. 8, pp. 858-866, August 1972.

34. J. P. Shen, W. Maly, and F. J. Ferguson, "Inductive Fault Analysis of MOS Integrated Circuits," *IEEE Design and Test of Computers*, vol. 2, pp. 13-26, December 1985.

35. D. Schmitt-Landsiedel, "Yield Analysis of CMOS ICs," *Proc. Of Gettering and Defect Engineering in Semiconductor Technology,"* vol. 57, pp. 327-336, 1997.

36. Z. Stamenkovic and N. Stojadinovic, "New defect size distribution function for estimation of chip critical area in integrated circuit yield models," Electron. Lett., vol. 28, no. 6, pp. 528-530, Mar. 1992.

37. C.H. Stapper, “Modeling of integrated circuit sensitivities,” IBM J. Res. Dev., vol. 27, no. 6, pp. 549-557, 1983.

38. C.H. Stapper, “The effects of wafer to wafer defect density variations on integrated circuit defect and fault distirbutions, “ IBM J. Res. Dev., vol. 29, no. 1, pp. 87-97, 1985.

39. C.H. Stapper, “Yield Model for fault clusters within integrated circuits,” IBM J. Res. Dev. Vol. 28, no. 5, pp. 636-640, 1984.

40. C.H. Stapper, “On Yield, fault distributions and clustering of particles,” IBM J. Res. Dev., vol. 30, no. 3, pp. 326-338, 1986.

41. C.H. Stapper, “Modeling of defects in integrated circuit photolithographic patterns,” *IBM-Journal of Research and Development*, vol.28, no.4, p.461-74, July 1984.

42. S. M. Sze, “VLSI Technology,” New York: *McGraw Hill Book Company*, 1983.

Chapter 3

DIGITAL CMOS FAULT MODELING

We begin with an overview of digital fault models. Different fault models are classified according to the level of abstraction. The merits and shortcomings of these models are reviewed. Special attention is paid on delay fault models and radiation induced soft errors.

1. OBJECTIVES OF FAULT MODELING

The exponential increase in the cost of functional testing has led to tests that are not functional in nature, but are aimed at detecting possible faulty conditions in ICs. The circuit under test (CUT) is analyzed for faulty conditions and tests are generated to detect the presence of such conditions. Like any other analysis, this fault analysis also requires a model (or abstraction) to represent the likely faults in ICs with an acceptable level of accuracy. This type of model is called the fault model and this type of testing is known as structural testing. The name structural test comes from two counts. First, the testing is carried out to validate the structural composition of the design rather than its function and, second, the test methodology has a structured basis, i.e., the fault model for test generation. In fact, the concept of structural testing dates back to the 1950s. In one of the first papers on the subject, Eldred proposed a methodology which tests whether or not all tubes and diodes within a gating structure are operating correctly [21]. However, structural testing gained popularity in the 1970s and the 1980s when structural design for test (DfT) methodologies like scan path and level sensitive scan design (LSSD) [20,27] emerged. These DfT methods became popular because their application could change distributed sequential logic elements into a big unified shift-register for testing purposes. As a result, the overall test complexity is reduced [106]. Owing to these techniques, the test generation and fault grading for complex digital circuits became a possibility.

Breuer and Friedman [11] described fault modeling as an activity concerned with the systematic and precise representation of physical faults in a form suitable for simulation and test generation. Such a representation usually involves the definition of abstract or logical faults that produce approximately the same erroneous behavior as the actual physical defects. Here, it is important to distinguish between a defect and a fault. A defect is physical in nature and a fault is its representation. Therefore, a fault can also be defined as follows: *A fault is the electrical impact of a physical defect at an appropriate level of abstraction.* A fault is often represented by its simulation model, which is termed as the fault model. Fault models have played a pivotal role in the success of structural testing whose goal is to test for the modeled faults. Structural testing has some notable advantages over functional testing. Foremost amongst them are:

- The effectiveness of the structural test is quantifiable. It is possible to ascertain the percentage of the modeled faults tested by a given test suite. This percentage is popularly known as the fault coverage. Thus, it allows the user to establish a relationship between the fault coverage of the test suite and the quality of tested ICs.
- Test generation for structural tests is considerably simpler compared to functional test generation for a complex CUT. Computer aided design (CAD) tools (e.g., automatic test pattern generator (ATPG) and fault simulator) ensure faster and effective test generation.
- In the case of functional testing, the choice of ATE is closely related to the CUT specifications, which may cause an undesirable dependence on a particular ATE for testing. However for structural testing, the choice of ATE is largely independent of the CUT specifications, which allows greater freedom in choosing an ATE.

The underlying assumption behind structural testing is that the design is essentially correct and its function on silicon has already been verified and characterized. The non-ideal manufacturing process introduces defects in the design (or IC). These defects cause faults, which result in erroneous IC behavior that needs to be tested. Moreover, fault models in structural testing are only for fault simulation purposes and they do not represent how likely one fault is compared to another. Structural testing further assumes time invariance of a fault. Time variant faults degrade over time, and therefore are difficult to test in the production environment. Similarly, a combination of environmental conditions may trigger a device to have a temporary fault that is very difficult to test in a production environment. Consequently, time variant and temporary faults are not considered for structured test generation. Lastly, it is assumed that a fault has a local impact. For example, a fault may cause the output of a NAND gate to be always logic high. The origin of this assumption lies in the fact that a fabrication process line is regularly monitored; hence, the global defects are controlled early in the production environment. As a result, the vast majority of defects which are to be tested for are local in nature.

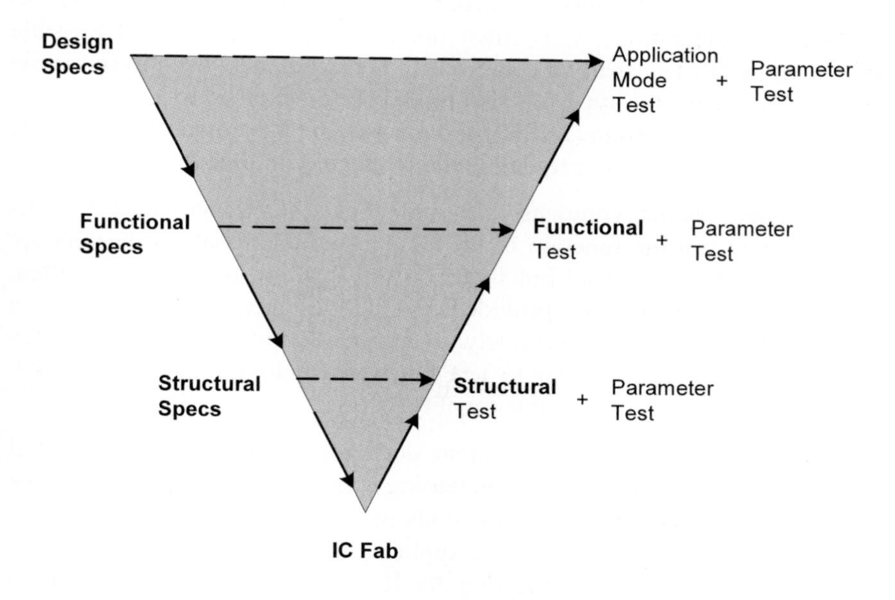

Figure 3-1. An idealized design and verification process [7].

2. LEVELS OF TESTING

It is prudent to mention here that the aims of structural testing are different from those of functional testing. Beenker et al. [7] suggested that test objectives should be mapped to various product development stages, and hence have different purposes. Different levels of testing and their significance are explained with the help of Figure 3-1. The product development stages range from requirement specification to functional specification, functional design, and structural design [7]. These stages are highlighted on the left hand side of the figure. The corresponding test levels, called verification tests, are shown on the right hand side of the figure. Each has a different objective, namely, assuring the corresponding specifications are faithfully implemented. At the end of the verification process, if the design has no flaws it is passed on to large scale production and the verification tests are replaced by the production tests.

As mentioned before, the objectives of structural testing include testing for the faithfulness of the structure of the IC with respect to the original functional specification. Fault models are required which represent production process defects. These faults are mapped onto the device under test (DUT) and test vectors are generated to cover such failures. Due to economic considerations, it is often not practical to test for all possible defects in the DUT. Usually a compromise is reached between the economic considerations and the type of fault model. DfT schemes (e.g., scan path, level sensitive scan design (LSSD), and macro test [9]) provide test solutions while retaining, or even improving fault coverage economically.

Functional testing contrasts with the above mentioned approach. Its objective is to test the function of the DUT. The functional test vectors are not based on a fault model but focus on the function of the DUT. Often, functional test vectors are provided by the designer. The quality of a functional test can not be objectively quantified. Despite these shortcomings functional vectors are popular to test functional marginalities (e.g., speed, minimum V_{DD}) and ensuring functionality.

The correct design of a device alone does not ensure that the device will work satisfactorily in the system. Increasing complexities of ICs and systems forced test professionals to find methods of characterizing the behavior of ICs in their real-world applications. Application mode testing emerged from this desire. During this type of testing, the ICs are tested with real-life data in a wide variety of environmental conditions that the DUT is supposed to encounter during its normal life cycle. In the fast-paced semiconductor industry, being good is not good enough. *"How good is good?"* is one of the questions answered by parametric testing. In other words, quantifying the performance of the DUT in terms of speed, AC and DC parameters, power consumption, and environmental (i.e., temperature and voltage) susceptibility are the objectives of parametric testing.

3. LEVELS OF FAULT MODELING

There are numerous ways to represent faults for fault simulation. In general, faults are categorized according to the level of abstraction. An appropriate level of abstraction essentially achieves a trade-off between the fault model's ability to accurately represent an actual physical defect and the speed of fault simulation. For example, behavior (function) level fault modeling is the fastest but is the least accurate. On the other hand, layout level inductive fault analysis is the most accurate but requires enormous computational resources. Most fault models can be classified according to the following levels of abstraction.

The various fault modeling levels are presented from a top-down perspective, proceeding from logic, to transistor, to layout level. Following this, the function level, or functional fault modeling is presented. Since this form of testing is used in complex SoCs, it is distinguished from the first three since it extends beyond pure digital logic testing. Next, delay faults and their various causes and models are discussed. Finally, temporary and operational faults are presented.

3.1 Logic Level Fault Modeling

Initial work on fault modeling was concentrated at the logic level. It was assumed that the faulty behavior due to defects can be mapped onto the incorrect Boolean function of basic gates in the CUT. Simple circuits and relatively large feature sizes justified this assumption. Moreover, fewer defects could cause non-logical faults. In the early days of ICs, the semiconductor industry was struggling to solve complex design and process related problems and paid little attention to IC testing. Yield of the early ICs was poor and was caused primarily by equipment and technological problems. Therefore, yield loss related to spot or lithographic defects was insignificant. Furthermore, limited knowledge about the origin and impact of defects on circuit behavior forced researchers to adopt many simplifying assumptions for testing and modeling purposes.

The implementation details of logic gates are not considered in fault modeling at the logic level. Fault modeling at the logic level has some notable advantages. The Boolean nature of the fault model allows usage of powerful Boolean algebra for deriving tests for complex digital circuits [2]. The gate level representation of the faulty behavior resulted in a technology independent fault model and test generation algorithms. Technology independent tests increased the ability to port designs to different technologies.

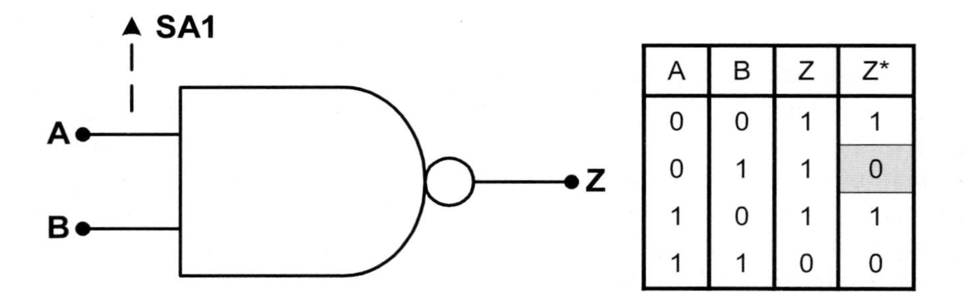

A	B	Z	Z*
0	0	1	1
0	1	1	0
1	0	1	1
1	1	0	0

Figure 3-2. The SAF model and its fault-free and faulty Boolean behavior.

3.1.1 Stuck-At Fault Model

The stuck-at fault (SAF) model is the most commonly used logic level fault model. Poage [68] was one of the first to propose the SAF model. It became a popular in the 1960s owing to its simplicity. It is widely used in academic research as well as in the industry. Its simplicity is derived from its logical behavior. SAFs are mapped onto the interconnects (or nets) between logic gates. Thus, they are also referred to as stuck-line (SL) faults [36,37]. Under the faulty condition, the affected line is assumed to have a permanent (stuck-at) logic 0 or 1 value that cannot be altered by input stimuli. Figure 3-2 illustrates a NAND gate and its truth table. Let us consider when line A has a stuck-at-1 (SA1) fault. The presence of the SA1 fault is detected by the faulty gate response when lines A and B are driven to logic 0 and 1, respectively. A fault is said to be detected when the expected output of the logic gate differs from the actual output. For example, the third and fourth columns of the table in Figure 3-2 illustrate the expected (fault-free, Z) and the actual (faulty, Z*) responses of the NAND gate. For the test vector {A = 0, B = 1}, the expected and actual responses differ and thus the fault is said to be detected.

The SAF model is widely used for many practical reasons. These reasons include the availability of CAD tools for test generation and the technology independence of the SAF model. Furthermore, many physical defects cause SAFs. With increasingly complex devices and smaller feature sizes, the likelihood that more than one SAF can occur at the same time in an IC has become significantly large. However, a large number of possible multiple SAFs force the test generation effort to be impractical. For example, assume that a circuit contains n lines. Each line is capable of having three distinct states; i.e, SA0, SA1 and fault-free. Therefore, there are $3^n - 1$ multiple SAFs possible. A typical IC may contain hundreds of thousands or more lines which may result in an enormously large number of possible faults. Therefore, it is a common practice to assume the occurrence of only single-stuck-line (SSL) faults in the IC. Hence, for an IC with n lines, only $2n$ SSL faults are possible. This single fault assumption is typically valid for two reasons. The first is that multiple faults in the same branch of a circuit are statistically much less likely than single faults. The second reason is that multiple faults are often easier to detect. Multiple faults typically result in more input vectors causing erroneous outputs, and thus are more likely to be detected.

Table 3-1. The SAF classes in a 2-input NAND gate.

A	B	Z	Fault Classes
1	1	0	A/0, B/0, Z/1
1	0	1	B/1, Z/0
0	1	1	A/1, Z/0
0	0	1	Redundant Test

3.1.2 Fault Equivalence, Dominance and Collapsing

A fault *f1* is considered to be equivalent to fault *f2* if their faulty behaviors for all possible input stimuli are indistinguishable from each other at primary outputs. Therefore, a stimulus detecting *f1* will also detect *f2*. Such faults can be bunched into an equivalent fault class. This point can be further illustrated from Figure 3-2. Consider once again a 2 input NAND gate. A SA0 fault at B (B/0) causes a SA1 fault at Z (Z/1) for all input conditions of the NAND gate. Therefore, both of these faults are indistinguishable from each other at the primary outputs. A prior knowledge of the fault classes in a network is useful in fault diagnosis. Furthermore, fault detection is simplified by using equivalent fault classes to reduce or collapse the set of faults into fault classes that need to be considered for test generation [56,84]. The fault resolution of a network depends on how widely equivalent faults are separated. Hence, knowledge of the equivalent fault classes is helpful in problems such as test point placement and logic partitioning to increase the fault resolution [33].

Poage [68] was the first to describe the concept of fault dominance. According to him, if all tests for a fault *f1* also detect *f2* but only a subset of tests for *f2* detect *f1*, then *f2* is said to dominate f1. However, Abraham [2] argued that it is safer to consider a test for *f1* which will ensure both faults are detected. Hence, *f1* is said to dominate *f2*. For example, in Table 3-1, A/1 does not cause Z/0 behavior for all input conditions of the NAND gate. A test for A/1 will also test for Z/0. The converse is not true because Z/0 is also tested by the test for B/1. Thus according to Abraham, the detection of A/1 will ensure the detection of Z/0 as well. Therefore, A/1 is said to dominate Z/0. On the other hand, according to Poage, Z/0 dominates A/1. However, it should be mentioned here that the difference between two approaches is in the definition and not in the process of fault collapsing. The concepts of fault equivalence and fault dominance allow us to collapse SAFs into fault classes. In general, for an n-input gate there will be $n + 1$ equivalent fault classes [2]. Table 3-1 shows these fault classes and test vectors needed for their detection for a 2-input NAND gate.

Table 3-2. Fault table for a 2-input NAND gate.

A	B	A/0	A/1	B/0	B/1	Z/0	Z/1
0	0	0	0	0	0	1	0
0	1	0	1	0	0	1	0
1	0	0	0	0	1	1	0
1	1	1	0	1	0	0	1

These concepts can be applied to larger circuits, but in general it is a complex computational problem. It is shown [33,40] that the problem of identifying fault equivalence in arbitrary networks belongs to the class of computationally difficult problems called NP complete. Nevertheless, a significant attention has been devoted to find equivalent fault classes in circuits bigger than a single logic gate [56,84,33,36,77,87,99].

Schertz and Metze [84] described fault collapsing as the process of combining faults by means of implication relationships derived from the network. They defined three stages of fault collapsing corresponding to three types of implication relationships. These stages can be explained as follows:

Consider a fault table T that contains a row for each possible input vector and each fault is represented by a column. Table 3-2 shows the resulting fault table for the 2 input NAND gate. The entry t_{ij} is 1 if and only if test i detects fault j. A column j dominates column k if for every 1 in column k there is also 1 in column j (same as definition as Poage). In Table 3-2, Z/0 dominates A/1 and B/1. If two columns of the table are identical then each dominates the other (A/0, B/0, Z/1). The first stage of the fault collapsing corresponds to identical columns in the fault table. The second stage of collapsing corresponds to the unequal columns, that is, the situations where one fault is more readily detected than the other. The third stage is concerned with the relationship between single and multiple faults. In order to illustrate these stages, let us consider the 2-input NAND gate of Figure 3-2 and its fault table shown in Table 3-2 once again. The faults A/0, B/0, and Z/1 are indistinguishable from each other and, hence, represent the first stage collapsing procedure. The dominance of Z/0 over A/1 and B/1 represents the second stage of fault collapsing. The third stage of fault collapsing is concerned with multiple faults. For example, Z/0 is indistinguishable from all input lines (A/1, B/1) having a multiple SA1 fault.

In the case of larger circuits having non-reconverging fanouts, the application of the foregoing analysis is rather straight forward. However, in the case of reconverging fanouts, the reconverging branch may interfere with the propagation of the fault to a primary output. Gounden and Hayes [33] further simplified the fault class identification problem for certain cases. They introduced the concepts of intrinsic and extrinsic fault equivalence. These concepts were used to derive some general conditions for fault equivalence and non-equivalence for a given network topology. They argued that every switching function can be realized by a two-level network. Therefore, for a CUT realized by a two-level network, it is possible to identify equivalent fault classes. For a given two-level network, the computational complexity is reduced to 19% of the original value [33]. Fault collapsing techniques have been extended to NMOS and CMOS circuits not only for stuck-at faults but also for transistor stuck-open (SOP) and stuck-on (SON) faults [26,87].

3.1.3 Mapping of Defects on Stuck-at Faults

In CMOS technology an m-input static logic gate is realized with m p-channel transistors and an equal number of n-channel transistors. The output of the gate is taken where p- and n channel transistor groups are connected to each other. Depending upon the Boolean state of the inputs, the output is driven either by p-channel transistor(s) to logic 1 or by n-channel transistor(s) to logic 0. The output is never driven by both types of transistors at the same time.

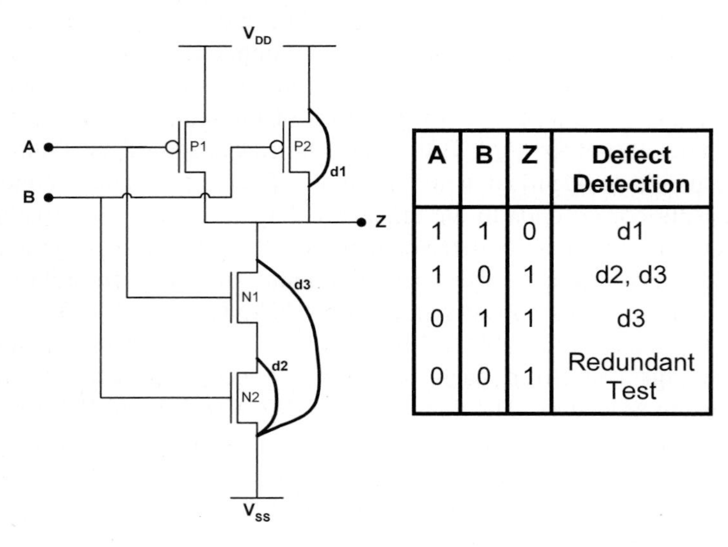

A	B	Z	Defect Detection
1	1	0	d1
1	0	1	d2, d3
0	1	1	d3
0	0	1	Redundant Test

Figure 3-3. Defects in a 2-input NAND gate and their detection.

Figure 3-3 illustrates a transistor-level schematic of a 2-input NAND gate in CMOS technology. Two p-channel and two n-channel transistors are needed to realize a 2-input NAND gate. The figure also illustrates some common bridging defects in the gate, and the table shows how they are detected by SAF test vectors. In this simplistic analysis we assume that the defect resistance is substantially low compared to the on resistance of a transistor. Therefore, in case of a conflict between defect and a transistor, the defect will override. Later, we will remove this assumption as we discuss lower-level fault models.

Defect *d1* causes a short between the output Z and the power supply V_{DD}. Assuming that the on resistance of N1 and N2 transistors is substantially high compared to defect resistance, the defect results in the Z/1 stuck-at fault. Needless to say, the test vector {A = 1, B = 1} will detect it. Similarly, defect *d3* causes the Z/0 stuck-at fault which can be detected by {A = 0, B = X}, or {A = X, B = 0}, test vectors.

Defect *d2* cannot be modeled by the SAF but it can be detected by SAF test vectors under the assumption that the resistance of the defect in series with the on-resistance of the N1 transistor is substantially smaller than the on-resistance of the P2 transistor. Test vector {A = 1, B = 0} causes transistors P2 and N1 to conduct. Therefore, in the presence of defect *d3*, output Z is driven from V_{DD} and V_{SS} simultaneously and has an intermediate voltage. This intermediate voltage is not the same as Z/1 or Z/0. However, this defect is detected provided that the following logic gate interprets this output level as logic 0 instead of logic 1. However, the interpretation by subsequent logic gate depends on several factors such as relative strengths of p- and n-channel transistor(s) in conflict, the defect resistance and the logic threshold of the subsequent logic gate. Semenov has investigated detection of such defects and concluded that scaling technology makes detection more difficult, and that lowered operating voltage (V_{DD}) and increased temperature make detection easier. Hence, during testing, the operating voltage should be decreased and the temperature increased to facilitate voltage detection of such faults [85].

3.1.4 Shortcomings of the Stuck-At Fault Model

Despite its simplicity and universal applicability, the SAF model has serious drawbacks in representing defects in CMOS technology. It can represent only a subset of all defects. Large numbers of defects that are not detected by a SA test set cause bad ICs to pass the test. In a study, Woodhall et al. [107] reported that open defects led to an escape rate of 1210 PPM when the CUT was tested with a 100% SAF test set. We illustrate some representative examples of faults that are not modeled by SAF.

Open Defects

The output of a CMOS logic gate retains its value when left in a high impedance state. Such a property has numerous applications in the data storage and discrete time signal processing areas. Furthermore, CMOS circuits offer high input impedance, hence, floating interconnects retain their previous logic value for a significantly long time. This sequential behavior of CMOS logic gates causes many open defects not to be detected by a SAF test set [103].

Consider once again the same 2-input NAND gate and its truth table illustrated in Figure 3-3. Some of the open defects affecting the operation of transistor P1 are not detected. The second test vector {A = 1, B = 0} drives the output to logic 1 through transistor P2. The third test vector {A = 0, B = 1} instead drives the output to logic 1 through transistor P1. Some of the open defects affecting transistor P1 cannot be detected by the third test vector (Figure 3-4). In the presence of these defects, the output is not driven to logic high in the third test vector but retains its logic high state from the second test vector. Therefore, these defects are not guaranteed to be detected by SA test vector set. However, these defects are detected if the order of test vectors is changed. For example, test vectors {T2, T1, T3} will detect these faults. A detailed treatment of open defects is presented in the next subsection. However, in general, for open defects two test vectors {T1, T2} are required. The first test vector T1 initializes the output to a specific logic level and T2 attempts to change the output state through a particular transistor-controlled path. For logic gates with higher complexity the SAF test vector set cannot guarantee detection of all open defects.

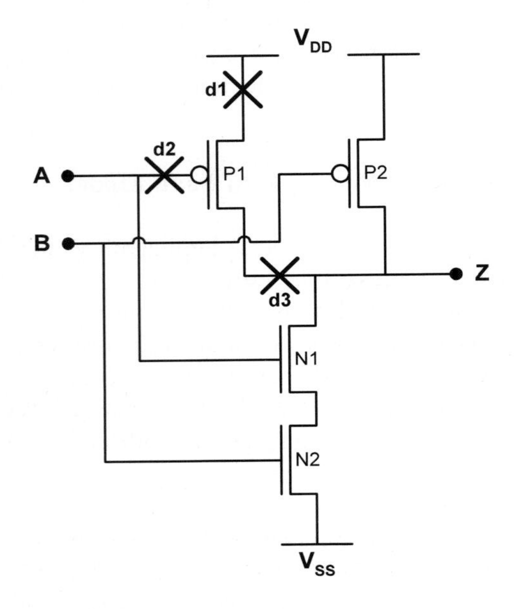

Figure 3-4. Undetected open defects in a 20input CMOS NAND gate by SAF test vectors.

Short Defects

A short defect is defined as an unintended connection between two or more otherwise unconnected nodes. Often they are referred to as bridging faults or simply as bridges. Shorts are the dominant cause of failures in modern CMOS processes. In the CMOS technology, shorts cannot be modeled as wired-OR or wired-AND logic. The circuit level issues (e.g., W/L ratio of driving transistors, defect resistance, logic thresholds of subsequent logic gates, Boolean input logic conditions, etc.) play an important role in their detection. Although a large number of shorts (shorts between V_{DD}/V_{SS} and input/output of logic gates) lead to SAFs, in general, the SAF model does not adequately represent all possible shorts in an IC. Hence, SAF test vectors do not ensure the detection of all shorts.

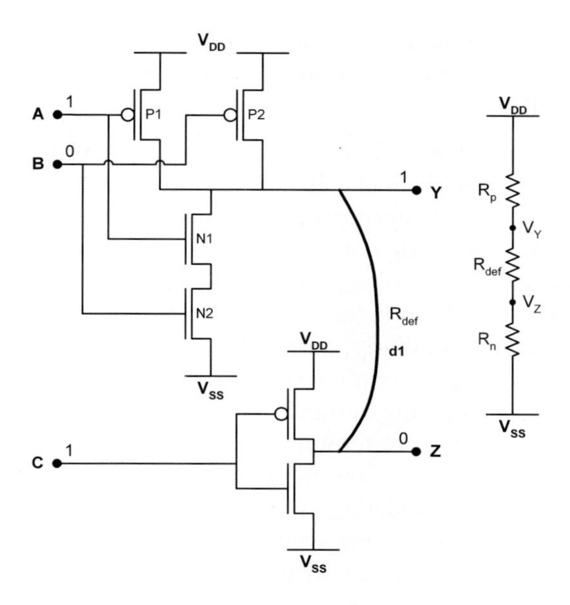

Figure 3-5. An external bridging defect not detected by SAF test vectors.

Shorts in ICs can be classified as internal bridges and external bridges. Internal bridges are those that affect the nodes within a logic gate. The shorts shown in Figure 3-3 are examples of this category. The external bridges are those that affect nodes within two or more logic gates. Figure 3-5 illustrates an external bridge and its electrical model. Besides the circuit level issues, the detection of external bridging faults also depends on exciting nodes Y and Z to opposite logic values. For a complex circuit, this is a non-trivial task. One must determine all potential locations for such bridging defects. Techniques like IFA can be useful in finding such locations.

3.2 Transistor Level Fault Modeling

The SAF model has limitations in representing defects of CMOS circuits. In general, there are many defects in CMOS circuits that may be represented by the SAF model and are detected by the SAF test set. There are other defects that are not modeled by the SAF model but are detected by chance by the SAF test set. However, there are still potentially many defects that are not modeled by the SAF model and are not detected by the SAF test set [2]. Therefore, we need transistor level fault models which represent faulty behavior with better accuracy. However, such fault models result in a significantly larger number of faults compared to that for the SAF model. Furthermore, a significant effort is directed towards a better understanding of defects and their influence on circuit behavior. The knowledge gained improved the fault models leading to efficient and effective tests, and better quality of tested ICs.

As described earlier, a static CMOS logic gate is constructed by a set of p-channel and a set of n-channel enhancement mode transistors between V_{DD} and V_{SS} terminals. An enhancement mode transistor in the absence of a gate to source voltage ($V_{GS} = 0$) does not conduct. It conducts only when an appropriate gate to source voltage ($V_{GS} > V_t$) is applied. For example, a p-channel transistor (e.g. P1 in Figure 3-3) conducts when its gate terminal (A) is logic 0 ($V_{GS} < V_{tp}$). Similarly, an n-channel transistor (e.g. N2 in Figure 3-3) conducts when its gate terminal (B) is logic 1 ($V_{GS} > V_{tn}$). Therefore for transistor level fault modeling, a transistor can be treated as a three terminal ideal switch having a control terminal or gate, which controls the flow of electrical signal (V_{DD} or V_{SS}) from the source terminal to the drain terminal. In transistor-level fault modeling, physical defects are mapped onto the functioning of these switches (transistors). For a given combination of input logic values, a CMOS logic gate may have one of the following states [73]:

- The output node is driven to V_{DD} via one or more paths provided by conducting p-channel transistors and no conducting path from output to V_{SS} exists through n-channel transistors.
- The output node is driven to V_{SS} via one or more paths provided by conducting n-channel transistors and no conducting path from output to V_{DD} exists through p-channel transistors.
- The output node is not driven to V_{DD} or V_{SS} via conducting transistors.
- The output node is driven by both V_{DD} and V_{SS} via conducting transistors.

In the first two cases, the output is logic high and low, respectively. In the third case, the output is in high impedance and its present logic state is the same as its previous logic state. In the fourth case, the output logic state is treated as indeterminate, since the actual voltage depends on the resistance ratio of the conducting paths of p- and n-channel transistors. However, logic gates are rarely designed to have a type of situation like the fourth case.

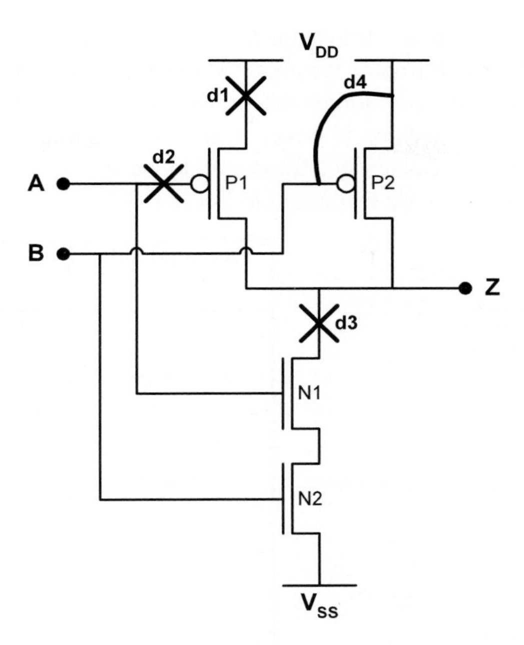

Figure 3-6. Open and bridging defects causing SOP behavior.

3.2.1 Transistor Stuck-Open Fault Model

The detection of transistor stuck-open (SOP) faults is a difficult problem and has received considerable attention in the past. In the presence of a SOP fault the affected transistor fails to transmit a logic value from its source terminal to its drain terminal. Therefore, the transistor can be treated as a switch which never closes and remains open despite all possible Boolean input conditions. As apparent from the name, such faults in enhancement mode transistors are primarily caused by open defects. However, short defects can also cause a transistor to have a SOP fault. Figure 3-6 illustrates some defects causing SOP faults in a logic gate. SOP faults are classified by their location as a fault at the source (S) or drain (D) terminal (S/D-line fault) or at the gate terminal (gate-line fault) of a transistor [45]. A S/D-line fault creates a break in the data transfer path and clearly causes a SOP fault in the transistor (defects *d1* and *d3*, Figure 3-6). A gate-line fault (defect *d2*, Figure 3-6) requires explanation. As stated above, in the CMOS technology enhancement-mode p- and n-channel transistors are used. Enhancement transistors have the property that in the absence of any gate voltage (V_{GS}) the transistor does not conduct. Only when the gate voltage exceeds the threshold voltage, V_t, does the transistor start to conduct. Therefore, a gate-line fault may cause a transistor to be in the non-conduction mode. A short defect between the gate terminal and the source terminal (defect *d4*, Figure 3-6) of the transistor forces the same voltage on both these terminals. As a result, the P2 transistor is never in the conduction mode, at the same time the same defect causes N2 transistor to be always in the conduction mode (SON fault which are discussed in Section 3.2.2).

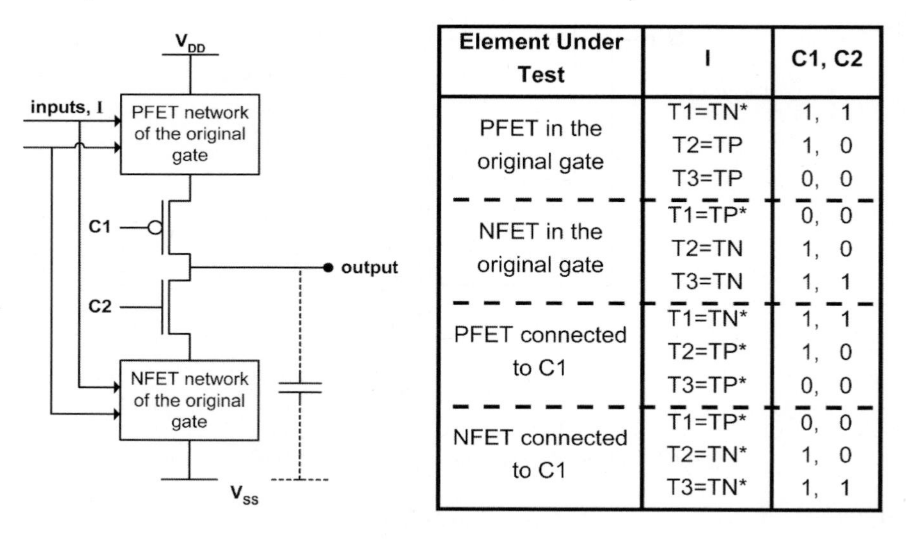

Element Under Test	I	C1, C2
PFET in the original gate	T1=TN*	1, 1
	T2=TP	1, 0
	T3=TP	0, 0
NFET in the original gate	T1=TP*	0, 0
	T2=TN	1, 0
	T3=TN	1, 1
PFET connected to C1	T1=TN*	1, 1
	T2=TP*	1, 0
	T3=TP*	0, 0
NFET connected to C1	T1=TP*	0, 0
	T2=TN*	1, 0
	T3=TN*	1, 1

Figure 3-7. Robust SOP fault detection scheme and the 3-pattern test procedure [74].

Wadsack demonstrated that in the presence of a SOP fault in a CMOS logic gate, the gate shows a memory effect under certain input conditions [103]. Therefore, such faults are not guaranteed to be detected by a SAF test set. In general, SOP fault detection requires a two pattern test sequence {T1, T2} [42]. The first test vector of the sequence, T1, is referred to as the initializing test vector and the second test vector of the sequence, T2, is referred to as the fault excitation test vector. Test vector T1 initializes the output to logic 0 (1) and T2 attempts to set the output to logic 1 (0) if the logic gate is fault-free. A failure to set the output to logic 1 (0) indicates the presence of a SOP fault.

Some SOP faults in static CMOS logic gates require only one test vector, T2. If there is only one path from output to V_{SS} (V_{DD}) and the SOP fault affects this path, it is not possible to set the output to V_{SS} (V_{DD}). For example, in the case of a 2-input NAND gate (Figure 3-4), test vector T2 {A = 1, B = 1} will detect SOP faults in the n-channel transistors. Effectively, such SOP faults disconnect all possible paths from the output to V_{SS} and cause a SA1 fault (Z/1) at the output of the NAND gate and fault detection requires only a single test vector [73].

Design for SOP Fault Testability

A 2-pattern test for SOP fault detection can be invalidated (i.e., may not detect the fault it was supposed to detect) by arbitrary circuit delays and glitches if patterns are not carefully selected. In fact, for some irredundant CMOS complex gates a robust test (which is not invalidated by arbitrary circuit delays) for SOP faults does not exist [74]. At the gate level, a robust test sequence is a sequence of test vectors in which each successive test vector differs from the previous test vector in only one bit position. However, it is difficult to generate a robust test for a given logic gate from primary inputs of an IC because even a single bit position change in a test vector may produce multiple changes at the faulty gate. As a remedy for robust SOP fault detection, DfT solutions for complex gates have been proposed [74]. In the first scheme, an addition of two transistors with two independent control lines to each complex gate was suggested. Three test vectors are needed to detect a SOP fault. Figure 3-7 shows the scheme and the test procedure. TN and TP are the fault evaluation patterns (T2). The pattern TN* (TP*) is any value of input I that would have established one or more paths from V_{SS} (V_{DD}) to the output node.

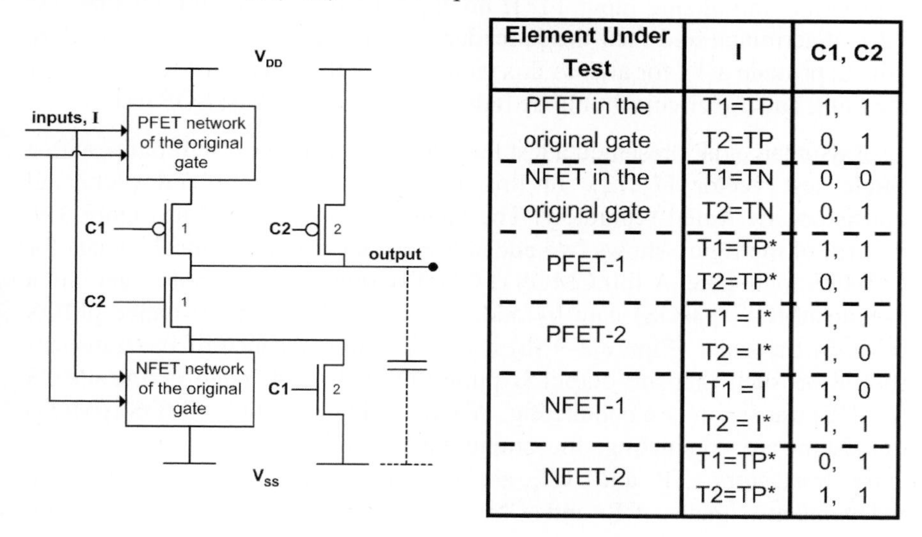

Element Under Test	I	C1, C2
PFET in the original gate	T1=TP	1, 1
	T2=TP	0, 1
NFET in the original gate	T1=TN	0, 0
	T2=TN	0, 1
PFET-1	T1=TP*	1, 1
	T2=TP*	0, 1
PFET-2	T1 = I*	1, 1
	T2 = I*	1, 0
NFET-1	T1 = I	1, 0
	T2 = I*	1, 1
NFET-2	T1=TP*	0, 1
	T2=TP*	1, 1

Figure 3-8. Robust SOP fault detection scheme and the 2-pattern test procedure [73].

In the above mentioned scheme an SOP fault needs a 3-pattern test, which may result in a longer test sequence and thus increased test costs.

Therefore, Reddy et al. [73] suggested a second scheme that requires a 2-pattern test; and the test will not be invalidated by arbitrary circuit delays. However, a total of four transistors for each complex gate are needed for implementation. The additional two transistors PFET-2 and NFET-2 are added between the output and V_{DD} (V_{SS}) in parallel to the PFET (NFET) network. The scheme and the test procedure are shown in Figure 3-8. To test a SOP fault in a PFET the initialization vector is provided through NFET-1 and NFET-2. During this time input-I is the evaluation test vector, however, its evaluation is blocked by non-conducting PFET-1. In the second test vector, T2, since only the control is changed and input-I stays the same, the test is not invalidated by the circuit delays. Similarly, other parts of a complex gate are tested for SOP faults. The test vector I* in Figure 3-8 signifies any arbitrary input test vector. Both of these DfT schemes could also detect SOP faults in added transistors. However, the practicality of these schemes is limited owing to high area overhead and performance degradation.

Reddy et al. [75] proposed a robust 2-pattern test for SOP faults in combinational circuits, if such a procedure existed, which will not be invalidated in the presence of arbitrary circuit delays. They assumed that T2 of the 2-pattern test is given and then provided a procedure to determine an appropriate initializing input T1. If no appropriate T1 is found, another test T2 is determined and then the procedure is repeated. In case the procedure fails to provide a T1 for all T2s this would imply that no robust test exists for the fault, and the circuit should be redesigned to have robust SOP test.

Rajsuman et al. presented a test technique for testing of SOP faults with a single test vector [71,72]. In this technique, n-channel and p-channel transistors are tested separately. The technique is illustrated in Figure 3-9. Part (a) of the figure shows two added transistors and two control signals for a CMOS logic gate. A full CMOS (FCMOS) logic gate is transformed into a pseudo nMOS (pMOS) gate by adding an extra high on resistance pMOS (nMOS) transistor (Figure 3-9 (b) and (c)). The resistance of the transistor should be such that the output is pulled high (low) if none of the nMOS (pMOS) transistors are conducting. When single or multiple nMOS (pMOS) transistors are conducting, the output voltage is close to V_{SS} (V_{DD}). Two extra transistors, TP and TN, are needed and are controlled by two independent signals, CP and CN, respectively. During normal circuit operation these transistors are switched-off (CP = 1,CN = 0). The testing of SOP faults in nMOS transistors is performed as follows: CP and CN are kept low. Inputs are applied to the nMOS part such that the output is pulled low through each possible Boolean combination of inputs. In the presence of a SOP fault in the nMOS part, the output is not pulled low for one or more input conditions. In fact, these inputs are the excitation vectors (T2) of the conventional 2-pattern test. Similarly, the pMOS part is tested by keeping CP and CN logic high.

In the presence of a SOP fault, the output of a pseudo nMOS (pMOS) logic gate shows a SA1 (SA0) behavior. Therefore, an automatic test pattern generating program (ATPG) can be used for test generation. Furthermore, a significant reduction in test generation and test application time is also expected. However, the scheme requires two transistors per logic gate. In addition, two control lines are required to control the transistors. These extra transistors themselves are untestable. Furthermore, extra transistors will cause an increase in parasitic capacitance which will have an impact on circuit performance. In a subsequent article, Jayasumana et al. [43] proposed a DfT solution which requires only one transistor and two control lines.

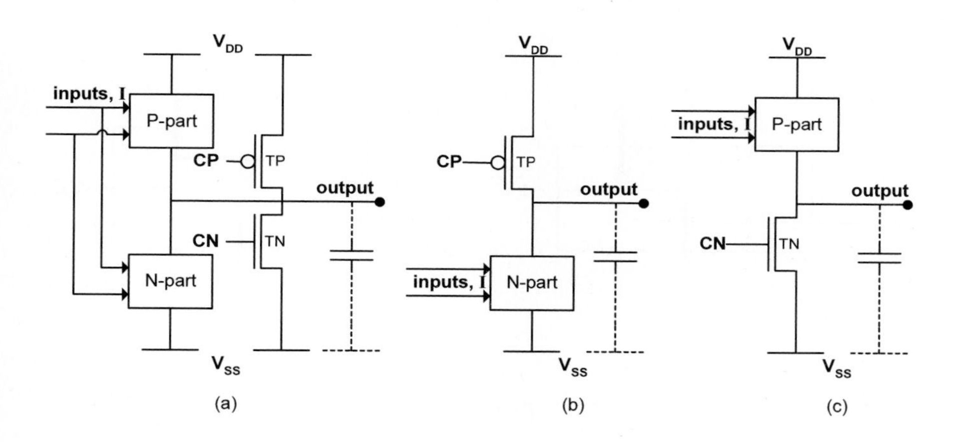

Figure 3-9. Single pattern SOP fault detection procedure.

Layout Rules for SOP Fault Detection

In the previous sub-section, we saw that robust 2-pattern test generation for SOP fault testability is difficult for a complex IC. Furthermore, SOP DfT schemes have area and performance penalties that restrict their application.

The layout of basic gates has a significant influence on the occurrence of open defects. A logic gate's layout can be modified such that the probability of a SOP fault is reduced or eliminated. Koeppe presented a set of layout rules to deal with SOP faults [45]. By application of these rules, SOP faults are either reduced or their detection is simplified. He argued that the SOP faults, in general, are caused by missing contacts, cracks in metal over oxide steps, and dust particles. For S/D-line faults, only faults in the parallel branches (p-channel transistors in NAND gates) of a basic logic gate (NAND, NOR) require 2-pattern test sequences. To detect such SOP faults, he suggested a reduction in the contact locations. Figure 3-10 illustrates the conventional and an alternative stick diagram for a 2-input NAND gate. In the alternative stick diagram (Figure 3-10 (b)) a contact is placed such that its absence affects all parallel branches together. Therefore, such a fault causes a SA (Z/0) fault and is detected by the SA test vector set. Similarly, for gate-line SOP faults, he suggested branchless and fixed order routing of signals inside the logic gates such that the chance of an open defect causing a single SOP transistor is reduced. In certain instances, it may not be possible to reduce contact locations. In such instances, placement of an extra contact reduces probability of a fault and improves the yield, often without area penalty.

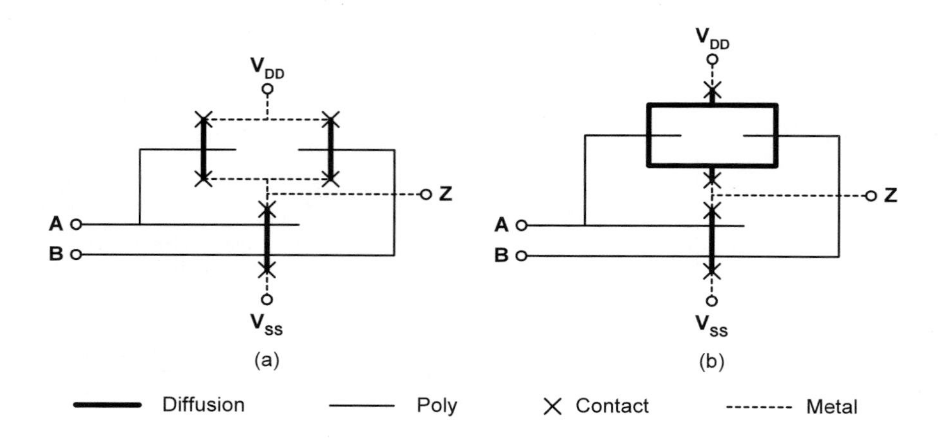

Figure 3-10. Stick diagrams of a 2-input NAND gate; (a) conventional stick diagram, (b) alternative stick diagram to avoid an open S/D line.

To investigate the effectiveness of the rules, Koeppe performed the fault simulation over original and modified layouts. The fault simulation results demonstrated that the SOP fault coverage of the SA test vector set increased substantially for the modified layout. The area overhead for the implementation of the rules was also low. It was expected to be between 0% and 20% depending upon the application of the rule and the original style of the cell layout. A small performance degradation was also expected since the parasitic capacitance of the transistor drains in the alternative layout was higher than in the original layout.

Some defects such as resistive contacts/vias may cause transistor delay to increase substantially. Such defects cause transistors to exhibit a SOP behavior at high frequency since increased delay result in timing failure. A relatively low frequency test may not be able to test the SOP behavior.

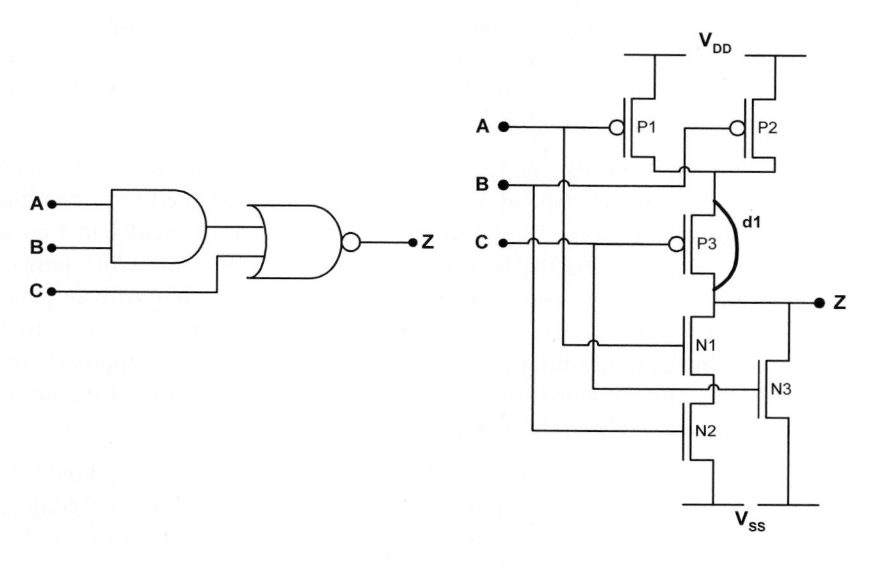

Figure 3-11. A SON fault in a CMOS complex gate.

3.2.2 Transistor Stuck-On Fault Model

A stuck-on (SON) fault forces a transistor into the conduction mode irrespective of the voltage on its gate terminal. Figure 3-11 illustrates a defect causing a transistor SON fault. The figure shows a 3-input AND-NOR complex gate. A bridging defect between source and drain of the P3 transistor causes a SON fault. SON faults cause state dependent degradation of Boolean output levels. Therefore, their detection depends on circuit level parameters. In order to detect the SON fault in transistor P3, the input test vector ABC is chosen as 001, 0X1 or X01. In the fault free case, the output should be logic low. However, due to the SON fault, a conflict is created between N3 and the p-channel transistors that causes a resistive ladder between V_{DD} and V_{SS}. The voltage on the output Z depends on the on resistance of the transistor N3, resistance of the defect *d1*, and the on resistance of conducting p-channel transistors. Often this voltage lies in the ambiguous region between logic 0 and logic 1. Hence, such faults are very difficult to detect by logic testing.

A unique property of CMOS circuits can be exploited to test for SON faults. The steady state current consumption (I_{DDQ} as it is popularly known) in CMOS circuits is low. A million-transistor IC implemented in 180 nm CMOS technology may have an I_{DDQ} value of less than 100 μA [39]. As we know, a SON fault at appropriate input logic conditions causes a resistive ladder between V_{DD} and V_{SS} nodes. The steady state current through this resistive path indicates a SON fault. For most practical situations, the difference between faulty and the fault-free I_{DDQ} is sufficiently large for an unambiguous fault detection [91].

However, the benefits of I_{DDQ} testing are beginning to erode with transistor scaling for high speed VLSIs. The set of rules used to shrink the dimensions of transistors, called general selective scaling results in higher MOSFET I_{OFF} currents. As the total chip leakage current approaches tens of mA range, the defect-free and defective I_{DDQ} distributions begin to overlap, hence reducing its effectiveness. New forms of I_{DDQ} are being devised which are effective in deep sub-micron technologies. Some of these methods are ΔI_{DDQ} and I_{CCQ}, current signatures which exploit differential measurement to cancel the increasing common-mode leakage current.

There have been efforts to detect SON faults with delay testing, however, with limited success. It can be shown that a SON fault affecting an n-channel transistor in a primary CMOS gate will cause an extra delay in the 0 → 1 output transition under certain input conditions. However, the same defect may speed up 1 → 0 output transitions under different input conditions. In a study of CMOS logic gates, Vierhaus et al. [102] found that in general SON faults can not be safely tested with delay testing but are effectively tested with I_{DDQ} testing.

3.3 Layout Level Fault Modeling

Layout level fault modeling is motivated by several factors. First and foremost amongst them is the inability of logic and transistor level fault models to represent physical defects with desired accuracy. Many defects (e.g., gate oxide defects) degrade the transistor behavior in a manner that cannot be mapped onto a transistor-level fault model. Similarly, bridging faults in interconnects cannot be mapped to higher level fault models. Higher packing densities and smaller feature sizes make such defects more likely in contemporary technologies. In other words, a large number of potential defects cannot be modeled by transistor or logic level fault models. The rising quality objective of 100 PPM or less necessitates that the layout information is exploited to generate better and more effective fault models. Often, all faults are assumed to be equally probable in logic and transistor level fault models. However, in reality this is rarely the case. Some faults are more likely than others. This information should be exploited not only for efficient, effective, and economic test generation but also for creation of defect insensitive layouts. This subject is treated at length in Chapter 2, Section 2.2.

3.4 Function Level Fault Modeling

Over the years, semiconductor technology has matured such that a VLSI chip today contains a variety of digital and analog functional blocks. The motivation behind this integration is to offer cheaper and more reliable system solutions. Very often a single IC package may contain all functional blocks of an entire micro-controller or DSP processor including memories and analog interfaces. These functional blocks or system components must share a common substrate and manufacturing process. This development has resulted in dramatic changes for testing. In spite of advances in CAD tools and CPU power, it is no longer possible to simulate faults in a complex IC at the transistor level of abstraction. Instead, a complex IC is divided into many functional modules or macros [7]. In many cases, it is possible to model (or map) the impact of defects on the function of the macro. Once this mapping is known test vectors can be generated. However, this mapping must be repeated for each transistor-level implementation.

Testing of semiconductor RAMs is a typical example of functional level fault modeling. However, RAM fault modeling, test algorithm development, and testing is a mature discipline by itself. A lot of attention has been paid on modeling [18,35,78] and testing of faults in RAM [62,83]. In Chapter 5 we address the defects and their detection strategies for RAMs. Hence, in this sub-section, we only address the function level fault modeling taking RAMs as a vehicle. Furthermore, for a tutorial overview an interested reader is referred to [1,16,32].

Thatte and Abraham [97] suggested a functional test procedure for semiconductor RAM testing. They argued that all RAM decoder and Read/Write logic faults can be mapped onto the RAM matrix as inter-cell coupling faults. An address decoder is combinational logic that selects a unique RAM cell for a given address. Assuming that under faulty conditions, the address decoder stays combinational it will behave in one of the following manners[1] :

- The decoder will not access the addressed cell. In addition, it may access another cell.

[1] Certain address decoder faults violate this assumption and we will discuss address decoder faults in greater detail in Chapter 5.

- The decoder will access multiple cells, including the addressed cell.

Both of these faulty situations can be viewed as coupling faults involving two or more RAM cells. Similarly, the impact of Read/Write logic faults is viewed as the SAF and/or coupling fault in the RAM matrix. On the basis of these arguments, the authors evolved efficient algorithms (complexity $O(n \cdot \log n)$) compared to more complex $O(n^2)$ methods prevalent in the 1970s. Similarly, there had been other attempts to model RAMs, PLAs [90], and microprocessors [10,98].

3.5 Delay Fault Models

The input-output relationship in digital circuits is Boolean in nature. The logic and transistor level fault models describe the steady state malfunctioning of the Boolean relationship, but cannot model the faulty delay behavior of a logic element. Timing (or delay) is also an important design parameter in the input-output relationship. An otherwise good IC may fail to perform correctly in a system if it fails to meet designed timing specifications. With increasing system complexities and higher operational frequencies, timing is becoming an important aspect of the design. Furthermore, rising quality expectations motivate testing for the correct temporal behavior, commonly known as delay testing [12].

Generally, pre-fabrication timing is verified at each successive level of design hierarchy. At each level the objective of the analysis is either to determine the maximum operational frequency at which circuit will behave correctly, or to guarantee that the circuit operates without any malfunction at a pre-specified clock rate [3]. Once a chip is fabricated, it still must be tested for a pre-specified clock frequency. A circuit is said to have a delay fault if the output of the circuit fails to reach its final value within pre-specified timing constraints. Often functional test vectors are used for speed testing. Alternatively, functional test vectors together with minimum V_{DD} is also used to enhance the delay coverage.

A timing or delay fault in an IC could be caused by a number of reasons that include subtle manufacturing process defects, transistor threshold voltage shifts, increased parasitic capacitance, improper timing design, etc. A substantial research effort has been directed towards delay fault testing [12,14,41,64,65,100,101]. Broadly, two fault models have been proposed for delay fault testing in the literature: (i) the gate delay fault model, and (ii) the path delay fault model.

3.5.1 Gate Delay Fault Model

Each gate in an IC is designed with a pre-specified nominal delay. However, under the gate delay fault model, the faulty gate may assume a considerably larger delay. The test complexity of the gate delay fault model is relatively small compared to that of the path delay fault model. This is because in a digital IC the number of paths can be exponential in the number of gates.

It appears that SA or SOP faults are special (limiting) cases of gate delay faults. For example, in the case of a SAF the logic gate output has an infinite delay for a class of input stimuli. Similarly, for a SOP fault the transistor has infinite delay. However, there is an important distinction between SAF or SOP faults and delay faults. Unlike SA or SOP faults, a gate delay fault does not necessarily cause the circuit to malfunction. In other words, a faulty gate may assume significantly larger delay than its nominal delay, and still the circuit could work within the timing constraints. Therefore, in general, an evaluation scheme for a delay fault test must not only compute whether or not a delay fault is detected but also calculate the size of the fault. The size of a delay fault is defined as the fault detection size (FDS) [66] or as the detection threshold [41] for a test, T. The FDS of a fault for the test T has the property that T is guaranteed to detect any fault at that site that is greater than the FDS. However, the best FDS achievable for any gate delay fault detecting test is the corresponding slack at the fault site [41,66]. The slack of a signal is defined as the difference between clock period and the propagation delay of the longest delay path through that signal. Therefore, the quality of the test set depends how small a delay fault can be tested by a test. Methods for designing tests that activate the longest sensitized path through every gate have been discussed in the literature [49,64].

Reducing the transistor and logic gate delay has been one of the primary motivations of technology scaling (Table 1-1). Figure 3-12 illustrates performance increase with each successive technology node. As can be seen from this figure, gate delays have been scaled to tens of pico-second in 130 nm CMOS technology. When logic gates delays are in pico-second regime, innocuous manufacturing deviations from the process recipe may cause undesirable delay effects. The propagation delay for a long channel transistor is given by following, simplified equation [70].

$$t_P = \frac{CV_{DD}}{\frac{1}{2}k_n'\left(\frac{W}{L}\right)(V_{DD} - V_t)^2} \tag{3.1}$$

As technology scales, the gate delay is reduced owing to capacitance, power supply and the transistor threshold voltage reduction. Figure 3-12 illustrates the decreasing gate delay. If the threshold voltage of a transistor is increased due to manufacturing anomaly it results in larger propagation delay which is illustrated by Equation 3.2.

$$t_P + \Delta t_P \cong \frac{CV_{DD}}{\frac{1}{2}k_n^{'}\left(\frac{W}{L}\right)\left(V_{DD} - (V_t + \Delta V_t)\right)^2} \tag{3.2}$$

Technology scaling has other undesirable consequences. As CMOS technology scales, the within-die and die-to-die threshold voltage variation of MOSFETS is increased (see Figure 3-13) for two basic reasons. The first reason is the increased variation of process technology parameters with scaling [59]. Typically, these parameters are the effective channel length (L_{eff}), the gate oxide thickness (T_{ox}), the channel doping (N_{ch}) and the depths of source and drain junctions (X_j). The second reason is the intrinsic random fluctuations of doped atoms in channel region. For example, devices with 90 nm channel length and channel doping approximately 5 x 10^{18} cm^{-3} have approximately 170 doping atoms in the channel depletion region. The statistical distribution of doping atoms in channel region has significant impact on threshold voltage variation in nanometer-scale MOSFETs [4].

3.5.2 Path Delay Fault Model

The major drawback of the gate delay fault model is that the interconnect delay is not considered. This was acceptable when the feature size was relatively large and the gate delay was relatively large compared to the interconnect delay. However, the scaling of process dimensions has changed this equilibrium. The transistor switching times have reduced dramatically due to smaller geometries. As line widths scale into the deep sub-micron regime and device switching speeds continue to improve, however, delays due to interconnects have not scaled. In fact, the interconnect resistance has increased due to smaller cross sectional area while the parallel plate capacitance is decrease due to the same reason. Furthermore, there is no significant decrease in interconnect *rc* time constant due to increased fringe capacitance. In addition, contemporary VLSIs have longer interconnect, therefore, interconnect delays have increased [6,8,82]. Thus, the gate delay fault model is restrictive in its application to finer geometries. Furthermore, the gate delay fault model cannot account for the cumulative effect of small delay variations along paths from primary inputs to primary outputs.

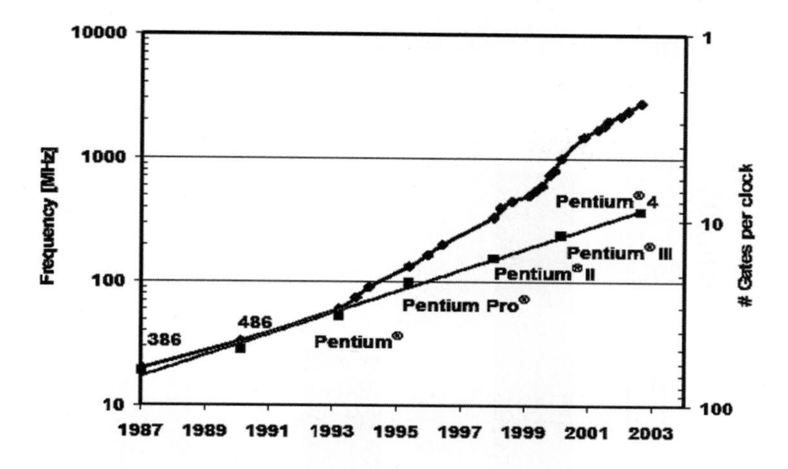

Figure 3-12. Frequency and gate delay with technology scaling [109].

The path delay fault model considers the cumulative delay of paths from primary inputs to primary outputs. The path delay fault model, in addition to single isolated failures, also considers distributed delay effects due to statistical process variations. A faulty situation may arise in spite of the fact that each individual component meets its individual delay specifications. A path delay test will detect both localized as well as distributed delay defects. Path delay faults may also provide a mechanism for monitoring process variations that may have significant impact on critical paths. Furthermore, they provide an ideal vehicle for speed-sorting since they have the most accurate description of the clock speed at which timing failures begin to occur [50,89].

However, the path delay fault model has the disadvantage that it is only practical to generate tests for a small number of the total paths in a given circuit. Hence, the path delay fault coverage tends to be low [101]. For all practical purposes, the delays in the longest and the shortest paths (critical paths) are considered. If these delays are within the clock cycle, the circuit is considered to be delay fault-free, otherwise it contains a path delay fault.

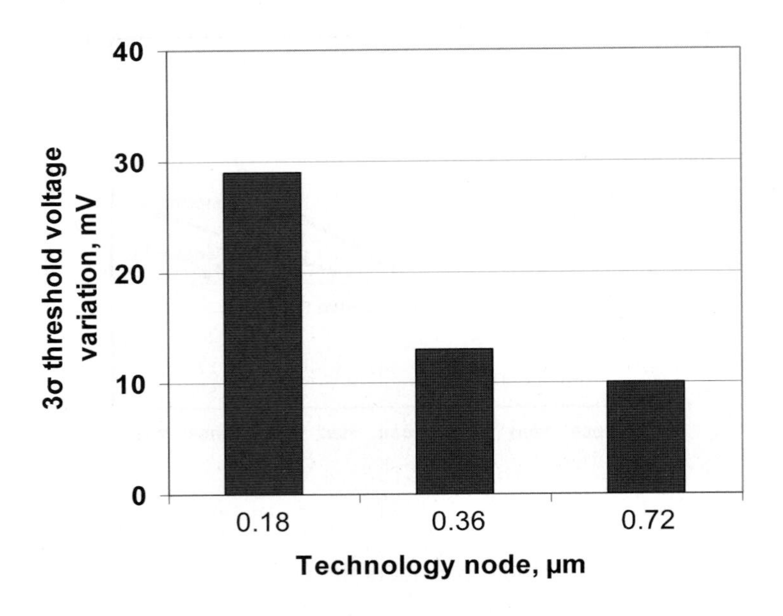

Figure 3-13. The spread of V_t increases with technology scaling [59].

3.5.3 Robust and Non-Robust Tests for Path Delay Faults

Delay fault testing assumes that the delay of a gate (or a path) depends on the transition propagated from the input to the output. However, the fault is independent of the vector that induces the given transition. Testing a delay fault requires a 2-pattern test, {V1, V2}. Similar to SOP testing, the first test vector, V1, is called initializing test vector and the second vector V2, is called the fault exciting test vector.

Irrespective of the fault mode (gate or path delay), the 2-pattern test may be categorized as robust or non-robust. A robust test detects the targeted delay faults irrespective of the presence of other delay faults in the circuit. Numerous classifications of robust path delay fault tests exist [12,47,88], e.g., hazard free robust tests, single/multiple input changing tests, and single/multiple path propagating tests. A necessary and sufficient set of path delay faults, known as primitive delay faults, must be tested to guarantee timing correctness of the circuit [44]. An important property of a path delay test is that the test must not be invalidated by variable delays of the fan-in signals of gates on the targeted path. On the other hand, a non-robust test detects the fault if no other delay faults affect the circuit. These faults are statically sensitizable. A fault is statically sensitizable if there exists at least one input vector which stabilizes all side inputs of the gates on the target path at non-controlling values [88]. In general, it is possible to determine whether or not a given 2-pattern test is robust by examining the logic structure of the circuit under test. The actual circuit delays are not important in this determination [65].

3.6 Leakage Fault Model

As mentioned before, static CMOS circuits have very low quiescent current (I_{DDQ}). Most manufacturing defects in CMOS ICs exhibit state dependent elevated I_{DDQ}. Therefore, I_{DDQ} testing is a powerful test method in manufacturing process defect detection. A defect-free MOS transistor has nearly infinite input impedance; hence, there should not be any current between gate and source, gate and drain, or gate and substrate (well). However, some defects, such as a gate oxide short, will cause leakage current between the gate and other nodes of the transistor. In general, a leakage fault may occur between any two nodes of a MOS transistor. Nigh and Maly [61] and Mao et al. [54] independently proposed a leakage fault model containing six types of faults for MOS transistors:

- F_{GS} – leakage fault between gate and source
- F_{GD} – leakage fault between gate and drain
- F_{SD} – leakage fault between source and drain
- F_{BS} – leakage fault between bulk and source
- F_{BD} – leakage fault between bulk and drain
- F_{BG} – leakage fault between bulk and gate

These faults include not only the gate oxide defect causing leakage but also the leakages between various diodes required to realize a MOS transistor. Furthermore, Nigh and Maly [61] suggested that well to substrate diode defects need not be considered explicitly since leakage or latchup caused by them is easily observable.

Leakage faults such as gate oxide shorts or pn-junction pinholes occur quite frequently in the CMOS process. Also, reduced geometries increase the electric field in MOS transistors, which may cause successive degradation of gate oxide, etc. Typically, small leakage faults do not cause a catastrophic failure of an IC. However, they are potential reliability hazards [91].

3.7 Temporary Faults

Unlike most other faults, temporary faults cannot be classified according to levels of fault modeling. Therefore, they should be treated separately. As their name suggests, temporary faults are not permanent in nature. A major portion of digital system malfunctions is caused by temporary faults. They are harder to detect because at the time of testing they are not reproduced. There are two types of temporary faults (i) transient faults or intermittent faults and, (ii) reliability faults.

3.7.1 Transient or Intermittent Faults

Transient or intermittent faults are non-recurring temporary faults. Typically they are caused by radiation or power supply fluctuations. Transient faults can also be caused by capacitive or inductive coupled disturbances and by an external electromagnetic field. They are not repairable because they do not cause physical damage to the hardware. Dynamic logic and memories are particularly susceptible to such faults. [80]

Soft Errors or Single Event Effects

A single event effect (SEE) is caused by a single energetic particle. Single event phenomena can be grouped into three different effects: Single event upset (soft error); Single event latchup (soft or hard error); and Single event burnout (hard failure).

1. Single Event Upset

Single event upset (SEU) is a radiation-induced error in microelectronic circuits caused when charged particles ionize the material through which they pass, leaving behind a wake of electron-hole pairs. SEUs are transient soft errors, and are non-destructive. A reset or rewriting of the device results in normal device behavior thereafter. An SEU may occur in analog, digital, or optical components, or may have effects in surrounding interface circuitry. SEUs typically appear as transient pulses in logic or support circuitry, or as bit flips in memory cells or registers. A multiple bit SEU is also possible in which a single ion hits two or more bits causing simultaneous errors. Multiple-bit upset poses a serious problem for bit error detection and correction (EDAC) circuits which may not be able to recover the data. A severe SEU is the single-event functional interrupt (SEFI) in which an SEU in the device's control circuitry places the device into a test mode, halt, or undefined state. The SEFI halts normal operations, and requires a power reset to recover [38].

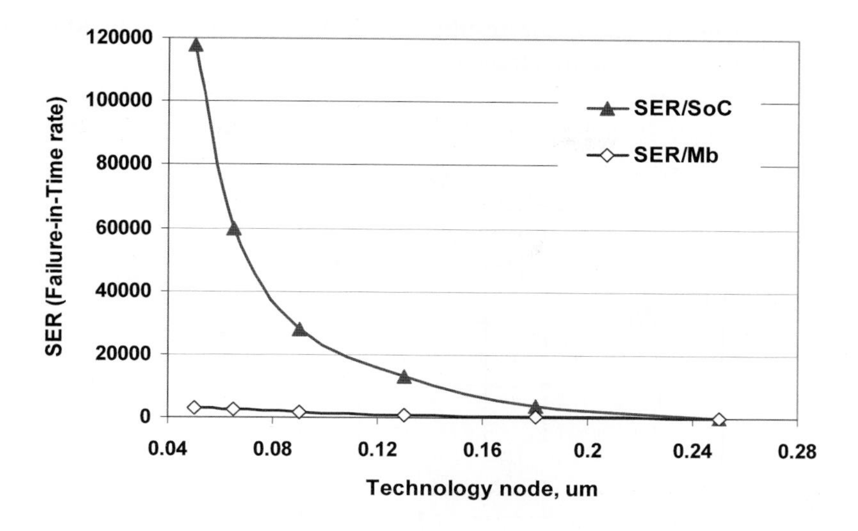

Figure 3-14. Alpha accelerated SER at nominal V_{DD} vs. technology scaling.

2. Single Event Latchup

Single event latchup (SEL) is a condition that causes loss of device functionality due to a single-event induced current state. SELs are hard errors, and are potentially may cause permanent damage. The SEL results in a high operating current, above device specifications. The latched condition can destroy the device, drag down the bus voltage, or damage the power supply. An SEL is cleared by a power off-on reset or power strobing of the device. If power is not removed quickly, catastrophic failure may occur due to excessive heating or metallization or bond wire failures. SEL is strongly temperature dependent: the threshold for latchup decreases at high temperature [38].

3. Single Event Burnout

Single event burnout (SEB) is a condition that can cause device destruction due to a high current state in a power transistor. SEB causes the device to fail permanently. SEBs include burnout of power MOSFETs, gate rupture, and noise in sensitive nodes. An SEB can be triggered in a power MOSFET biased in the OFF state (i.e., blocking a high drain-source voltage) when a heavy ion passing through deposits enough charge to turn the device on. SEB susceptibility has been shown to decrease with increasing temperature.

A power MOSFET may undergo single-event gate rupture (SEGR), which is the formation of a conducting path (i.e., localized dielectric breakdown) in the gate oxide resulting in a destructive burnout. SEB can also occur in bipolar junction transistors (BJTs) [38].

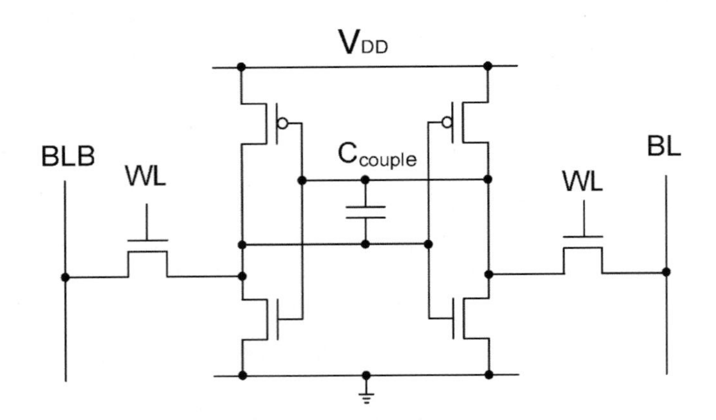

Figure 3-15. Schematic of the SRAM with the improved SEU immunity.

Impact of Scaling

The effect of technology scaling on soft error rate (SER) was widely investigated using the alpha accelerated SER measurements. It was shown that SER is exponentially increased with V_{DD} scaling. The strong dependence on V_{DD} is well known, as the critical charge of a storage cell is a direct function of the supply voltage. Note, that storage-cell node capacitance primarily consists of the gate capacitance, junction capacitance and local interconnect parasitic capacitance. The size and areas of last two components are reduced with technology scaling and typically can not be compensated by the increase of gate capacitance due to the reduction of gate dielectric thickness. The reduction of node capacitance results in an increased sensitivity to soft upsets, since the critical charge of the cell is directly related to the node capacitance. Figure 3-14 shows the FIT rate in SRAM cells for different CMOS technologies and nominal V_{DD} specified for each technology [76]. It can be seen that whereas there is a linear and moderate increase of SER on a per Mbit basis, the system SER significantly increases with the number of SRAM bits embedded in the chips. The trend of exponential increasing of the accelerated alpha SER in SRAM cells with decreasing technology node has been also reported in [23].

Since soft errors are a critical issue for deep submicron CMOS technologies, the different process and design techniques were proposed to improve SER immunity. For example, epitaxial (EPI) substrates with a lightly doped EPI layer on a heavily doped substrate can be easily integrated into standard CMOS process flow. If an alpha particle hits the silicon, the substrate doping leads to rapid charge recombination and reduction of SER. The EPI layer should be as thin as possible without degrading the MOSFET and connecting the EPI substrate with the well tubs at the same time. The typical EPI thickness is 2.5-3.0 μm [23]. The triple-well technology has been also used in deep-submicron VLSI technology to provide a complete electrical isolation for NMOS devices in a p-type substrate and to improve the overall SRAM SER performance [110]. It was reported that SRAM cells implemented in triple-well CMOS technology has soft error immunity lower by ~2x than SRAM cells implemented in conventional CMOS technology without n-well. This is because the quantity of charges reaching a memory cell is reduced since some of the charges are absorbed in the n-well.

The design techniques, which are typically used for radiation hardening ICs, are based on increased node capacitance in latches, flip-flops and SRAM cells. For example, Y.Z. Xu et al. proposed to add a backend MIM capacitor on the top of the metal one layer to increase the storage capacitor of the nodes (see Figure 3-15). As a result, the SER FIT rate was reduced by 80% [108]. Similar design solutions were proposed in [76,60]. For example, the "robust SRAM" (rSRAM) cell with two extra stacked capacitors was developed in [76], it was shown that rSRAM is 250 times more robust that a conventional SRAM cell. The combining of *rc* network with the 10 transistor SRAM cell, allowed to develop a memory chip with a dose rate upset threshold more than $2x10^{12}$ rad(Si)/s implemented in silicon-on-sapphire (SOS) technology [60]. The developed radiation hardened SRAM chip is suitable for military and space applications.

Crosstalk, and Power Supply Noise

An undesirable capacitive or inductive coupling from a neighboring wire to a sensitive circuit node introduces a noise that is often called cross talk. This coupling may result in data dependent delay faults, or intermittent faults. The impact of the cross talk is severe if the victim line or node is floating. A floating line is unable to recover from the impact of the cross talk and in some limiting cases may result in erroneous condition. On the other hand, driven nodes or wires, are more robust against crosstalk.

3.7.2 Reliability Faults

Reliability faults often occur in nature and appear at regular intervals. These faults are caused by circuit parameter degradations, aging [58], or soft defects. This degradation is progressive until a permanent failure occurs.

These faults can also occur due to design sensitivity to environmental conditions like ambient temperature, humidity, and vibrations. The frequency of their occurrence depends on how effective an IC (system) is protected against environmental conditions through cooling and shielding.

4. CONCLUSIONS

The functional testing of complex digital ICs is prohibitively expensive and does not ensure that the IC is fault-free. Structural tests that target faulty circuit behavior provide an alternative. The effectiveness of a structural test is quantifiable in terms of the covered faults. Thus, it allows the user to establish a relationship between the test coverage and the quality of the tested devices. The test generation for structural test is considerably simpler due to availability of CAD tools. However, the structural test requires a fault model that represents likely manufacturing process defects with an acceptable accuracy and provides an objective basis for the structural test generation. A number of fault models are available and most of them are classified according to the level of abstraction. Gate level (SAF), transistor level (SOP, SON) and function level are some of the examples of the abstraction levels. The level of abstraction is essentially a compromise between the fault model's ability to represent actual defects and the speed of processing of the fault in a fault simulation environment.

The conventional fault modeling approaches do not consider the likely or realistic faults in a given layout of a circuit. The layout of a circuit has a significant impact on the faulty circuit behavior. The IFA takes into account the circuit layout and the defect data from the manufacturing site to generate a list of realistic faults. The word realistic signifies that each fault has a physical basis (i.e., defect). In this manner the circuit layout dependent fault models are evolved. Many reported experiments illustrate the effectiveness of the method in realistic fault model generation with success.

References

1. M. S. Abadir, and H. K. Reghbati, "Functional Testing of Semiconductor Random Access Memories," ACM Computing Surveys, 15(3), pp. 175-198, Sept. 1983.

2. J. A. Abraham, "Fault Modeling in VLSI," VLSI Testing, vol. 5, pp. 1-27, 1986.

3. V. D. Agrawal, "Synchronous Path Analysis in MOS Circuit Simulator," Proceedings of the 19th Design Automation Conference, pp. 629-635, 1982.

4. A. Asenov, A.R. Brown, J.H. Davies, S. Kaya, G. Slavcheva, "Simulation of Intrinsic Parameter Fluctuations in Deca-nanometer and Nanometer-scale MOSFETs," IEEE Trans. on Electron Devices, vol. 50, No. 9, pp. 1837-1852, 2003.

5. P. Banerjee, and J. A. Abraham, "Characterization and Testing of Physical Failures in MOS Logic Circuits," IEEE Design and Test of Computers, vol. 1, pp. 76-86, August 1984.

6. H. B. Bakoglu, and J. D. Meindel, "Optimal Interconnection Circuits for VLSI," IEEE Transactions on Electron Devices, vol. ED-32, no. 5, pp. 903-909, May 1985.

7. F. P. M. Beenker, K. J. E. van Eerdewijk, R. B. W. Gerritsen, F. N. Peacock, and M. van der Star, "Macro Testing, Unifying IC and Board Test," IEEE Design and Test of Computers, vol. 3, pp. 26-32, December 1986.

8. S. Bothra, B. Rogers, M. Kellem, and C. M. Osburn, "Analysis of the Effects of Scaling on Interconnect Delay in ULSI Circuits," IEEE Transactions on Electron Devices, vol. ED-40, no. 3, pp. 591-597, March 1993.

9. H. Bouwmeester, S. Oostdijk, F. Bouwmann, R. Stans, L. Thijssen, and F. Beenker, "Minimizing test time by exploiting parallelism in macro test," Proceedings of the IEEE International Test Conference, pp. 451-460, 1993.

10. D. S. Brahme, and J. A. Abraham, "Functional Testing of Microprocessors," IEEE Transactions on Computers, vol. C-33, pp. 475-485, 1984.

11. M. A. Breuer, and A. D. Friedman, "Diagnosis and Reliable Design of Digital Systems," Woodland Hills, California: Computer Science Press, 1976.

12. T. J. Chakraborty, V. D. Agrawal, and M. L. Bushnell, "Delay Fault Models and Test Generation for Random Logic Sequential Circuits," Proceedings of the 29th Design Automation Conference, pp. 165-172, 1992.

13. R. Chandramouli, "On Testing Stuck-Open Faults," Proceedings of the 13th Annual International Symposium on Fault Tolerant Computing Systems, pp. 258- 265, 1983.

14. K. T. Cheng, "Transition Fault Simulation for Sequential Circuits," Proceedings of the IEEE International Test Conference, pp. 723-731, 1992.

15. K. T. Chang, and H.C. Chen, "Classification and Identification of Nonrobust Untestabel path Delay Faults," IEEE Transactions on CAD, vol. 15, pp. 845-853, August 1996.

16. B. F. Cockburn, "Tutorial on Semiconductor Memory Testing," Journal of Electronic Testing: Theory and Applications, vol. 5, no. 4, pp. 321-336, November 1994.

17. H. Cox, and J. Rajaski, "Stuck-Open and Transition Fault Testing in CMOS Complex Gates," Proceedings of the IEEE International Test Conference, pp. 688-694, 1988.

18. R. Dekker, F. Beenker, and L. Thijssen, "Fault modeling and Test Algorithm Development for Static Random Access Memories," Proceedings of the IEEE International Test Conference, pp. 343-352, 1988.

19. C. Di, and J. A. G. Jess, "On Accurate Modeling and Efficient Simulation of CMOS Open Faults," Proceedings of the IEEE International Test Conference, pp. 875-882, 1993.

20. E. B. Eichelberger, and T. W. Williams, "A Logic Design Structure for LSI Testability," Journal of Design Automation and Fault Tolerant Computing, vol. 2, no. 2, pp. 165-178, May 1978.

21. R. D. Eldred, "Test Routines Based on Symbolic Logical Statements," Journal of ACM, vol. 6, no.1, pp. 33-36, January 1959.

22. Y. M. El-Ziq, and R. J. Cloutier, "Functional-Level Test Generation for Stuck-Open Faults in CMOS VLSI," Proceedings of the IEEE International Test Conference, pp. 536-546, 1981.

23. A. Eto, M. Hidaka, Y. Okuyama, K. Kimura, and M. Hosono, "Impact of neutron flux on soft errors in MOS memories," IEEE International Electron Devices Meeting, pp. 367-370, 1998.

24. F. J. Ferguson, and J. P. Shen, "Extraction and Simulation of Realistic CMOS Faults using Inductive Fault Analysis," Proceedings of the IEEE International Test Conference, pp. 475-484, 1988.

25. A. V. Ferris-Prabhu, "Introduction to Semiconductor Device Yield Modeling," Boston: Artech House, 1992.

26. M. L. Flottes, C. Landrault, and S. Pravossoudovitch, "Fault Modeling and Fault Equivalence in CMOS Technology," Journal of Electronic Testing: Theory and Applications, vol. 2, no.3, pp. 229-241, August 1991.

27. S. Funatsu, N. Wakatsuki, and T. Arima, "Test Generation Systems in Japan," Proceedings of 12th Design Automation Symposium, pp. 114-122, 1975.

28. J. Galiay, Y. Crouzet, and M. Vergniault, "Physical versus Logical Fault Models in MOS LSI Circuits: Impact on Their Testability," IEEE Transaction on Computers, vol. C-29, no. 6, pp. 527-531, June 1980.

29. S. K. Gandhi, "VLSI Fabrication Principles", John Wiley and Sons, 1983.

30. D. Gaitonde, and D. H. H. Walker, "Test Quality and Yield Analysis Using the DEFAM Defect to Fault Mapper," Proceedings of the International Conference on Computer Aided Design, pp. 202-205, 1993.

31. D. S. Gardner, J. D. Meindel, and K. C. Saraswat, "Interconnection and Electromigration Scaling Theory," IEEE Transactions on Electron Devices, vol. ED-34, no. 3, pp. 633-643, March 1987.

32. A. J. van de Goor, "Testing Semiconductor Memories: Theory and Practices," John Wiley and Sons, 1991.

33. A. Goundan, and J. P. Hayes, "Identification of Equivalent Faults in Logic Networks," IEEE Transactions on Computers, vol. c-29, no. 11, pp. 978-985, November 1980.

34. R. J. A. Harvey, A. M. D. Richardson, E. M. J. Bruls, and K. Baker, "Analogue Fault Simulation Based on Layout Dependent Fault Models," Proceedings of IEEE International Test Conference, pp. 641-649, 1994.

35. J. P. Hayes, "Detection of Pattern-Sensitive Faults in Random Access Memories," IEEE Transactions on Computers, vol. C-24, no.2, pp. 150-157, February 1975.

36. J. P. Hayes, "Fault Modeling for Digital Integrated Circuits," IEEE Transactions on Computer-Aided Design of Circuits and Systems, CAD-3, pp. 200-207, 1984.

37. J. P. Hayes, "Fault Modeling," IEEE Design & Test of Computers, vol. 2, pp. 88-95, April 1985.

38. K. Holbert, "Single Event Effects" http://www.eas.asu.edu/~holbert/eee460/see.html, 2005.

39. T. Hook, L. Wissel, D. Mazgaj, "Estimation of I_{DDQ} for early chip and technology design decisions" Proceedings of the IEEE Custom Integrated Circuits Conference, pp. 627-630, 2003.

40. O. H. Ibarra, and S. K. Sahni, "Polynomial Complete Fault Detection Problems," IEEE Transactions on Computers, vol. c-24, no. 3, pp. 242-249, March 1975.

41. V. S. Iyenger et al., "On Computing the Sizes of Detected Delay Faults," IEEE Transactions on CAD, vol. 9, no. 3, 299-312, 1990.

42. S. K. Jain, and V. D. Agrawal, "Modeling and Test Generation Algorithm for MOS Circuits," IEEE Transactions on Computers, vol. 34, no. 5, pp. 426-43, May 1985.

43. A. P. Jayasumana, Y. K. Malaiya, and R. Rajsuman, "Design of CMOS Circuits for Stuck-Open Fault Testability," IEEE Journal of Solid-State Circuits, vol. 26, no. 1, pp. 58-61, January 1991.

44. W. Ke, and P. R. Menon, "Synthesis of Delay Verifiable Combinational Circuits," IEEE Transactions on Computers, vol. 44, pp. 213-222, February 1995.

45. S. Koeppe, "Optimum Layout to Avoid CMOS Stuck-Open Fault," Proceedings of the 24th ACM/IEEE Design Automation Conference, pp. 829-835, 1987.

46. F. C. M. Kuijstermans, M. Sachdev, and L. Thijssen, "Defect Oriented Test Methodology for Complex Mixed-Signal Circuits," Proceedings of the European Design and Test Conference, pp. 18-23, 1995.

47. W. K. Lam, A. Saldanha, R. K. Brayton, and A. L. Sangiovanni-Vincentelli, "Delay Fault Coverage and Performance Trade-offs," Proceedings of the 30th Design Automation Conference, pp. 446-452, 1993.

48. K. J. Lee, and M. A. Breuer, "On the Charge Sharing Problem in CMOS Stuck- Open Fault Testing," Proceedings of the IEEE International Test Conference, pp. 417- 425, 1990.

49. A. K. Majhi, J. Jacob, L. M. Patnaik and V. D. Agrawal, "On Test Coverage of Path Delay Faults," Proceedings of the 9th International Conference on VLSI Design, pp. 418-421, 1996.

50. Y. K. Malaiya, and R. Narayanaswamy, "Modeling and Testing for Timing Faults in Synchronous Sequential Circuits," IEEE Design and Test of Computers, vol. 1, no. 4, pp. 62-74, November 1984.

51. W. Maly, F. J. Ferguson, and J. P. Shen, "Systematic Characterization of Physical Defects for Fault Analysis of MOS IC Cells," Proceedings of the International Test Conference, 390-399, 1984.

52. W. Maly, A. J. Strojwas, and S. W. Director, "VLSI Yield Prediction and Estimation: A Unified Framework," IEEE Transactions on Computer Aided Design, vol. CAD- 5, no. 1, pp. 114-130, January 1986.

53. W. Maly, W. R. Moore, and A. J. Strojwas, "Yield Loss Mechanisms and Defect Tolerance," SRC-CMU Research Center for Computer Aided Design, Dept. of Electrical and Computer Engineering, Carnegie Mellon University, Pittsburgh, PA 15213.

54. W. Mao, R. Gulati, D. K. Goel, and M. D. Ciletti, "QUIETEST: A Quiescent Current Testing Methodology for Detecting Leakage Faults," Proceedings of the International Conference on CAD, pp. 280-283, 1990.

55. P. Mazumder, and K. Chakraborty, "Testing and Testable Design of High-Density Random-Access Memories," Boston: Kluwer Academic Publishers, 1996.

56. E. J. McCluskey, and F. W. Clegg, "Fault Equivalence in Combinational Logic Networks," IEEE Transactions on Computers, vol. c-20, no. 11, pp. 1286-1293, November 1971.

57. A. Meixner, and W. Maly, "Fault Modeling for the Testing of Mixed Integrated Circuits," Proceedings of the IEEE International Test Conference, pp. 564-572, 1991.

58. S. Mourad, and E. J. McCluskey, "Fault Models," Testing and Diagnosis of VLSI and ULSI, Boston: Kluwer Academic Publishers, pp. 49-68, 1989.

59. S. Narendra, D. Antoniadis and V. De, "Impact of using Adaptive Body Bias to Compensate die-to-die VT Variation on Within-die VT Variation," IEEE Int. Symp. on Low Power Electronics and Design, pp. 229-232, 1999.

60. A. Y. Nikiforov, and I. V. Poljakov, "Test CMOS/SOS RAM for transient radiation upset comparative research and failure analysis," IEEE Transactions on Nuclear Science, vol. 42 , No. 6 , pp. 2138-2142, 1995.

61. P. Nigh, and W. Maly, "Test Generation for Current Testing," IEEE Design and Test of Computers, pp. 26-38, February 1990.

62. C. A. Papachristou, and N. B. Sahgal, "An Improved Method for Detecting Functional Faults in Semiconductor Random Access Memories," IEEE Transactions on Computers, vol. C-34, no.2, pp. 110-116, February 1985.

63. E. S. Park, B. Underwood, T. W. Williams, and M. R. Mercer, "Delay Testing Quality in Timing-Optimized Designs," Proceedings of the IEEE International Test Conference, pp. 879-905, 1991.

64. E. S. Park, and M. R. Mercer, "An Efficient Delay Test Generation System for Combinational Logic Circuits," IEEE Transactions on CAD, vol. 11, pp. 926-938, July 1992.

65. A. Pierzynska, and S. Pilarski, "Non-Robust versus Robust," Proceedings of the IEEE International Test Conference, pp. 123-131, 1995.

66. A. K. Pramanick, and S. M. Reddy, "On the Computation of the Ranges of Detected Delay Fault Sizes," IEEE International Conference on CAD, pp. 126-129, 1989.

67. I. Pramanick, and A.K. Pramanick, "Parallel Delay Fault Coverage and Test Quality Evaluation," Proceedings of the IEEE International Test Conference, pp. 113-122, 1995.

68. J. F. Poage, "Derivation of Optimum Tests to Detect Faults in Combinational Circuits," Proceedings of the Symposium on Mathematical Theory of Automata, pp. 483-528, 1963.

69. B. Prince, "Semiconductor Memories," John Wiley and Sons, 1991.

70. J. M. Rabaey, A. Chandrakasan, B. Nikolic, "Digital Integrated Circuits – A Design Perspective," Printice Hall, ISBN 0-13-090996-3, 2003.

71. R. Rajsuman, A. P. Jayasumana, and Y. K. Malaiya, "CMOS Stuck-Open Fault Detection Using Single Test Patterns," Proceedings of the 26th ACM/IEEE Design Automa¬tion Conference, pp. 714-717, 1989.

72. R. Rajsuman, A. P. Jayasumana, and Y. K. Malaiya, "CMOS Open-Fault Detection in the Presence of Glitches and Timing Skews," IEEE Journal of Solid-State Circuits, vol. 24, no. 4, pp. 1129-1136, August 1989.

73. S. M. Reddy, and S. Kundu, "Fault Detection and Design for Testability of CMOS Logic Circuits," Testing and Diagnosis of VLSI and ULSI, pp. 69-91, 1989.

74. S. M. Reddy, M. K. Reddy, and J. G. Kuhl, "On Testable Design for CMOS Logic Circuits," Proceedings of the IEEE International Test Conference, pp. 435-445, 1983.

75. S. M. Reddy, M. K. Reddy, and V. D. Agrawal, "Robust Test for Stuck-Open Faults in CMOS Combinational Logic Circuits," Proceedings of the 14th International Symposium on Fault Tolerant Computing, pp. 44-49, 1984.

76. P. Roche, F. Jacquet, C. Caillat, J. P. Schellkopf, "An alpha immune and ultra low neutron SER high density SRAM," IEEE Reliability Physics Symposium, pp. 671-672, 2004.

77. B. K. Roy, "Diagnosis and Fault Equivalence in Combinational Circuits," IEEE Transactions on Computers, vol. c-23, no. 9, pp. 955-963, September 1974.

78. M. Sachdev and M. Verstraelen, "Development of a Fault Model and Test Algorithms for Embedded DRAMs," Proceedings of the IEEE International Test Conference, pp. 815-824, 1993.

79. M. Sachdev, "Defect Oriented Analog Testing: Strengths and Weaknesses," Proceedings of the 20th European Solid State Circuits Conference, pp. 224-227, 1994.

80. M. Sachdev, "A Defect Oriented Testability Methodology for Analog Circuits," Journal of Electronic Testing: Theory and Applications, vol. 6, pp. 265-276, June 1995.

81. M. Sachdev, "Reducing the CMOS RAM Test Complexity with I_{DDQ} and Voltage Testing," Journal of Electronic Testing: Theory and Applications, vol. 6, no. 2, pp. 191-202, April 1995.

82. K. C. Saraswat, and F. Mohammadi, "Effect of Scaling of Interconnections on the Time Delay of VLSI Circuits," IEEE Transactions on Electron Devices, vol. ED- 29, no. 4, pp. 645-650, April 1982.

83. J. Savir, W. H. McAnney, and S. R. Vecchio, "Testing for Coupled Cells in Random Access Memories," Proceedings of the IEEE International Test Conference, pp. 439- 451, 1989.

84. D. R. Schertz, and G. Metze, "A New Representation for Faults in Combinational Digital Circuits," IEEE Transactions on Computers, vol. c-21, no. 8, pp. 858-866, August 1972.

85. O. Semenov, "Impact of Technology Scaling on Bridging Fault Modeling and Detection in CMOS Circuits," M.A.Sc. Thesis, University of Waterloo, Waterloo, ON, Canada, 2001.

86. J. P. Shen, W. Maly, and F. J. Ferguson, "Inductive Fault Analysis of MOS Integrated Circuits," IEEE Design and Test of Computers, vol. 2, pp. 13-26, December 1985.

87. H. C. Shih, and J. A. Abraham, "Fault Collapsing Techniques for MOS VLSI Circuits," Proceedings of the Fault Tolerant Computing Symposium, pp. 370-375, 1986.

88. M. Sivaraman, and A. J. Strojwas, "Test Vector Generation for Parametric Path Delay Faults," Proceedings of the IEEE International Test Conference, pp. 132-138, 1995.

89. G. L. Smith, "Model for Delay Faults Based upon Paths," Proceedings of the IEEE International Test Conference, pp. 342-349, 1985.

90. J. E. Smith, "Detection of Faults in Programmable Logic Arrays," IEEE Transactions on Computers, vol. C-28, pp. 845-853, 1979.

91. J. M. Soden, C. F. Hawkins, R. K. Gulati, and W. Mao, "I_{DDQ} Testing: A Review," Journal of Electronic Testing: Theory and Applications, vol. 3, pp. 291-303, November 1992.

92. M. Soma, "An Experimental Approach to Analog Fault Models," Proceedings of the Custom Integrated Circuits Conference, pp. 13.6.1-13.6.4, 1991.

93. M. Soma, "A Design for Test Methodology for Active Analog Filters," Proceedings of the IEEE International Test Conference, pp. 183-192, 1990.

94. M. Soma, "Fault Modeling and Test Generation for Sample and Hold Circuit," Proceedings of the International Symposium on Circuits and Systems, pp. 2072- 2075, 1991.

95. D. S. Suk and S. M. Reddy, "A March Test for Functional Faults in Semiconductor Random Access Memories," IEEE Transactions on Computers, vol. C-30, no.12, pp. 982-985, Dec. 1981.

96. S. M. Sze, "VLSI Technology," New York: McGraw Hill Book Company, 1983.

97. S. M. Thatte, and J. A. Abraham, "Testing of Semiconductor Random Access Memories," Proceedings of International Conference on Fault Tolerant Computing, pp. 81-87, 1977.

98. S. M. Thatte, and J. A. Abraham, "Test Generation for Microprocessors," IEEE Transactions on Computers, vol. C-29, pp. 429-441, 1980.

99. K. To, "Fault Folding for Irredundant and Redundant Combinational Circuits," IEEE Transactions on Computers, vol. C-22, no. 11, pp. 1008-1015, November 1973.

100. B. Underwood, W. O. Law, S. Kang, and H. Konuk, "Fastpath: A Path-delay Test Generator for Standard Scan Designs," Proceedings of the IEEE International Test Conference, pp. 154-163, 1994.

101. P. Varma, "On Path Delay testing in a Standard Scan Environment," Proceedings of the IEEE International Test Conference, pp. 164-173, 1993.

102. H. T. Vierhaus, W. Meyer, and U. Glaser, "CMOS Bridges and Resistive Faults: I_{DDQ} versus Delay Effects," Proceedings of the IEEE International Test Conference, pp. 83-91, 1993.

103. R. L. Wadsack, "Fault Modeling and Logic Simulation of CMOS and MOS Integrated Circuits," Bell Systems Technical Journal, vol. 57, no.5, pp. 1449-1474, May-June 1978.

104. S. H. Walker, and S. W. Director, "VLASIC: A Catastrophic Fault Yield Simulator Integrated Circuits," IEEE Transactions on Computer Aided Design of Integrated Circuits and Systems, vol. CAD-5, pp. 541-556, October 1986.

105. H. Walker, "VLASIC System User Manual Release 1.3," SRC-CMU Research Center for Computer Aided Design, Dept. of Electrical and Computer Engineering, Carnegie Mellon University, Pittsburgh, PA 15213.

106. T. W. Williams, and K. P. Parker, "Design for Testability – A Survey," Proceedings of the IEEE, vol. 71, no. 1, pp. 98-113, January 1983.

107. B. W. Woodhall, B. D. Newman and A. G. Sammuli, "Empirical Results on Undetected CMOS Stuck-Open Failures," Proceedings of the IEEE International Test Conference, pp. 166-170, 1987.

108. Y. Z. Xu, H. Puchner, A. Chatila, O. Pohland, B. Bruggeman, B. Jin, D. Radaelli, and S. Daniel, "Process impact on SRAM alpha-particle SEU performance," IEEE Reliability Physics Symposium, pp. 294-299, 2004.

109. R. Yung, S. Rusu, K. Shoemaker, "Future Trend of Microprocessor Design: Challenges and Realities," Invited presentation, ESSCIRC 2002.

110. J. F. Ziegler, and H. Puchner, "SER – Histoty, Trends and Challenges. A guide for designing with memory ICs," Cypress Semiconductor Corp., 2004.

Chapter 4

DEFECTS IN LOGIC CIRCUITS AND THEIR TEST IMPLICATIONS

A substantial amount of research was carried out since the 1980s to verify the validity of various fault models. This chapter summarizes some of the key research work done in this area. During the same time frame, quiescent current measurement technique also known as I_{DDQ} testing became popular owing to its ability to uncover defects in CMOS circuits. Studies were conducted over the relative effectiveness of Boolean (logic) and I_{DDQ} methods of testing for defect detection. Salient features and observations of these studies are reproduced.

1. INTRODUCTION

In the previous two chapters, we studied causes of defects and corresponding fault models. In addition, we discussed the basics of defect-oriented test methodology (IFA). In this chapter we shall discuss important defect-oriented studies conducted on logic circuits. The focus is on the pragmatic side of the concepts discussed in the previous chapter. This material provides a complementary treatment for the concepts presented in the previous chapter.

Logic circuits include combinational logic gates (INVERTER, NAND, NOR, etc.) as well as sequential circuits (flip-flops, scan chains, etc.). Standard cells are basic building blocks used to implement a logic function. Owing to the simplicity of standard cell logic gates and complexity of carrying out the defect-oriented analysis, the early experiments with IFA were conducted over standard cells. Since the SAF model had gained wide acceptance for testing and test generation, researchers typically wanted to verify from IFA studies, "*Do stuck-at faults represent manufacturing defects* [4]?"

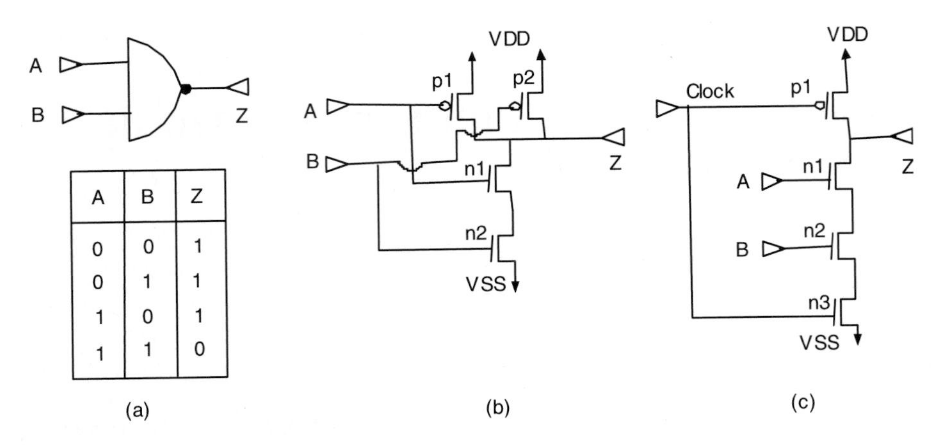

Figure 4-1. A two input NAND gate and its static CMOS (b), and dynamic CMOS (c), implementations.

The complementary metal oxide semiconductor (CMOS) technology is the most popular technology today. Logic functions in CMOS technology can be implemented with static logic gates or dynamic logic gates. Figure 4-1 illustrates a 2-input NAND gate in static as well as dynamic CMOS configurations. In the static implementation (Figure 4-1(b)), the output Z of the logic gate retains its state so long as the inputs are unchanged. However, in dynamic CMOS (Figure 4-1(c)), the output Z of the logic gate has the correct logic value at a certain instant of the clock. When the clock is low the output is pre-charged to logic 1 (pre-charge phase). At the instant clock goes high, the output response is evaluated depending upon the states of inputs (evaluation phase). The correct output is only available in the evaluation phase. In the case of a dynamic NAND, if both inputs are high, the output makes a transition to logic 0, otherwise it retains the pre-charged value of 1. However, if output stays high it is not driven by any source, and its value is due to the pre-charged output capacitance. Very high input impedance of MOS transistors helps dynamic logic to retain the pre-charged value under appropriate input conditions.

Most logic circuits use static CMOS technology for its better noise margin, robust implementation, and low power consumption. In the static CMOS implementation, a combinational logic gate has two distinct and equal parts: PMOS transistor(s), and NMOS transistor(s). The number of

PMOS transistors is equal to the number of NMOS transistors. Figure 4-1(b) shows PMOS and NMOS transistors arranged in a complementary manner with respect to each other. For example, if all PMOS transistors are in parallel, then all NMOS transistors are in series (NAND gate). Following the same principle complex logic gates can be formed such as AND-OR-INVERT, etc.

2. STUCK-AT FAULTS AND MANUFACTURING DEFECTS

Chapter 3 (Sections 3.1.3 and 3.1.4), touched upon the shortcomings of the stuck-at fault model in representing shorts and open defects in CMOS circuits. In this section we expand upon defects and their detection strategies in standard cell logic gates. Understanding the effect of physical failures on digital systems is essential to design test for them and to design circuitry to detect and tolerate them [3].

Table 4-1. Failure modes observed by Galiay et al. [13].

Defect	%
Short between metallization	39
Open metallization	14
Short between diffusions	14
Open diffusions	6
Short between metallization and substrate	2
Unobservable	10
Insignificant	15

2.1 Study by Galiay, Crouzet, and Vergniault

Galiay et al. [13] studied the physical origin of failures. Their study was concluded over an NMOS 4-bit microprocessor chip. They performed failure analysis of 43 failed devices. Table 4-1 illustrates the observed failure modes of the IC. It is clear from the table that opens and shorts in metallization and diffusion were the primary causes of failures. These failures were easily detected and causes of the failures were established. A total of 10% failures did result in logical faults that were detected. However, no conclusive failure mode could be established. Similarly, 15% of failures were caused by large scale imperfection like scratch, etc., those were easily detected and were considered insignificant for test purposes. Figure 4-2 depicts some observed failures in an NMOS logic gate identified by Galiay et al. [13]. These failures identify two broad issues in the modeling of such failures.

1. All failures can not be modeled by stuck-at faults. For example in Figure 4-2(a), short-1 and open-3 can, respectively, be modeled by SA1 at input e and by SA0 at input e (or input f or both). On the other hand, short-2 and open-4 can not be modeled by any SAF because they cause modification of the function realized by the gate. For the same reason, the short between outputs of logic gates can not be modeled by SAF.
2. The actual topology of the circuit is often different from the logical representation of the circuit. Some connections in the logic circuit do not map onto the actual topology, and vice versa. Figure 4-2(b) illustrates the electrical and logic diagrams of a complex gate. For example, short-2 which is physically possible cannot be represented on the logic diagram. Similarly, short-1 in logic diagram has no physical meaning.

In order to detect failures at the logic level, Galiay et al. [13] defined conduction path as a path between output and V_{SS} if all transistors in the path are in the conduction mode. In NMOS technology, if no conduction path exists, then the load transistor pulls the output (S) to logic high. An open defect in this switch-like network may remove one or more conduction paths. Detection of open defects in NMOS technology is simpler compared to open defect detection in CMOS [48]. Readers will recall (Chapter 3) that CMOS stuck-open faults, in general, require two test vectors, T1 and T2, for detection. In the NMOS technology, the load transistor always provides the first initializing vector (T1) for the detection of open defects in the switching network. In order to detect a particular open defect, a conduction path should be uniquely activated. The logic gate contains the above mentioned open defect if the output of the gate remains at logic high displaying a SA1 fault.

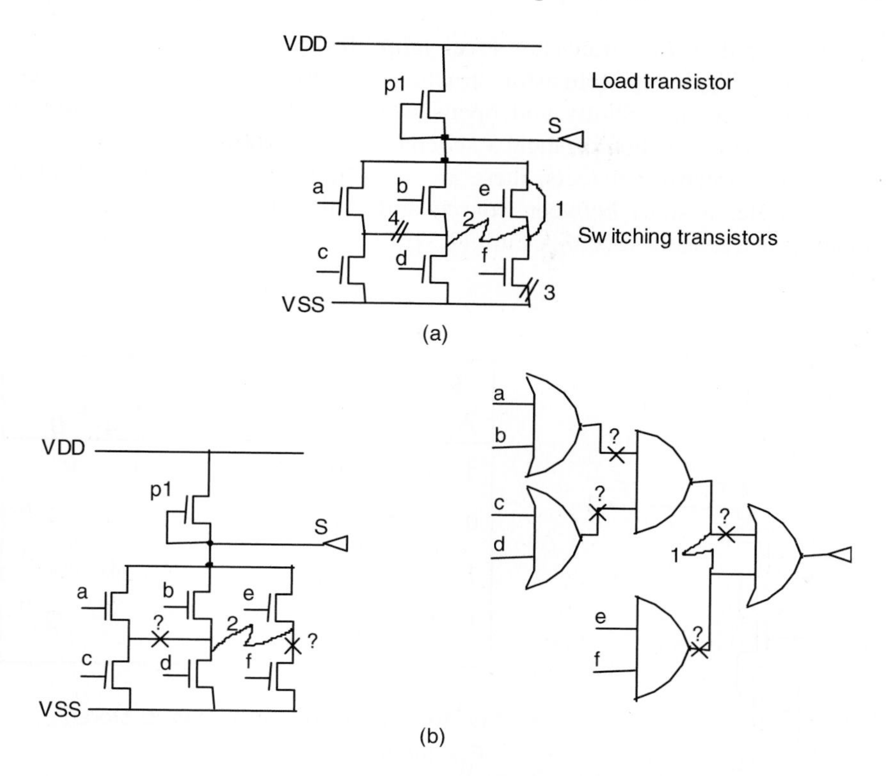

Figure 4-2. Failure examples in a MOS gate (a); Relationship between electrical and logic diagram (b) [13].

A short between two nodes inside a logic gate creates one or more conduction paths in the gate. Two conditions are required to detect a short between nodes *i* and *j*: (1) activation of at least one conduction path between *i* (*j*) and output, and activation of at least one conduction path between *j* (*i*) and V_{SS}; and (2) blocking of all conduction paths in the logic gate. This short pulls the output down to logic low which is different to the output in fault-free condition. In general, there may be more than one test vectors that will detect a given short. However, this detection holds only for low resistive shorts. In the case of high resistive shorts the output level of the gate may not degrade enough so as to be interpreted as a fault by the subsequent logic gate.

2.2 Study by Banerjee and Abraham

Transistor level fault models represent physical failures better than the logic level fault models. Several test generation algorithms have been reported for detecting transistor level faults for combinational circuits [6-8,10,17,18]. Typically, transistor structure is converted into an equivalent logic gate structure. Shorts and opens were modeled as stuck-at faults at logic gate level. Although such a scheme is fairly successful in modeling most of the transistor defects, there are some defects that are not modeled. For example, a short between source and drain of transistor can not be modeled as transistor level SA fault [2,3].

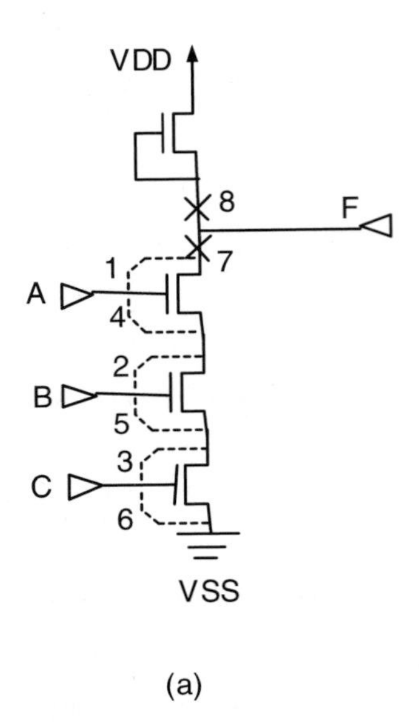

Inputs			Outputs					
A	B	C	F_0	F_1	F_2	F_3	F_4	F_5
1	1	1	0	1	1	1	1	0
0	1	1	1	0	1	1	1	Q^n
1	0	1	1	1	0	1	1	Q^n
1	1	0	1	1	1	0	1	Q^n

O^n means previous logic state is stored

F_0 is the normal fault-free output

Fault classes and faults: $F_1 = \{1\}$; $F_2 = \{2\}$; $F_3 = \{3\}$; $F_4 = \{4,5,6,7\}$; $F_5 = \{8\}$.

(b)

Figure 4-3. NMOS NAND gate (a); and tests for various failures (b) [3].

Modeling of MOS circuits as a network of simple transistors (switches) began in the late 1970s and early 1980s. Bryant presented an overview of switch level modeling and test pattern generation [6]. Switch level modeling offers several advantages. The switch level model is a close representation of schematic. Furthermore, it also models many important phenomena associated with MOS circuits, such as bidirectionality of signal flow, dynamic charge storage, resistance ratios. In some sense, switch-level modeling and simulation is an excellent trade-off between accuracy of circuit level and speed of logic level modeling and simulation.

Banerjee and Abraham [3] characterized physical failures of simple NMOS and CMOS circuits and translated them into logic level faults. They choose five logic levels to represent various voltage ranges in logic gates:

1. (0): Hard zero
2. (0*): Soft zero
3. (I): Indeterminate, near the logic threshold
4. (1*): Soft one
5. (1): Hard one

Logic level 1 represents a hard one and logic level 1* represents a soft one which is recognized as logic one by a fanout logic gate but can not drive an NMOS pass transistor to the fully on state. Logic level I represents an indeterminate level near the logic threshold of an inverter. Such a level may be interpreted as 0, or 1, or I (i.e., the output of the gate is also indeterminate) by the following gate. Logic level hard 0 is always interpreted as logic 0. Finally, logic 0* corresponds to a soft zero, i.e., if it is applied to a dynamic latch, (NMOS pass gate and output capacitance), it can discharge any stored charge on the drain in a time comparable to the propagation delay of the transistor switch, provided source is grounded [2,3].

2.2.1 NMOS Logic Gates

Figure 4-3(a) depicts a three input NMOS NAND gate with various failures locations marked. Fault 7 refers to an open anywhere in the conducting path from output terminal to the ground due to open in interconnect or due to a missing contact. Fault 1 refers to a short between the gate and the drain of a transistor. Similarly, Faults 2-6 represent shorts between terminals of transistors. Fault 8 represents an open in the path from output to V_{DD}.

Floating gate failures (such as a defect on input A, B or C) can be modeled in many ways. For enhancement mode NMOS switching transistors, the presence or absence of trapped charge in the gate (thin) oxide will cause transistor to be either stuck-on or stuck-off. This behavior is equivalent to a SA1 or SA0 fault on the gate terminal. However, in most cases, if the failure is permanent, the stored charge will eventually leak away through the leakage path from the gate terminal to the substrate [3]. Therefore, the substrate and gate voltage will remain the same and the enhancement mode transistor will always be off. As a result such defects in NMOS NAND gate can be represented as fault 7 of Figure 4-3(a). On the other hand, for depletion mode NMOS load transistor, the same failure will appear as low supply of charging current resulting in a long charging time. The floating gate will not give any logical error and will cause timing errors.

Figure 4-3(b) illustrates fault-free (F0) and fault class responses (F1 - F5) of the input test vectors in a tabular form. If two defects produce the same faulty response, they are put together in a fault class. Qn represents the high impedance state where previous logic state is stored. If all inputs are driven by logic gates and are logic high, then the output of the NAND will be 1* for defects 1, 2, and 3. However, if inputs are driven through pass transistors, then the output of NAND will be I which may be interpreted as a 0 or a 1 by the following logic gate. Therefore, this is not a reliable test for the above mentioned defects; nevertheless, it is a good test for defects 4, 5, 6, and 7. It is visible from the table that the stuck-at test set (111, 011, 101, 110) detects all above mentioned fault classes. Similarly, for a three input NMOS NOR gate, a particular sequence of stuck-at test set (100, 010, 001, 000) detects all fault classes similar to those in the illustrated NAND gate.

2.2.2 CMOS Logic Gates

Defects in CMOS circuits have similar fault behavior compared to NMOS circuits except for a class of open defects. This class of open defects in CMOS logic gates cause logic gates to have memory like behavior for certain input conditions (Chapter 3). Figure 4-4(a) illustrates a three input NAND gate with possible failure locations and Figure 4-4(b) depicts the fault detection table. Readers should notice that although in CMOS implementation, compared to NMOS implementation, the number of transistors has increased from four to six, however, the number of fault classes has more than doubled. In general, it is more difficult to test a CMOS logic gate compared to a NMOS logic gate for likely defects.

As it appears the short defects are detected by the stuck-at test set, however, open defects require a particular sequence for detection. Therefore, stuck-at test set is not enough and a test vector sequence of 111, 011, 111, 101, 111, 110 is needed to detect all modeled defects. In general, test vectors for an arbitrary CMOS logic gate may be generated in a manner explained below. First, all primitive sub-networks in the pull-down network are identified. Defects in pull-down sub-networks are detected by the following test sequence [3]:

1. Apply test vector producing logic high at the gates of N channel transistors in the sensitized N channel sub-network. This test detects open defects at all three terminals of the transistor in question, and source/drain short in the transistor. This test will also detect shorts between gate and drain of corresponding P channel transistor.

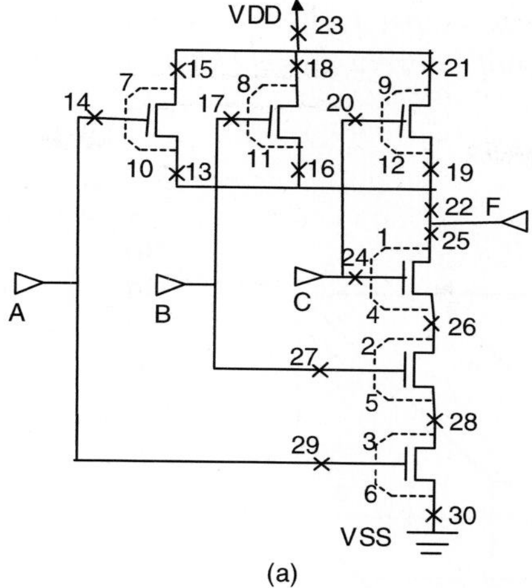

(a)

Inputs			Outputs												
A	B	C	F_0	F_1	F_2	F_3	F_4	F_5	F_6	F_7	F_8	F_9	F_{10}	F_{11}	F_{12}
1	1	1	0	I	I	I	1	0	0	0	0	0	0	Q^n	0
0	1	1	0	1	1	I	1	0	1	1	Q^n	1	1	1	Q^n
1	1	1	0	I	I	I	1	0	0	0	0	0	0	Q^n	0
1	0	1	1	1	I	1	1	1	0	1	1	Q^n	1	1	Q^n
1	1	1	0	I	I	I	1	0	0	0	0	0	0	Q^n	0
1	1	0	1	I	1	1	1	1	1	0	1	1	Q^n	1	Q^n

Fault classes and faults: $F_1 = \{1,12\}$; $F_2 = \{2,11\}$; $F_3 = \{3,10\}$; $F_4 = \{4,5,6\}$; $F_5 = \{7\}$; $F_6 = \{8\}$; $F_7 = \{9\}$; $F_8 = \{13,14,15\}$; $F_9 = \{16,17,18\}$; $F_{10} = \{19,20,21\}$; $F_{11} = \{24,25,26,27,28,29,30\}$; $F_{12} = \{22,23\}$.

(b)

Figure 4-4. CMOS NAND gate (a); and tests for various failures (b) [3].

2. Next, a test vector producing a logic low at the gate of N channel transistors in N channel sub-network and logic high on the gates of other N channel transistors is applied. In this way the N channel transistor whose gate is set at logic low is tested for shorts between gate and drain. This test also detects the source/drain short and open defects of all three terminals of the corresponding P channel transistor in the pull-up network. This procedure is repeated for every transistor in the N channel sub-network of the logic gate.

The above mentioned test procedure is a typical example of transistor level test generation. Here, it is pertinent to mention that defects considered in this analysis are simplistic, and zero and infinite impedances are assumed for shorts and open defects, respectively. We shall see in subsequent studies that this assumption in not entirely correct and actual defect detection is more difficult.

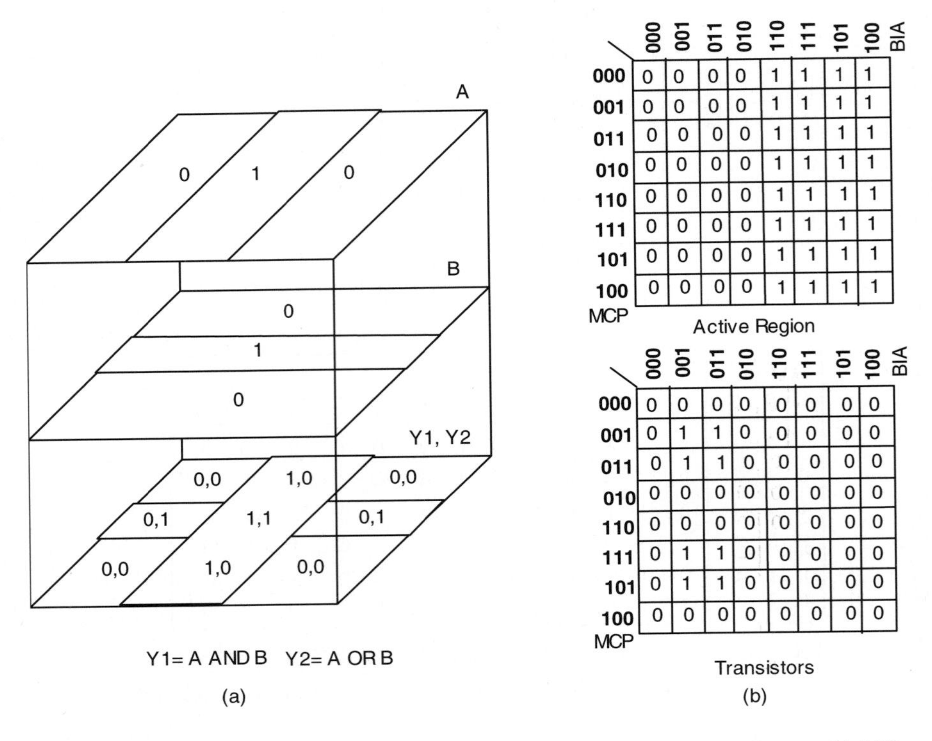

Figure 4-5. An example of topological operation (a); examples of Karnaugh maps (b) [42].

2.3 Study by Maly, Ferguson and Shen

Maly, Ferguson and Shen [23] analyzed the impact of physical defects over NMOS and CMOS cells and developed a systematic methodology [42] for such an analysis. The main difference between their study and the studies reported earlier was the determination of physical origin of the defect and subsequent modeling at the appropriate level of abstraction for the fault simulation purpose. A MOS technology has a set of layers to be processed through masks. Each mask discriminates between areas to be processed and not to be processed on a given layer. The processed area is assigned the value one and unprocessed area is assigned zero. In this way, it is possible to make a set of Karnaugh maps for the active area, poly, or transistor, etc. In other words, an ideal signature for a given process is determined.

Table 4-2. Results of the defect simulation at the circuit level (a); and at the logic level (b) [42].

	Extra Metal	Extra Diff.	Extra Poly.	Miss. Metal	Miss. Diff.	Miss. Poly.
simulated	110	64	130	138	145	147
Single SA faults	7	1	2	0	7	1
Source/drain shorts	0	2	0	0	0	7
Source/drain opens	0	0	0	0	8	0
Bridging faults	4	0	12	0	0	0
Floating lines & gates	0	0	1	16	1	17
Mixed faults	0	0	0	0	1	1
Power faults	0	0	0	5	0	0
Total with logical significance: 93	11	3	15	21	17	26

(a)

	Extra Metal	Extra Diff.	Extra Poly.	Miss. Metal	Miss. Diff.	Miss. Poly.
Total: 93	11	3	15	21	17	26
Can be modeled as SA0/SA1: 65	7	3	2	15	13	25
Can be modeled as Single cell I/O SA0/SA1: 31	5	1	2	12	8	3

(b)

A major cause of defects is the improper processing of layers, or improper interaction among different layers. There could be several reasons for improper processing. However, it is important to put the improper processing step onto Karnaugh maps and establish defect and fault relationship. In this way realistic faults can be obtained. Once a fault and its impact on the circuit are known a test could be developed for it. This process is explained with the help of Figure 4-5. The figure illustrates two layers, A and B. The area to be processed on these layers is marked with 1 and rest of the layer with 0, respectively. If a logic AND operation is performed over layers A and B, it results in an area that is common in both layers. This layer is depicted as Y1 (coded as 1,1). Similarly, if a logic OR function is performed over A and B, it results in area depicted on layer Y2 (coded as 1,0 OR 0,1). In a similar fashion more complex Boolean relationships can be summarized among different layers using Karnaugh maps. Figure 4-5(b) illustrates Karnaugh maps for active region and transistors, respectively. The axes of Karnaugh map represent different masks. For example, M, C, and P represent metal, CVD SiO_2, and poly-silicon, respectively. Similarly, B, I, and A represent thin (gate) oxide, transistor implant, and thick (field) oxide, respectively. The electrical equivalence between different combinations representing the shapes of layers (or processing steps) can also be determined using Karnaugh maps in a similar manner. In this fashion, defects at the processing level of abstractions are translated to the device level of abstraction using a table look-up approach. Such a table can be developed along with the Boolean functions for equivalence classes. The impact of a defect at device level can be determined using a circuit level simulator. Although this is a time consuming process, in many cases a higher level of simulator may be used without a significant loss of accuracy.

A full adder was analyzed through this procedure in an NMOS technology. The presence of extra material and absence of material was considered on metal, diffusion, and poly-silicon layers. Therefore, the fault model included six types of spot defects. Based on this fault model, a total of 734 spot defects in the NMOS full adder were generated. These defects were analyzed. Results of this analysis at the circuit level are shown in Table 4-2(a). This table includes only the defects that can be translated into logic level faults. The first category represents single SA faults. Such faults are typically shorts between a circuit node and V_{DD} (V_{SS}). Similarly, category four represents bridging faults which short two (or more) internal circuit nodes. Categories 2 and 3, source/drain shorts and opens, represent transistor level faults and are self explanatory. Mixed faults are multiple faults, which have more than one previously described categories. For example, a missing diffusion may cause a particular transistor gate to be at SA0 and other transistor to be stuck-open. However, such faults are less likely and only one fault is reported. Finally, all defects which create an open circuit on V_{SS} or V_{DD} are classified as power faults.

Table 4-2(b) illustrates the results of logic level fault simulation for a number of logic level significant faults (93). The single SA0/1 row gives the number of defects of each type that can be modeled as a single line SA0 or SA1. While performing this analysis two basic assumptions are made:

1. The circuit level schematic is known. A logic level schematic with best correspondence to the circuit level schematic is evolved such that circuit level faults can be mapped at logic level.
2. Floating lines and gates are assumed to remain at a fixed value (0 or 1). This allows a majority of the defects resulting in floating lines and floating gates to be modeled as single SAF.

The second row illustrates the number of faults that can be modeled as a single input or output of the cell SAF which is same as the traditional logic level SAF model. This fault model is able to model approximately 35% of the analyzed spot defects. Furthermore, out of 80 possible SAF in the logic schematic of the adder only 37 occurred in this simulation experiment. However, it should be remembered that the defect densities of the six types may vary significantly from process to process. Furthermore, these types do not represent the complete defect spectrum.

2.4 Gate Oxide Shorts: Study by Hawkins and Soden

Gate of a MOS transistor is isolated from its channel by a thin layer of an insulating oxide (SiO_2). For a typical 130 nm technology, the gate oxide thickness is about 3 nm and each successive scaled generation requires further reduction in this thickness. Growing of thin oxide is a very critical and sensitive process step of VLSI manufacturing. For high performance ICs, thin oxide quality and reliability are a major concern [38].

A gate oxide short is an electrical connection through the thin oxide between the gate and any of the other three ports of a MOS transistor (Figure 4-6). Typically, the resistance of such a short is a few kΩ, therefore, such shorts are not detected by structural and functional voltage testing. However, such defects give rise to elevated I_{DDQ} when excited logically. In one of the first studies of its kind conducted in Sandia National laboratories, gate oxide shorts were found to be the major reliability concern. In a high voltage screen experiment conducted over more than 5,000 1k CMOS static RAMs, 687 devices failed due to Time Dependent Dielectric Breakdown (TDDB) of the gate oxide and 254 (37%) of these devices failed the I_{DDQ} test but passed the functional test. The other 433 devices failed the I_{DDQ} as well as functional tests [14,43].

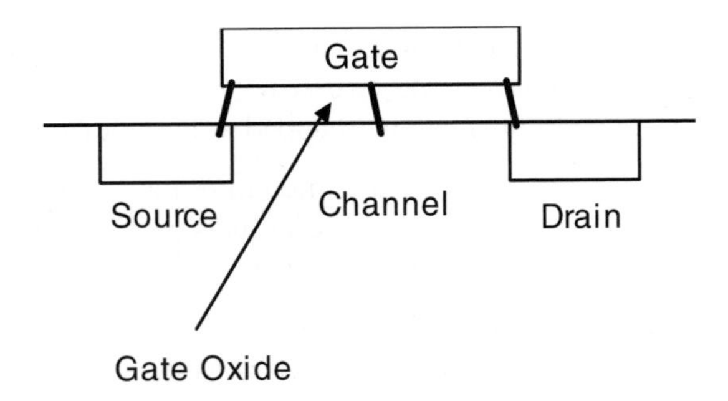

Figure 4-6. Structure of a MOS transistor and possible gate oxide short locations.

A gate oxide short may result from many reasons. Shorts between the gate and source or drain are often caused by electrostatic discharge or electrical over-stress. Furthermore, gate oxide shorts may also occur during the fabrication process. In addition, these shorts can occur later when defects in the oxide results in oxide break-down because of electrical field and thermal stress. The delayed occurrence of the gate oxide short is often referred to as TDDB. There are two dominant causes of TDDB: (i) Defect or contaminant based failures- silicon surface impurities or imperfections cause a local thinning of thin oxide. A local thinning of gate oxide thickness results in higher electric field across the spot that further damages the thin spot. This positive feedback process continues with time until a complete break-down occurs. (ii) Hot electron based failures- Hot electrons cause damage to the gate oxide resulting in trapped charge in the gate oxide which attracts more hot electrons causing further damage. This process also continues until it results in a complete breakdown of the oxide [15,38,43].

Hawkins and Soden used electrostatic discharge (ESD) and laser techniques to create gate oxide shorts to study the properties of such defects. ESD creates higher electric field near the edge of the gate and typically produces gate oxide shorts between gate and source/drain of a transistor. For an *n*-doped poly and n^+ diffusion such defects cause ohmic connections with values ranging from 800Ω to 4kΩ. Contrary to ESD, the laser technique could create gate oxide defects in any region. Hence, it was used to study defects between *n*-doped poly and *p*-well. Such defects formed a *pn*-junction between the terminals. Similarly, gate oxide shorts between *n*-doped poly and source/ drain in PMOS transistors formed a *pn*-junction and gate oxide shorts between poly and *n*-type substrate formed a resistor. Unless a gate oxide defect results in a complete oxide breakdown, a gate oxide defect typically does not prevent transistor from performing its logical operation. However, its performance (rise or fall time) is degraded substantially.

Hawkins and Soden reported an average of 29% reduction in transistor transconductance by a gate oxide defect. As reported earlier, such defects result in abnormally high I_{DDQ} and, therefore, are easily detected by I_{DDQ} measurements.

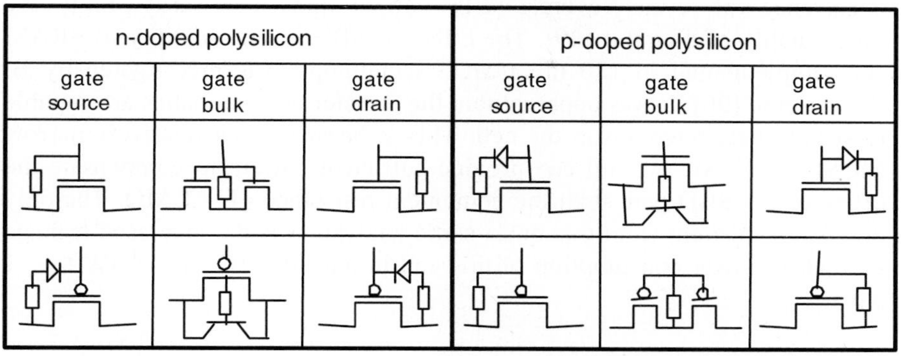

n-doped polysilicon			p-doped polysilicon		
gate source	gate bulk	gate drain	gate source	gate bulk	gate drain

Figure 4-7. A generalized gate oxide defect model [39].

A substantial research effort has been devoted to the modeling of gate oxide defects [39,46]. In general, such defects are either ohmic or rectifying depending upon the polarity (p-type or n-type) of the two shorted nodes. Figure 4-7 illustrates a generalized gate oxide defect model for MOS transistors.

The gate oxide thickness is decreased with each technology generation in order to increase the drive current and to control the short channel effects. The experimental measurements of time-to-breakdown of ultra-thin gate oxides with thickness less than 4 nm show that the conventional E and 1/E time dependent dielectric breakdown models (TDDB) cannot provide the necessary accuracy for calculation and prediction. Hence, starting from about the 180-nm CMOS technology (gate oxide thickness of about 26–31 Å), a new TDDB model was proposed [45]. Experiments show that the generation rate of stress-induced leakage current (SILC) and charge to breakdown (Q_{BD}) in ultra-thin oxides is controlled by gate voltage rather than the electric field. This model includes the gate-oxide thickness (T_{ox}) and the gate voltage (V_G) [28]:

$$T_{BD} = T_0.\exp[\gamma(\alpha T + (E_a/kT_j) - V_G)] \qquad (4.1)$$

where γ is the acceleration factor, E_a is the activation energy, α is the oxide thickness acceleration factor, T_0 is a constant for a given technology, and T_j is the average junction temperature. With the continuous gate oxide downscaling and the consequent operating voltage reduction, a new mode of gate dielectric breakdown was observed. It was named "soft" breakdown (SBD) due to its relatively low conductivity [32]. Several researchers investigated the effect of SBD on single MOSFET and predicted that occurrence of even several SBDs might not necessarily limit the functionality of a circuit [29]. The effect of SBD on the stability of SRAM cells implemented in 130 nm CMOS technology was investigated by B. Kaczer et al [20]. It was apparent that the transfer characteristics are capable to sustain data retention in the cell. This is because even relatively narrow MOSFETs in SRAM cell can provide sufficient current to compensate the effect of the SBD path with the equivalent resistance of ~ 1 MΩ. The only drawback apparent from this quasi-static analysis was the increased leakage current or power consumption of this cell in one state (from ~ 1 nW to ~ 1 μW).

3. IFA EXPERIMENTS ON STANDARD CELLS

In another study, Ferguson and Shen [11] extracted and simulated CMOS faults using the IFA technique. A CAD tool, FXT, was developed with a capability of automatic defect insertion/extraction for a reasonably large layout. This tool was used to analyze five circuits from a commercial CMOS standard cell library. The five circuits were, (i) a single bit half adder cell, (ii) a single bit full adder cell, (iii) a counter, (iv) another counter, and (v) a 4x4 multiplier. They sprinkled more than ten million defects in two counters which caused approximately 500,000 faults. Similarly, over 20 million defects were sprinkled in the multiplier which caused approximately 1 million faults. Approximately 1/20th of the defects caused faults that conform to the fact that most of the defects are too small to cause a fault. The majority of extracted faults could be modeled as bridging faults, break faults or a combination of these two. For example in the 4x4 multiplier, the bridging and break faults amount to 48% and 42%, respectively. Almost all remaining faults were transistor SON faults which can also be represented as bridging faults between source and drain of transistors. Similarly, a transistor SOP fault is equivalent to a break fault. Therefore, almost all faults could be represented as bridging or break faults or a combination of the two. The only two categories that were not equivalent to above mentioned categories were new transistors and exceptions. A new transistor in the layout is created by a lithographic spot on poly or diffusion mask. Less than 0.7% of faults fell into this category.

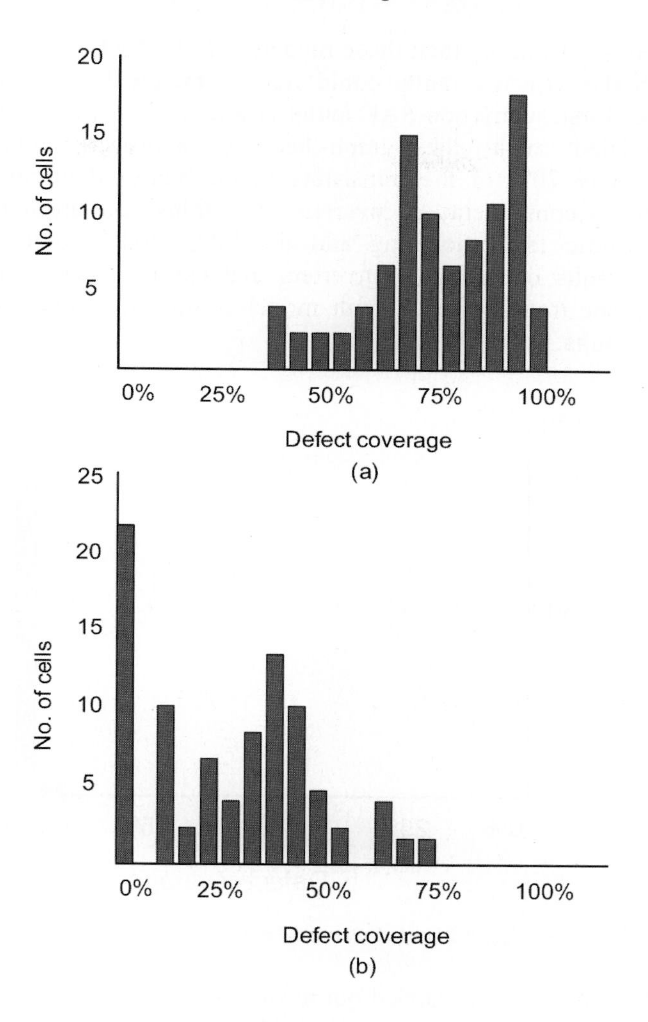

Figure 4-8. Voltage based coverage of a standard cell library with low resistive bridging defects (a); and 2kΩ resistive defects (b) [31].

The SAF model performed rather poorly in modeling the extracted faults [11]. In the case of the 4x4 multiplier, only 44% of the bridging faults could be modeled by SAFs. For non-bridging faults only 50% of the faults could be modeled as SAFs. Hence, for the multiplier, less than 50% of all extracted faults could be modeled as SAFs. A similar comparison was carried out with graph-theoretic (transistor-level) fault models. It was estimated that only 57% of the extracted faults could be modeled as graph-theoretic fault model.

Though this was higher than those modeled by the SAF model, the majority of non-SAF extracted faults could not be modeled. Two reasons were attributed. First, many non-SAF faults bridge input nodes together and are not modeled with the graph-theoretic approach either. Second, approximately 70% of the transistors in the analyzed circuits were pass transistors or components of inverters. Pass transistors are not modeled in graph-theoretic fault modeling and the SAF model could model most transistor faults occurring in inverters that caused change in the logical function. The transistor SOP fault model could represent only 1% of the extracted faults.

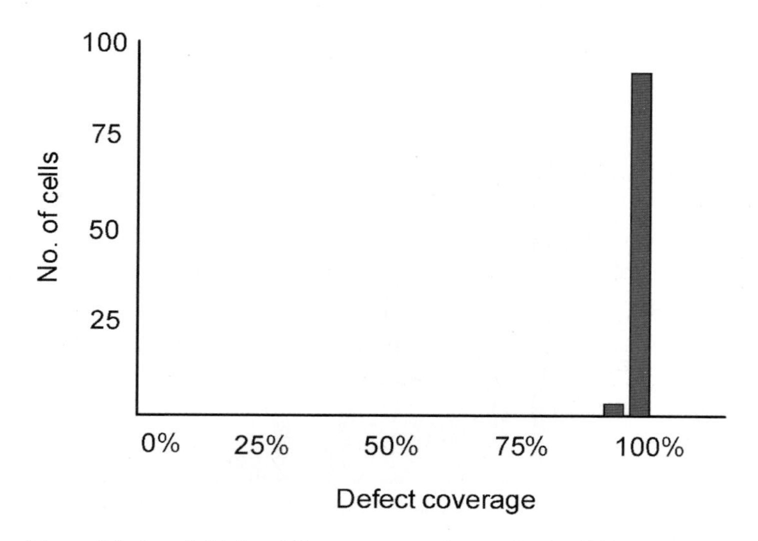

Figure 4-9. I_{DDQ} *bridging defect coverage of a standard cell library [31].*

Further analysis was carried out to find out how well SA and exhaustive test sets detect the extracted faults [11]. Simulation time was reduced by simulating only extracted bridging faults in counters. The SA test set could only detect from 73% to 89% of the circuit's bridging faults. Even under the unrealistic assumption that all non-bridging faults are detected by SA test set, the 100% SA test set could detect between 87% and 95% of extracted faults. The fault coverage of extracted bridging faults by the exhaustive test set was relatively high. The exhaustive test set detected between 89% and 99.9% of the extracted bridging faults. As a test solution for better fault coverage quiescent current monitoring (I_{DDQ}) was suggested. It was implied that I_{DDQ} will provide the best test set for bridging fault detection.

Peters and Oostdijk [31] analyzed the Philips standard cell library using six types of leakage faults. Their analysis was restricted to bridging defects in standard cells. They did not consider open defects, assuming that the probability of occurrence of open defects is relatively small. The bridging fault coverage of the library by voltage and I_{DDQ} tests is illustrated in Figure 4-8 and Figure 4-9, respectively. Figure 4-8(a) depicts that for low resistive defects (typically less than 10Ω) no cell has 100% coverage for the modeled faults by voltage testing. Most cells have voltage coverage between 60% - 90% for the modeled faults. Figure 4-8(b) illustrates the voltage fault coverage when the bridging fault resistance is $2k\Omega$. As expected, the voltage fault coverage declines rapidly as the defect resistance is increased. A large number of cells (22) were reported with zero fault coverage.

The I_{DDQ} coverage of the leakage fault model is 100% for all combinational cells for a threshold current of 10 μA. Sequential cells do not have 100% I_{DDQ} fault coverage. It is known that a class of bridging defects in latches and flip-flops do not give rise to elevated I_{DDQ}. Later in this chapter, we shall discuss the defects in sequential circuits and suggest measures to improve the defect coverage.

An analysis of the results stresses that I_{DDQ} testing is a necessity for reliability since it detects bridging and transistor short defects. Even assuming zero ohm defect resistance, the defect coverage of voltage vectors can not be expected to be higher than about 75%. An analysis of the detailed results has shown that especially defects on serial transistors and defects between inputs are hard to detect using a voltage test. These defects are easily detectable using I_{DDQ} testing. Also, many GOS defects can not be detected using voltage vectors, which could be a possible explanation for early life failures.

For minimizing the overlap between detection by voltage or I_{DDQ} vectors, the use of a critical resistance is very important. Defects with resistance below the critical resistance will be detected by voltage testing. However, detection of defects above the critical resistance is not assured. Introduction of a threshold resistance value for voltage testing will assure the detection of defects up to a particular resistance value will be detected. In other words, based on defect resistance statistics for a given process, one can quantify how I_{DDQ} testing will improve the defect detection.

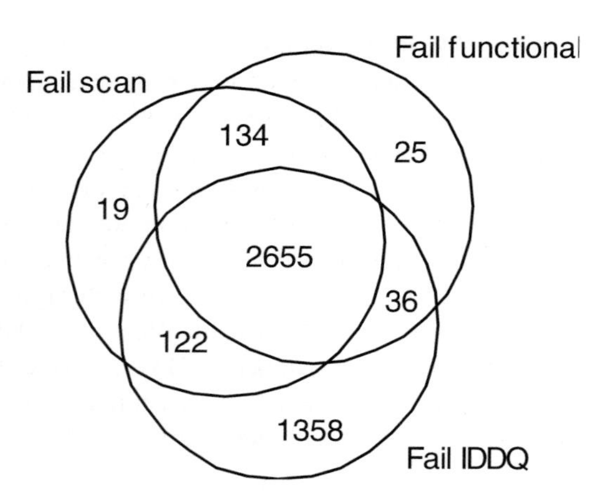

Figure 4-10. Distribution of failing die in each test class reported by Maxwell et al. [24].

4. I_{DDQ} VERSUS VOLTAGE TESTING

Logic testing, structural as well functional, has been the cornerstone for IC testing. In the last decade, I_{DDQ} testing was increasingly used as a quality improving supplement to the logic testing. A number of studies were reported on the relative effectiveness of both of these test techniques [1,24,30,44]. Some aspects are discussed below briefly to give an idea to the reader.

Table 4-3. Reject rate for various tests reported by Maxwell et al. [24].

Reject Rates (%)	**Scan and Functional Tests**			
	Neither	Noscan/ Func	Scan/ Nofunc	Both
Without IDDQ	16.46	6.36	6.04	5.80
	0.80	0.09	0.11	0.00

Perry [30] reported a three year study of CMOS ASICs. I_{DDQ} testing was implemented to reduce the early life failures. A set of 13 ASICs were analyzed in this study and typical SA fault coverage of devices was more than 99%. It was demonstrated in the study that with the implementation of I_{DDQ} the rejection rate was reduced at least by a factor of four. Similarly, Maxwell et al. [24] conducted a study of three wafer lots containing 26,415 die (excluding parts which failed initial continuity and parametric tests). The distribution of failing die in each of the test class is illustrated in Figure 4-10. Most defective chips (2,655) were identified by all tests. A large number of defective chips (1,358) were detected only by I_{DDQ} test and 25 and 19 failures were only detected by functional and scan tests, respectively. Table 4-3 shows the reject rates which would result for various combinations of the tests. If no testing is done, the reject rate would be 16.5% while if just functional and I_{DDQ} tests were performed, the rate would be 0.09% (900 PPM). If only the I_{DDQ} test was to be performed the reject rate would be 0.80%.

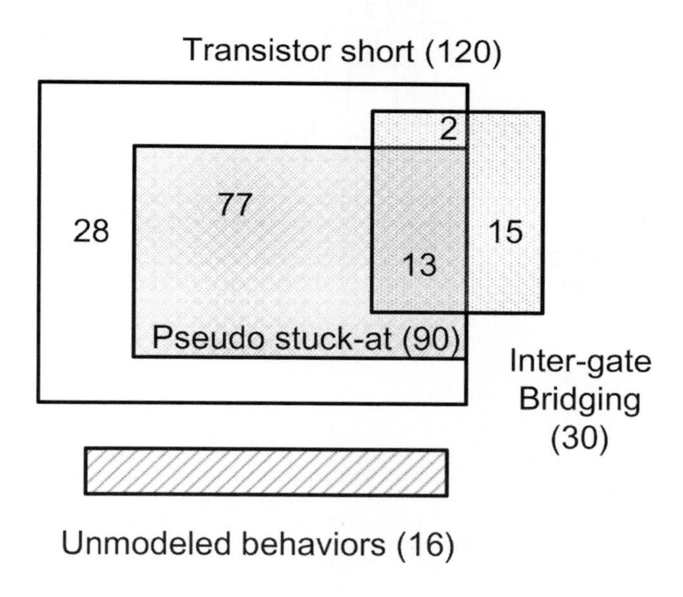

Figure 4-11. Overlap of faulty behaviours observed by Aitken [1].

Aitken [1] investigated the potential of I_{DDQ} testing in defect diagnosis. He showed that using both inter and intra-gate shorts as fault models and measuring the current under different steady state input conditions, it is possible to diagnose a defect location and/or determine its cause. This hypothesis was applied to ASICs and out of 151 parts in the sample, diagnosis was obtained for 135. In many cases, the predicted defects were confined to a single standard cell. The transistor short model (transistor leakage fault model) could diagnose 120 of the diagnosed defects. The input SA fault model was second with 90, however, all of them were also detected by transistor short fault model. Inter-gate bridge fault model could diagnose only 30 of these out of which 15 were only detected by this fault model. A total of 16 failed devices could not be modeled by any of the above mentioned fault models. The success of the transistor short model may be due to the fact that the I_{DDQ} test is the only one that specifically targets those faults. Furthermore, these results are biased towards I_{DDQ} test since failing parts had passed all tests except I_{DDQ} tests.

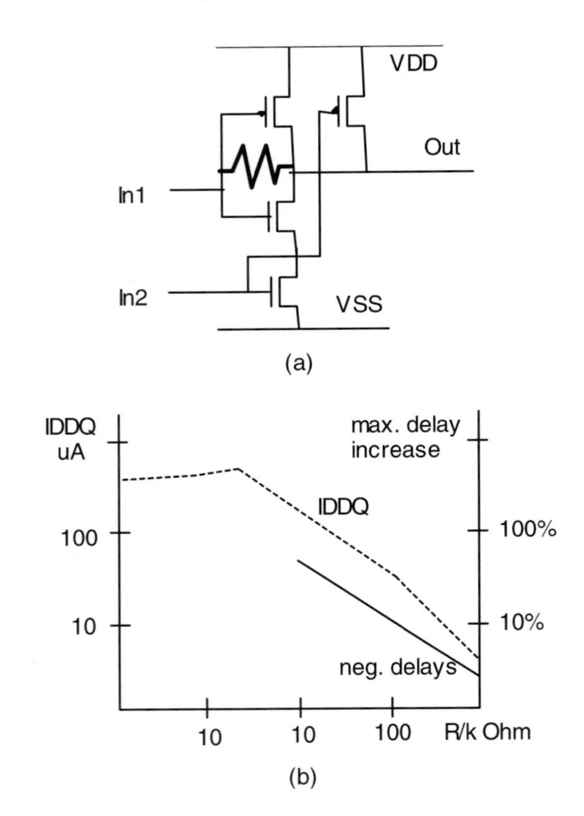

Figure 4-12. A bridging fault improving logic delay of a two input NAND gate [47].

Subtle defects such as resistive transistor faults usually change the transistor transfer characteristics. Such changes in the transfer characteristics may cause an increased transistor delay or an increased I_{DDQ} current.

Vierhaus et al. [47] carried out extensive simulations to quantify the impact of resistive stuck-on, stuck-open, and bridging faults on delays and I_{DDQ} for typical CMOS logic gates. The results of their analysis were more or less predictable for I_{DDQ} testing. All resistive transistor faults gave rise to state dependent elevated I_{DDQ}. However, their analysis also highlighted certain faults that reduced the logic delay. It may be recalled that bridging faults typically degrade the switching characteristics and increase the transistor/logic delay. This phenomenon can be further explained with the help of Figure 4-12. This figure illustrates a two input NAND gate with a bridging fault between an input (In1) and output (Out). Such a fault may be caused either by low a resistive poly-silicon bridge or a high resistive oxide defect through the thin oxide of the particular transistor (gate-drain gate oxide defect). Typically, a low resistive bridge causes a feedback condition resulting in a functional fault. A strong input driver, via the bridge, directly drives the faulty gate's output node. In the case of a weak driver, the output of faulty gate switches at an intermediate voltage (usually close to $V_{DD}/2$), depending upon the input. As the fault resistance is increased (>10kΩ), the fault causes positive delay and/or SAF. However, as the fault resistance is increased further above 10 kΩ, a positive feedback effect is created.

The resistive transistor stuck-on and stuck-open faults behave similarly. A low resistive stuck-on fault in a PMOS transistor of a NAND gate will most likely cause a SAF behavior. As the resistance of the faulty transistor rises, it causes delay fault in the NAND gate. A stuck-on fault in an NMOS transistor usually does not result in a SAF on the output of the NAND gate. The fault supported transitions result in smaller delays. All these faulty conditions are detected by I_{DDQ} measurements. The faulty behavior of CMOS gates under non-ideal stuck-open conditions (transistor source/drain open) with approximate resistance value of 50kΩ results in gross delay fault. Such defects are not detected by I_{DDQ} measurements.

5. DEFECTS IN SEQUENTIAL CIRCUITS

We mentioned how scan path, LSSD, and their derivatives became popular because their application could change distributed sequential logic elements into a big unified shift-register for testing purposes [9,12]. As a result, the overall test complexity is reduced [49]. Owing to these techniques, test generation and fault grading for complex digital circuits became a possibility.

From our discussion it is abundantly clear that I_{DDQ} is the best method for bridging defect detection provided back ground leakage current is low. It must be mentioned that in nano-metric technologies keeping the background leakage low has become a non-trivial task. However, there are several I_{DDQ} based test techniques such as delta I_{DDQ} to enhance the defect detection capability of I_{DDQ} testing. An interested reader is referred to recent research on this topic.

It is important to note that certain bridging defects in sequential circuits do not give rise to elevated I_{DDQ}. On the other hand, the voltage detection of such defects depends on transistor level parameters (e.g., width and length) of the affected transistors and resistance of the defect. As VLSI complexity is growing, the number of flip-flops is also growing therefore, we should pay attention to this class of defects that are difficult to detect with I_{DDQ} or voltage means.

Unlike combinational circuits, the controllability condition in sequential circuits (e.g., flip-flops) does not ensure that a bridging fault (short defect) is detected by I_{DDQ} testing. Many low resistance shorts in sequential circuits do not cause an elevated I_{DDQ}. The voltage detection of such defects depends on transistor level parameters (e.g., width and length) of the affected transistors and resistance of the defect. Therefore, in this section, we concentrate on defects in sequential circuits and proposed some design for testability solutions.

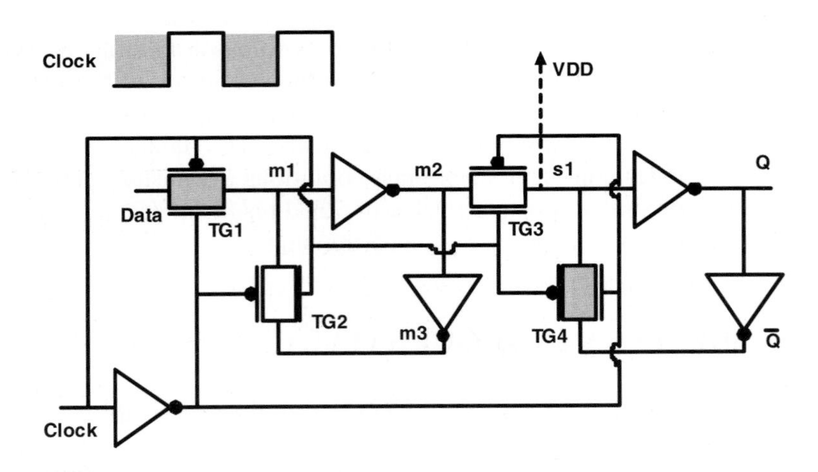

Figure 4-13. A bridging defect between s1 and V_{DD} in a typical CMOS flip-flop is not detected by I_{DDQ}.

5.1 Undetected Defects

Lee and Breuer [22] highlighted a certain class of bridging faults in sequential circuits that are not detected by I_{DDQ}. Rodriguez et al. [34], carried out inductive fault analysis of a scan flip-flop to find out the relative effectiveness of I_{DDQ} and voltage test methods for realistic bridging faults detection. Conclusions of their analysis are: (a) For zero resistance bridging faults, 8% can not be detected by the I_{DDQ}. However, these can be detected by output voltage measurements. (b) For bridging faults with resistance above 2kΩ, I_{DDQ} detected all defects but only some were voltage detectable.

The above analysis demonstrates that all bridging faults in the analyzed flip-flop are detectable either by voltage or I_{DDQ} test method. Nevertheless, it should be stressed that the detection of such bridging faults by I_{DDQ} or voltage test method strongly depends on the circuit level parameters (e.g., W and L) of transistors in the flip-flop and the resistance of the defect. In a paper, Metra et al. [26] showed that for CMOS flip-flops implemented with NAND gates and/or with pass gates, neither I_{DDQ} testing, nor voltage testing, nor did the combination of the two achieve the complete bridging fault coverage. Their study has two broad conclusions: (a) Irrespective of flip-flop implementation, bridging fault coverage in flip-flops is low, and (b) Circuit level parameters have an influence on bridging fault detection. For voltage detection the logic thresholds of intermediate logic gates driven by the flip-flop and the satisfaction of observability conditions will also play an important role before faults can be detected at the primary outputs of the device under test (DUT).

Figure 4-13 shows a typical flip-flop implementation in CMOS technology. It is a single phase clock, master-slave flip-flop. While Clock is at logic low level, transmission gates TG1 and TG4 are conducting. At the same time TG2 and TG3 are in the non-conducting state. Hence in this clock state, the master latch of the flip-flop accepts the new data from the data input (Data) while the slave-latch retains the old data. At the rising edge of the clock the master-latch no longer accepts the input data and transfers the current data to the slave-latch. In this fashion the master-slave operation of the flip-flop is realized.

Extra material defects in metallization layers, gate oxide shorts were identified as major contributors to overall defect mechanisms in CMOS ICs. These defects are collectively known as shorts or bridging faults. In nano-metric technology with copper dual damascene metallization, the open defects in vias and contacts are on the rise. Most of these bridging faults, under appropriate steady state input stimuli conditions, give rise to an abnormally high current. Such defects under specific steady-state input stimuli conditions create a DC current path between V_{DD} and V_{SS} and can be detected by the I_{DDQ}. However, the current through the DC path should be higher than the various leakage currents (e.g., transistor leakage) in the IC so as to give the indication of a defect. On the other hand, a defect which can not cause an elevated quiescent current between V_{DD} and V_{SS} in any of the steady-state input stimuli conditions can not be detected by the I_{DDQ}. A low resistive bridging defect between node s1 and V_{DD} (or V_{SS}) in Figure 4-13 can not cause high quiescent current and therefore is not detected by I_{DDQ} [35,37].

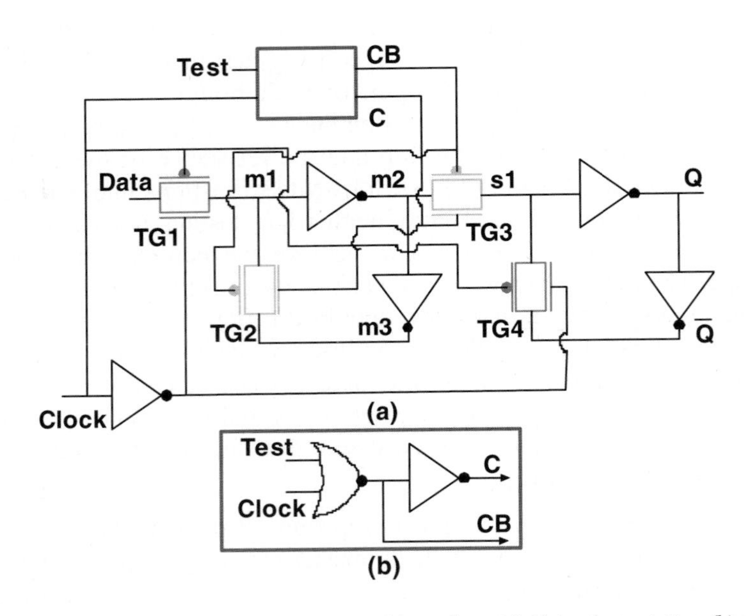

Figure 4-14. Defect detection concept (a); and test block implementation (b).

The problem of bridging fault detection in CMOS flip-flops with I_{DDQ} is generic in nature. For flip-flops implemented with pass gates, the non-detection of bridging faults is explained as below. CMOS flip-flops are made economically using switches or TGs. These switches are alternately closed or opened to ensure the master-slave operation of the flip-flop (Figure 4-13). The reason for non-detection of this bridging fault by I_{DDQ} is the bidirectional nature of switches. At the rising transition of Clock, TG1 and TG4, which were conducting, stop conducting; and TG2 and TG3, which were not conducting, start to conduct. Now the node m2 starts to drive node s1 which till this moment was driven by node Q through TG4. The input of node m2, itself is going through a transitory phase (since TG1 is turning off and TG2 is turning on), therefore, the node m2 has a limited driving capability. In a defect-free case, positive feedback via a pair of back to back inverters allows the flip-flop to ride through this transitory phase. Now, due to the bridging fault, the node s1 is constantly driven to V_{DD} (or V_{SS}) level. In the case of a low resistive bridging fault, the voltage driving strength through the defect is much stronger than that of m2 and as a result overrides the master latch. This operation is similar to the write operation carried on a SRAM cell. Therefore, in steady-state no current exists and the defect is not detected by I_{DDQ}.

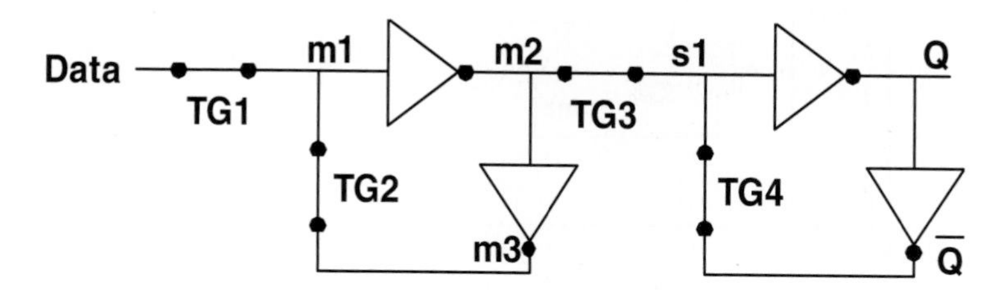

Figure 4-15. The equivalent circuit of the flip-flop.

5.2 Defect Detection Technique

As explained in the previous section, a bridging fault is I_{DDQ} detected if a logical conflict between faulty nodes can be created and sustained in the steady-state. In this particular case, the conflict can be kept alive as long as the master-latch of the flip-flop is not over-written by the drive through the defect. This can be achieved in one of the following ways [35,36]:

- Initialize the master-latch with appropriate data value keeping Clock low. Use an external control signal, Test, which breaks the feedback loop in the master-latch. Finally change Clock to logic high so that TG3 starts conducting.
- Maintain the appropriate data value at Data input of the flip-flop. Use an external control signal, Test, and Clock, make all TGs conducting at the same time.

The first solution can be implemented with an additional TG in the feedback path of the master-latch, an inverter and a control input, Test. However, the latter solution of making the flip-flop transparent is a better one since all nodes are driven to either logic high or low. A flip-flop can be made transparent locally if an additional test block is added to it such that in test mode, all four TGs of the flip-flop are made to conduct at the same time. This will ensure, that nodes m1, m2, and s1 are always driven by Data input.

Figure 4-14(a) shows the concept, and Figure 4-14(b) gives a possible implementation of extra test logic required in a flip-flop. Clock and Clock' signals to TG2 and TG3 are routed through the test logic. A control signal, Test, is needed to switch between normal and test modes.

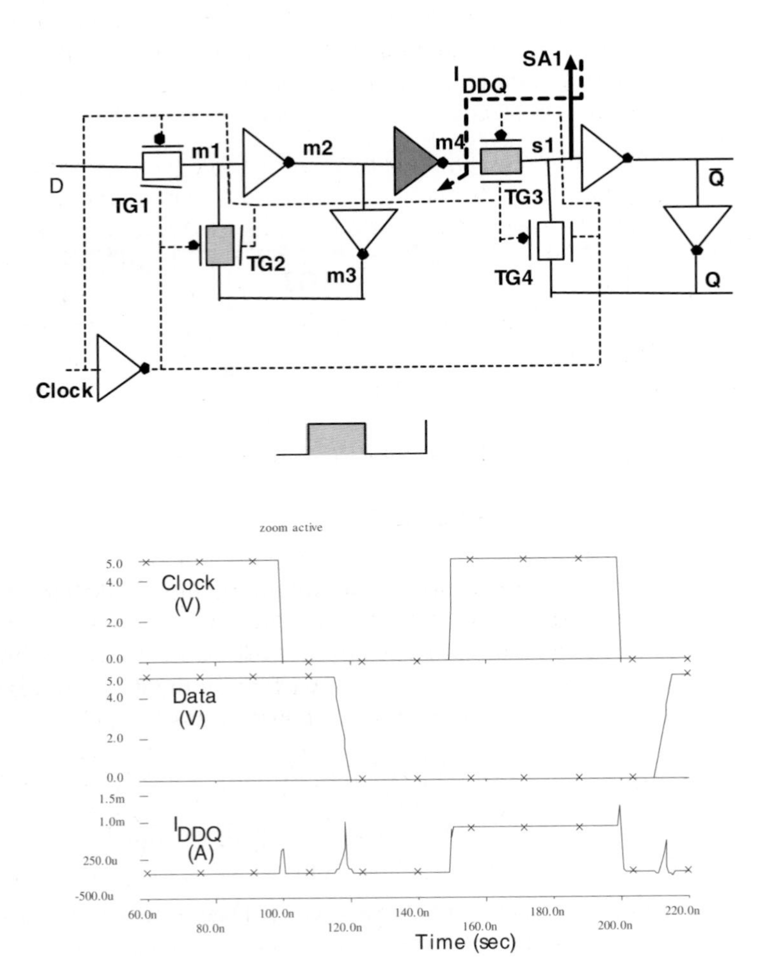

Figure 4-16. *I_{DDQ} testable flip-flop and simulation results showing the defect detection [32].*

The scheme can be implemented with a NOR gate and an inverter. This extra test logic is added to the flip-flop such that when Test is low, the flip-flop behaves exactly as a normal flip-flop. Outputs of the test logic, C and CB, are Clock and Clock'. However, in the test mode (Test=1), C and CB are held at logic levels high and low, respectively, irrespective of the logic level of Clock. This ensures that TG2 and TG3 are conducting during the test mode. Now if Clock is at logic low, TG1 and TG4 are also conducting. Therefore, in the test mode the flip-flop is transparent. Figure 4-15 shows the test mode equivalent circuit of the flip-flop. Since in test mode all nodes within the flip-flop are driven simultaneously, any bridging defect between flip-flop node and V_{DD} (V_{SS}) node can be detected by I_{DDQ} as well as by voltage measurement.

Input data values can be setup to detect various likely bridging defects by I_{DDQ} as well as by voltage measurement. Assuming that s1 is shorted to V_{DD}, Data is also kept high. In this condition there will be a driven logic conflict between nodes s1 and V_{DD} that will give rise to a large quiescent current. Similarly, logic low Data will detect the bridging defect between node s1 and V_{SS}. The cost of this local implementation is six extra transistors and the test signal routing to each desired flip-flop. A similar solution can be applied to two-phase and level-sensitive flip-flops.

5.3 I_{DDQ} Testable Flip-flop

Shorts or bridging faults in the slave latch can be made I_{DDQ} detectable if the master latch is prevented from being over-written due to a fault in the slave latch. Since the master latch is not over-written, the data stored in master sustain a logical conflict with the fault in the slave. For example, a logical conflict is created and sustained between the fault and node s1 while TG3 is conducting (Figure 4-16). The prevention of master latch over-writing is achieved if TG3 is made unidirectional such that it transfers data only from the master to the slave latch. TG3 is made unidirectional either by putting an additional inverter just before it or by replacing TG3 by a clocked-inverter. A schematic of the flip-flop configuration is shown in Figure 4-16. The additional inverter improves fault detection by I_{DDQ} test method and also improves various timing aspects of the flip-flop. Since the master latch is effectively isolated from the slave latch, setup and hold times of the flip-flop are improved.

5.4 Defects and Scan Chains

Hard to detect bridging defects in flip-flops can be tested efficiently by I_{DDQ} measurements if flip-flops are organized in scan chains and simple DfT solutions are implemented. Figure 4-17 illustrates a transparent scan chain. The concept of flip-flop transparency (Figure 4-15) can be extended to the entire scan chain. Flip-flops can be organized to operate in three different operational modes. These modes are normal functional mode, scan mode, and transparent scan mode.

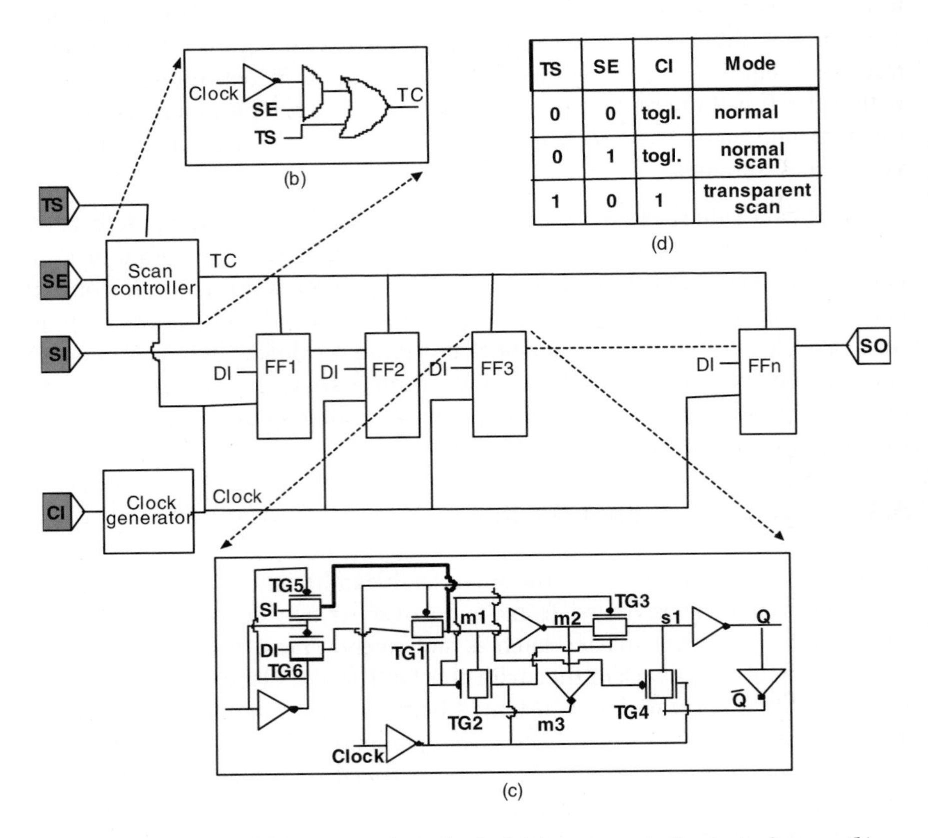

TS	SE	CI	Mode
0	0	togl.	normal
0	1	togl.	normal scan
1	0	1	transparent scan

Figure 4-17. Transparent scan with single clock (a), scan controller logic diagram (b), modified scan flip-flop to facilitate single clock test mode operation (c), and flip-flop operational modes (d).

The basic idea exploits the normal mode non-clock signals (e.g., TC/SE) as clock signals in the test mode to reduce test and implementation complexity. Since these are non-clocked signals in the normal mode, they are not timing critical. Moreover, normal mode clock signal is kept unchanged so that normal mode performance is not affected. It can be argued that this is a better, cost effective approach compared to building additional control for the normal mode clock signal [35,36].

An implementation of the idea is shown in Figure 4-17. Figure 4-17(a) illustrates a scan chain, clock generator, and an additional block, scan controller that controls the modes of operation of flip-flops in the scan chain. Figure 4-17(b) depicts the logic schematic of the scan controller. This implementation requires 3 logic gates. Figure 4-17(c) illustrates the logic schematic of the scan flip-flip. The flip-flop requires no extra logic gates. The only modification needed in the flip- flop is to reroute the front-end multiplexer outputs. The output of TG5 which normally goes to input of TG1, is now connected to the output of TG2. This change is shown by a bold line in Figure 4-17(c). The normal mode data path of the scan flip-flop is not changed to keep the normal mode set-up and hold times of the flip-flop unchanged. Furthermore, the clock path of the flip-flop is also not changed. Thus, the normal mode operation of the flip-flop remains unaffected. Essentially, in the scan mode, TG1 is replaced by TG5, which is clocked by TC. Rest of the TGs (TG2-TG4) are clocked by the Clock. The signal TC can be controlled asynchronously at logic high as well as logic low. Moreover, it can act as a clock, depending upon the input decoding conditions. These conditions are shown in Figure 4-17(d). In the transparent scan mode, TC is kept at logic high by virtue of logic high on input TS. This ensures that TG5 is conducting. Clock is held at logic high so that TG2 and TG3 are conducting. Hence, the flip-flop becomes transparent in this mode. Therefore, the complete scan chain can be quickly tested.

Mercer and Agrawal [25] investigated the optimization of clock and scan enable signals. Their motivation was to save routing of an extra signal to two-phase clock flip-flops. In a two-phase clock scan flip-flop, two clock signals, and a scan enable signal are typically needed. These control signals, can have $2^3 = 8$ control states. However, not all control states are used in a scan flip-flop, and some control states can be decoded locally. A simple decoding logic can be a part of a flip-flop. As a result, only two control signals are required for each flip-flop. Alternatively, similar to the clock generator modification discussed above, clock generator may be modified to reduce the area required for local decoding of signals. In their scheme, the scan chain transparency (flush test) is not possible.

Subsequently, Bhavsar implemented this technique with very few MOS transistors [5]. He proposed a modification of latches in single latch design style of LSSD such that in scan mode the latches function as dynamic master-slave flip-flops. This technique requires very little area overhead for scan implementation in single latch design style. However, it has some disadvantages. Scan chains are dynamic in nature; therefore, extra care must be taken while propagating data through them. Secondly, the flush test is not possible in this methodology. Finally, in this technique, a static latch is converted into a dynamic master/slave flip-flop. As a result, the feedback paths are turned off during the shift test. Therefore, testing defects in the feedback paths require an additional hold test.

Table 4-4. CMOS IC defect classes and their detection techniques [16].

Defect Classes	Description	Test Method	100% Detection
Bridge Type -1	Transistor node, inter-logic gate, logic gate to power bus, power bus to bus	I_{DDQ}, Boolean	Yes, No
Bridge Type -2	Layout identified bridges	I_{DDQ}, Boolean	Yes
Bridge Type -3	Sequential intra-nodal	I_{DDQ}, Boolean	Yes
Open Type -1	Transistor-on	I_{DDQ}, Boolean	Yes, No
Open Type -2	Transistor pair-on	I_{DDQ}, Boolean	Yes, Yes
Open Type -3	Transistor pair-on/off	I_{DDQ}, Boolean	Yes, No
Open Type -4	Sequential	I_{DDQ}, Boolean	Yes
Open Type -5	Transistor-off (memory effect)	I_{DDQ}, Boolean	No, No
Open Type -6	Delay	I_{DDQ}, Boolean	No, No
Parametric Delay	R_{via}, V_t, $\Delta(W/L)$	I_{DDQ}, Boolean	No, No

6. DEFECT CLASSES AND THEIR TESTING

Fault models represent defective behaviors with limited accuracy. We saw in previous sections that there is no single test method suitable for testing all possible defect types. Furthermore, tests should focus on defects rather than on faults if the objective is to guarantee a low escape rate (low PPM). Hence, it is of vital importance to categorize types of defects in CMOS circuits and outline their detection strategies. Keeping this objective in mind Hawkins et al. [16] identified defect classes. Defects can be segregated into different types of bridges and opens. Table 4-4 illustrates these types.

It is apparent from the table that several defect classes are 100% detectable considering that appropriate test vectors are either available or can be generated. I_{DDQ} is the most effective test for bridging defect classes with Boolean testing also achieving 100% coverage for Type-2 and Type-3 bridges assuming that the defect resistance is lower than 1kΩ. If the defect resistance increases beyond 1kΩ, the Boolean coverage of defects is reduced.

An open defect causing a transistor stuck-on behavior (Open Type-1) is 100% detectable using pseudo stuck at fault (PSAF) patterns. Boolean detectability is difficult since stuck-on behavior causes degraded voltage levels at logic output. Open Type-2 defects are 100% detectable by I_{DDQ} tests as well as by Boolean testing with PSAF patterns. If an open defect causes a transistor-pair on/off then Boolean test will detect it and I_{DDQ} will not detect it. Open defects in sequential circuits will be detected by Boolean as well as I_{DDQ} tests. In order to detect Open Type-5 a 2-pattern sequence is needed and simple SAF testing or I_{DDQ} testing will not be effective. Marginal opens (Open Type-6) are hard to detect by Boolean or I_{DDQ} testing. Similarly, parametric faults are hard to detect by both methods. In order to detect such faults delay fault testing or at speed testing seems to be promising alternatives.

7. APPLICATION OF IFA IN NANO-METRIC TECHNOLOGIES

As we move to nano-metric regime, IFA must deal with circuit complexity and varying defect mechanisms. Most of the studies reported in this chapter so far were on modest complexity circuits by modern standards. Can we utilize IFA to analyze a complete microprocessor or an ASIC? The short answer is yes. However, before we get into details, it is pertinent to discuss the progress made towards this objective in recent years.

IFA is based on the assumption that the probability of a defect occurring at a particular site is a function of the local layout geometry and the distribution of mechanisms observed for the manufacturing process. Traditionally, IFA focused on layout geometry and defect distribution, and it ignored the testability of a fault. However, if the faults identified using IFA are highly testable, it means they are easily covered by conventional tests for stuck-at faults, then using an IFA based approach will not yield a significant DPM (Defects per Million) improvement over a standard stack-at fault model. For nano-metric CMOS technologies, the challenge for effective IFA tools is to identify faults that are both most probable and relatively difficult to be detected using stuck-at fault vectors [41].

In deep submicron CMOS technologies product yields are high due to clean processes, good design techniques, and years of experience. Even for complex VLSIs it is not uncommon to have yield better than 95%. Note, that IFA has the statistical nature. It means that seemingly identical circuit structures do not always produce identical results. The problem with IFA is that in order to obtain realistic yield loss estimation, we need suitably large sample size. It means that we should simulate one million or more faulty circuits to obtain realistic yield loss. When we examine the time that it takes to perform large number of fault simulation, most of the time is spent processing circuits in which there are no defects. On the other hand, one can speed up the IFA process by simulating fewer circuits, but then we have a smaller sample set and some faults will disappear from the list.

This problem can be overcome by artificially increasing the defect densities in the IFA experiments. As described in Chapter 2, the probability of defect between two parallel conductors is proportional to the defect density and the distance between two lines. It means that if we scale D0 we directly scale the probability of all fault occurrences. In other words, if we reduce the number of circuits that we simulate but scale the defect densities up accordingly, we will get the same results much more quickly, since we are not simulating so many fault-free circuits. A. Platts et al. show that by scaling defect densities upwards by two or three orders of magnitude and scaling the number of devices simulated downwards in the same ratio we obtain accurate IFA results in a relatively short time [33]. Zachariah and Chakravarty proposed a methodology of extracting bridging faults for large million transistor circuits [50]. They created partially overlapping segments of a layout. Computation of weighted critical area (WCA) was one of the key factors in their computation. WCA is computed over different defect sizes and weighed with their respective probabilities of occurrence. The higher the WCA for two nets, higher is the probability of occurrence of a bridging defect. Even though authors restricted themselves to just the bridging defects, ability to handle large designs was no mean achievement.

Perhaps the most ambitious recent study was carried out by Krishnaswamy et al., who successfully performed IFA on the entire layout of the Pentium 4 microprocessor [19]. It was a 32 bit microprocessor with over 40 million transistors implemented in a 0.18 μm CMOS process. The layout database of this large industrial circuit was divided into several layout hierarchies (B1, B2 ... B30), which for the most part correspond to RTL hierarchies. They were motivated to seek answers to: *(i) Where bridging defects are most likely to occur in the chip, and why? (ii) What kinds of failures are caused by these defects, and (iii) What changes to be brought in test strategies to screen these defects.*

Krishnaswamy et al used the IFA tool developed by Zachariah and Chakravarty [50]. It was determined the IFA tools ran most efficiently on circuits of a few hundreds of thousand of transistors. The data at the full chip level says that roughly 80% of all bridges will occur between a signal node and V_{DD} or V_{SS}, while only 20% occur between non supply nodes. In general, these data suggest that although the mechanism for inducing stuck fault may have changed with technology scaling, it is still a very viable fault model to continue testing against. IFA also found that bridging faults are more likely to occur on global signals (70%) as opposed to the leaf level signals (30%). Analysis of leaf level weighted critical area (WCA) showed that 50% of leaf level faults were devoted to RAM or ROM arrays, 10% came from synthesized logic, and roughly 30% from hand drawn high speed data paths. At the leaf level WCA, the data also indicated that the stuck fault model is still useful for providing high defect coverage. The using a commercial sequential ATPG tool provided the stuck fault coverage greater than 95%. However, ATPG tool provided disappointing coverage of 37% and 28% on non supply bridges in two of these blocks. It was apparent that the functional test vectors are better suited for testing hard to detect bridges.

The design of the clock network is becoming more critical with the progressive increase in chip area and operating frequencies,. It is generally assumed that an incorrect clock signal leads to a catastrophic failure of the whole system which, consequently can be easily detected during manufacturing testing. However, recently it was shown that this assumption is no longer true for high-speed synchronous systems [27]. The probability of clock distribution faults in the Intel Itanium microprocessor was estimated by IFA. It was found that this probability is two orders of magnitude higher than other most likely microprocessor faults. Only a small percentage of these faults results in a catastrophic failure of the CPU, while the majority result in a local delay failure which will compromise the microprocessor operation and result in an unacceptable decrease in quality and reliability.

8. CONCLUSIONS

The popularity of stuck-at faults led to several studies to determine whether the SAF model really represents manufacturing defects. Results of various studies are mixed. Galiay et al. found that only a subset of defects resulted in logical faults [13]. The study of Banerjee and Abraham concluded that the transistor level fault models represent transistor defects with fair amount of accuracy [2,3]. However, their study assumed zero and infinite impedances for shorts and opens, respectively. With more realistic defect impedances their fault coverage would have been lower. A MOS technology is a collection of a set of layers to be processed through masks and each mask discriminates between areas to be processed or not processed on a given layer. A defect on a mask or a dust particle on a wafer may result in the improper processing of any layer. Defects are abstracted at the device level by performing logical operations on the masks of different layers. In this way realistic faults are generated. Maly et. al. proposed a methodology for the same and carried out an analysis for NMOS and CMOS circuits [23,42]. The conventional SAF model could represent only 35% of all defects that could occur and out of 80 possible SAFs in a given circuit only 37 actually occurred. Similarly, studies conducted on standard cells by Ferguson and Shen, and Peters and Oostdijk, respectively, demonstrated that the SAF model is a poor abstraction of realistic defects. Their studies concluded I_{DDQ} is far more effective in defect detection.

The realization of gate oxide is one of the most critical process steps. Each successive scaled generation has still thinner gate oxide and the quality of gate oxide often determines the product quality and reliability in the field. Hawkins and Soden conducted an experiment on SRAMs where 37% of faulty devices (poor gate oxide) passed all functional tests but failed I_{DDQ} tests. Several subsequent studies were conducted on I_{DDQ} and logic testing. The results of these studies were overwhelmingly in favor of I_{DDQ} testing.

Flip-flops are indispensable building block in digital ICs. However, most of the present static flip-flop configurations suffer from poor coverage of bridging faults by the I_{DDQ} test technique. Bridging faults are among the most prevalent faults in ICs. Considering that a complex digital IC may contain several thousand flip-flops and I_{DDQ} is emerging as the quality improving complement to logic testing, such I_{DDQ} escapes will have a severe quality impact. Researchers have attributed this poor coverage to the architecture of flip-flops where the controllability condition is not sufficient for bridging fault detection with the I_{DDQ} test technique. In this chapter, several solutions of testing defects in flip-flops and scan chains were suggested. By virtue of flip-flop and scan path transparency defects can be effectively and efficiently detected. Flip-flop or scan path transparency requires an independent control of Clock and Clock' signals.

As we scale the technology to nano-metric dimensions, application of IFA becomes extremely expensive. Complexity of analysis, diverse defect mechanisms, and large chip size do contribute to it. However, researchers analyzed the layout of the entire Pentium microprocessor and concluded that 80% of all bridges occurred between signal node and V_{DD} or V_{SS}. Therefore, these shorts could be modeled as the SA faults and most of them (95%) were detected using SA based test pattern generation. However, 20% of bridging faults were between signal nodes and could not be modeled as the SA faults and most of these could not be detected by the SA based test patterns.

References

1. R.C. Aitkens, "A Comparison of Defect Models for Fault Location with I_{DDQ} Measurements," Proceedings of International Test Conference, 1992, pp. 778-787.

2. P. Banerjee and J.A. Abraham, "Fault Characterization of VLSI Circuits," Proceedings of IEEE International Conference on Circuits and Computers, September 1982, pp. 564-568.

3. P. Banerjee and J.A. Abraham, "Characterization and Testing of Physical Failures in MOS Logic Circuits," IEEE Design & Test of Computers, vol. 1, pp. 76-86, August 1984.

4. C.C. Beh, K.H. Arya, C.E. Radke and K.E. Torku, "Do Stuck Fault Models Reflect Manufacturing Defects?" Proceedings of International Test Conference, 1982, pp. 35-42.

5. D. Bhavsar, "A New Economical Implementation for Scannable Flip-Flops in MOS," IEEE Design & Test, vol. 3, pp. 52-56, June 1986.

6. R. Bryant, "A Survey of Switch-Level Algorithms," IEEE Design & Test of Computers, vol. 4, pp. 26-40, August 1987.

7. R. Chandramouli, "On Testing Stuck-Open Faults," Proceedings of 13th Annual International Symposium on Fault Tolerant Computing Systems, June 1983, pp. 258-265.

8. K.W. Chiang and Z.G. Vranesic, "Test Generation For MOS Complex Gate Networks," Proceedings of 12th Annual International Symposium on Fault Tolerant Computing Systems, June 1982, pp. 149-157.

9. E.B. Eichelberger and T.W. Williams, ''A Logic Design Structure for LSI Testability,'' Journal of Design Automation and Fault Tolerant Computing, vol. 2, no. 2, pp. 165-178, May 1978.

10. Y.M. El-Ziq and R.J. Cloutier, "Functional Level Test Generation for Stuck-Open Faults in CMOS VLSI," Proceedings International Test Conference, 1981, pp. 536-546.

11. F.J. Ferguson and J.P. Shen, ''Extraction and Simulation of Realistic CMOS Faults using Inductive Fault Analysis," Proceedings of International Test Conference, 1988, pp. 475-484.

12. S. Funatsu, N. Wakatsuki and T. Arima, ''Test Generation Systems in Japan,'' Proceedings of 12th Design Automation Conference, 1975, pp. 114-122.

13. J. Galiay, Y. Crouzet and M. Vergniault, ''Physical versus Logical Fault Models in MOS LSI Circuits: Impact on Their Testability,'' IEEE Transaction on Computers, vol. C-29, no. 6, pp. 527-531, June 1980.

14. C.F. Hawkins and J.M. Soden, "Electrical Characteristics and Testing Considerations for Gate Oxide Shorts in CMOS ICs," Proceedings of International Test Conference, 1985, pp. 544-555.

15. C.F. Hawkins and J.M. Soden, "Reliability and Electrical Properties of Gate Oxide Shorts in CMOS ICs," Proceedings of International Test Conference, 1986, pp. 443-451.

16. C. F. Hawkins, J. M. Soden, A. Righter, and J. Ferguson, "Defect Classes – An Overdue Paradigm for CMOS IC Testing," Proceedings of International Test Conference, 1994, pp. 413-425.

17. S.K. Jain and V.D. Agrawal, "Modeling and Test Generation Algorithm for MOS Circuits," IEEE Transactions on Computers, vol. 34, no. 5, pp. 426-43, May 1985.

18. N.K. Jha and S. Kundu, Testing and Reliable Design of CMOS Circuits, Boston: Kluwer Academic Publishers, 1990.

19. Krishnaswamy V., Ma A.B., Vishakantaiah P., "A study of bridging defect probabilities on a Pentium (TM) 4 CPU," IEEE Int. Test Conf., 2001, pp. 688-695.

20. B.Kaczer, R.Degraeve, E. Augendre, M. Jurczak, and G. Groeseneken, "Experimental verification of SRAM cell functionality after hard and soft gate oxide breakdowns," European Solid-State Device Research Conf. (ESSDERC), pp. 75-78, 2003.

21. B. Kruseman, R. van Veen, K. van Kaam, "The Future of delta I_{DDQ} testing," in Proc. of ITC, pp. 101-110, 2001.

22. K.J. Lee and M.A. Breuer, ''Design and Test Rules for CMOS Circuits to Facilitate I_{DDQ} Testing of Bridging Faults,'' IEEE Transactions on Computer-Aided Design, vol. 11, no.5, pp. 659-669, May 1992.

23. W. Maly, F.J. Ferguson and J.P. Shen, ''Systematic Characterization of Physical Defects for Fault Analysis of MOS IC Cells,'' Proceedings of International Test Conference, 1984, 390-399.

24. P. C. Maxwell, R. C. Aitken, V. Johansen and I. Chiang, "The Effectiveness of I_{DDQ}, Functional and Scan Tests: How Many Fault Coverages Do We Need?," Proceedings of International Test Conference, 1992, pp. 168-177.

25. M.R. Mercer, and V.D. Agrawal, "A Novel Clocking Technique for VLSI Circuits Testability," IEEE Journal of Solid State Circuits, vol. sc-19, no. 2, pp. 207-212, April 1984.

26. C. Metra, M. Favalli, P. Olivo, and B. Ricco, "Testing of Resistive Bridging Faults in CMOS Flip-Flop," Proceedings of European Test Conference, 1993, pp. 530- 531.

27. Metra C., Di Francescantonio S., Mak TM., "Implications of Clock Distribution Faults and Issues with Screening them during Manufacturing Testing," IEEE Trans. on Computers, Vol. 53, No. 5, pp. 531-546, 2004.

28. F. Monsieur, E. Vincent, D. Roy, S. Bruyere, G. Pananakakis, and G. Ghibaudo, "Time to breakdown and voltage to breakdown modeling for ultra-thin oxides (Tox < 32 Å)," in Proc. IEEE Int. Reliability Workshop, 2001, pp. 20-25.

29. K. Okada, H. Kubo, A. Ishinaga, and K. Yoneda, "A concept of gate oxide lifetime limited by "B-mode" stress induced leakage currents in direct tunneling regime," Symp. VLSI Technol. Dig., pp. 57-58, 1999.

30. R. Perry, "I_{DDQ} Testing in CMOS Digital ASIC's - Putting It All Together," Proceedings of IEEE International Test Conference, 1992, pp. 151-157.

31. F. Peters and S. Oostdijk, "Realistic Defect Coverages of Voltage and Current Tests," Proc. of IEEE International Workshop on I_{DDQ} Testing, 1996, pp. 4-8.

32. T. Pompl, H. Wurzer, M. Kerber, R.C.W. Wilkins, I. Eisele, "Influence of soft breakdown on NMOSFET device characteristics", Proc. IRPS, pp. 82-87, 1999.

33. A. Platts and D. Taylor, "Rapid Inductive Fault Analysis for High-Yield Circuits," Microelectronics Journal, vol. 33, No. 3, p 279-284, 2002.

34. R. Rodriguez-Montanes, J. Figueras and R. Rubio, ''Current vs. Logic Testability of Bridges in Scan Chains,'' Proceedings of European Test Conference, 1993, pp. 392-396.

35. M. Sachdev, ''Transforming Sequential Logic for Voltage and I_{DDQ} Testing,'' Proceedings of European Design and Test Conference, 1994, pp. 361-365.

36. M. Sachdev, ''Testting Defects in Scan Chains,'' IEEE Design & Test of Computers, vol. 12, pp. 45-51, December 1995.

37. M. Sachdev, "I_{DDQ} and Voltage Testable CMOS Flip-flop Configurations," Proceedings of International Test Conference, 1995, pp. 534-543.

38. K. F. Schuegraf and C. Hu, "Reliability of thin SiO2," Proceedings of IEE, vol. 9, pp. 989-1004, September 1994.

39. J. Segura, C. Benito, A Rubio and C.F. Hawkins, "A Detailed Analysis of GOS Defects in MOS Transistors: Testing Implications at Circuit Level," Proceedings of International Test Conference, 1995, pp. 544-551.

40. O. Semenov, A. Vassighi and M. Sachdev, "Leakage current in sub-quarter micron MOSFET: A perspective on stressed delta I_{DDQ} testing," Journal of Electronic Testing (JETTA), vol. 19, No.3, pp. 341-352, 2003.

41. S. Sengupta, S. Kundu, S. Chakravarty, P. Parvathala, R. Galivanche, G. Kosonocky, M. Rodgers, TM. Mak, "Defect-Based Test: A Key Enabler for Successful Migration to Structural Test," Intel Tech. Journal, Q1, pp 1-14, 1999. http://developer.intel.com/technology/itj/q11999/pdf/defect_based.pdf

42. J.P. Shen, W. Maly and F.J. Ferguson, ''Inductive Fault Analysis of MOS Integrated Circuits,'' IEEE Design & Test of Computers, vol. 2, pp. 13-26, December 1985.

43. J. M. Soden and C.F. Hawkins, "Test Considerations for Gate Oxide Shorts in CMOS ICs," IEEE Design & Test of Computers, vol. 3, pp. 56-64, August 1986.

44. T. Storey, W. Maly, J. Andrews, and M. Miske, "Stuck Fault and Current Testing Comparison Using CMOS Chip Test," Proceedings of International Test Conference, 1991, pp. 311-318.

45. J.H. Suehle, "Ultra thin gate oxide reliability: physical models, statistics, and characterization," IEEE Trans. Electron Devices, vol. 49, pp. 958-971, June 2002.

46. M. Syrzycki, "Modeling of Gate Oxide Shorts in MOS Transistors," IEEE Transactions on Computer Aided Design, vol. 8, no. 3, pp. 193-202, March 1989.

47. H.T. Vierhaus, W. Meyer, and U. Glaser, "CMOS Bridges and Resistive Faults: I_{DDQ} versus Delay Effects," Proceedings of International Test Conference, 1993, pp. 83-91.

48. R.L. Wadsack, "Fault Modeling and Logic Simulation of CMOS and MOS Integrated Circuits," Bell Systems Technical Journal, vol. 57, no.5, pp. 1449-1474, May-June 1978.

49. T.W. Williams and K.P. Parker, ''Design for Testability--A Survey,'' Proceedings of the IEEE, vol. 71, no. 1, pp. 98-113, January 1983.

50. S. T. Zacariah and S. Chakarvarty, "A Scalable and Efficient Methodology to Extract Two Node Bridges from Large Industrial Circuits," Proceedings of IEEE International Test Conference, pp. 750-759, November 2000.

Chapter 5

TESTING DEFECTS AND PARAMETRIC VARIATIONS IN RAMS

RAMs are integral building blocks of modern ICs and systems. As far as the testing is concerned, RAMs suffer from quantitative issues of digital testing along with the qualitative issues of analog testing. Sheer number of transistors, extremely high packing density, mixed-signal nature of the design and sensitivity to process variations make RAM testing expensive and crucial. In this chapter, we review the fault models for semiconductor RAMs, and describe the algorithmic and DFT means to test defects in them. As SRAMs are scaled in sub 130 nm regime, their cell stability is being compromised with scaling. Special DFT techniques to detect the SRAM stability are also discussed.

1. INTRODUCTION

Semiconductor random access memories (RAMs) probably represent the biggest product segment of the semiconductor industry. The intense R&D directed toward RAMs has resulted in several orders of increase in the capacity of RAM chips in the last two decades. RAMs have played a significant role in the electronic revolution that pervades our lives. The performance of modern computers, communication networks and systems heavily depends on the ability to store and retrieve massive amounts of data quickly and inexpensively. Furthermore, RAMs have found their way into diverse applications like aerospace, automobiles, banking and consumer electronics.

The ever increasing demand for higher density RAMs is matched by test quality and reliability expectations. On one hand, development of high density, large RAMs puts a severe strain on testing. On the other hand, system reliability and economics have forced a merger of RAMs with CPU or DSP cores on the same substrate. This merger has resulted in dramatic changes for embedded RAMs which must be fabricated using a process developed primarily for standard logic. This leads to new challenges in the design, manufacturing, and testing of embedded RAMs.

Embedded RAMs are special in many ways. Not only they are almost analog devices, operating in a noisy digital environment but they are also harder to test owing to system-limited controllability and observability. In addition, RAMs must be designed with the layout density reaching the limits for the available technology. Even though most embedded RAMs are tested using built in self test (BIST), BIST is only a part of the total test procedure. BIST is largely a go/no go test, hence other diagnostic test procedures such as access through scan chains must be used if required. Such procedures, when expanded into the individual vectors, can easily exceed the tester's maximum pattern depth [9]. High packing density, standard manufacturing process implementation and the analog nature of operation make embedded RAMs susceptible to catastrophic as well as non-catastrophic defects. Non-catastrophic or soft defects, as they are popularly known, are too numerous to be ignored [5,83]. Present and future manufacturing of single chip systems will be strongly related to the design of embedded memories. However, the quality and reliability of such systems will depend on our ability to test them with sufficient defect coverage.

In this chapter, we discuss the impact of defects on RAMs and their test strategies. This chapter is divided as follows: The chapter begins with a brief overview of the conventional memory fault models and algorithms. Subsequently, defect-oriented fault models and algorithms for SRAMs and DRAMs are described. The DRAM fault model is evolved considering the catastrophic defects as well as abstract coupling faults. Algorithms are developed to cover these fault models. The SRAM and DRAM fault models are validated with the manufacturing test results. Certain classes of address decoder defects need special attention since these defects are unlikely to be detected by most march algorithms.

SRAMs are the most popular means to realize embedded random access storage. However, the increasing process spreads in modern semiconductor processes and subtle manufacturing defects lead to a growing number of SRAM cells with marginal stability. Research shows that such marginal cells are not detected through test algorithms, and require dedicated DfT techniques. Understanding the factors affecting the static noise margin (SNM) of SRAM cells and modeling the stability is essential in developing effective stability tests. We will present an extensive SRAM SNM sensitivity analysis, discuss the stability fault modeling and overview the techniques providing extended stability fault coverage and reduced test time.

2. TRADITIONAL RAM FAULT MODELS

RAMs require a special treatment as far as their testing is concerned. Test techniques of the digital domain are not sufficient to cover many defect/fault mechanisms that are likely to occur in RAMs. Their special test requirements have been recognized and addressed by several researchers. A variety of test algorithms, ranging from complexity $O(n)$ to $O(n^2)$ have been evolved. On one hand, these algorithms include simple algorithms, like MSCAN and ATS, that cover only stuck-at faults [4,28] and on the other, algorithms covering complex pattern- sensitive faults have been proposed [12-14,19,81]. Van de Goor [16] and Mazumder and Chakraborty [48] gave excellent overviews covering the theory and practical aspects of semiconductor memory test.

The evolution of RAM test algorithms is closely related to their fault model development. Abadir et al. [1] segregated RAM test algorithms according to their respective capabilities to detect various failures possible in RAMs. They segregated memory fault models (and hence the test algorithms) into three broad categories, listed below according to their order of complexity.

2.1 Stuck-at Fault Model

Stuck-at faults are often caused by shorts and opens in RAM cells and address decoders. As a result, one or more cells have a fixed logic state that cannot be overwritten by the write operation. In this fault model, it is assumed that SAFs adequately represent the faulty behavior of the given RAM. Simple test procedures like MSCAN [4], ATS [28], MATS [55] were developed to cover stuck-at faults in memories. Nair proved that MATS covers all stuck-at faults in RAMs independently of the address decoder design [55]. The complexity of these algorithms is linear with respect to the number of memory addresses.

2.2 Coupling Fault Model

An important type of fault that can cause a RAM to function incorrectly is the cell coupling fault [1]. Coupling faults occur because of the mutual capacitance between cells or the current leakage from one cell to another [59, 82]. Savir et al. [70] defined coupling between a pair of cells such that a $0 \rightarrow 1$ transition in one cell causes a $0 \rightarrow 1$ transition in another cell only for some fixed values of other cells in the neighborhood. Let G denote some pattern in other cells of the memory. For example, let $g1$ and $g2$ be two cells. A pattern such that $g1 = 1$ and $g2 = 0$ is denoted by $G = g1g'2$. When G is void, it is called a general 2-coupling fault between cell i and cell j, on which the other $n-2$ cells in the memory have no influence [70]. When the content of G is limited to a single bit it is called a 3-coupling fault. Because of the enormous complexity of a fault model for more than 1 bit in G, only 2-coupling and restricted 3-coupling faults have been investigated [59].

Galloping 1's and 0's or GALPAT was proposed by Breuer and Friedman [4] to cover coupling faults. The major disadvantage of this algorithm is its length, which is $O(n^2)$. This makes it impractical for large memories [1]. Nair et al. [55] proposed algorithms of complexity $30n$ and $n + 32n\log_2 n$ to cover coupling faults. Furthermore, Suk and Reddy proposed two algorithms of complexity $14n$ and $16n$ to cover all 2-coupling faults with some restrictions. A detailed account of these algorithms is given in [82]. Papachristou and Sahgal developed two test procedures of complexity $36n$ and $24n\log_2 n$ [59]. The two procedures put together have a similar capability as that of Nair et al. and GALPAT, but they require shorter test application time.

2.3 Pattern Sensitivity Fault Model

A RAM cell is said to suffer from a pattern sensitive fault (PSF) if its content is influenced by a particular pattern in the array. Hayes demonstrated that testing for unrestricted pattern sensitive faults in large RAMs is impractical. He introduced the concept of a neighborhood. Rather than considering each write or read operation on a cell C_i to be capable of affecting or being affected by the state of every cell in the memory array, M_r, he assumed that these operations can only involve a certain set of cells N_i called the neighborhood of C_i [19]. Hayes [20] and Suk and Reddy [81] tackled the problem of single PSFs in two dimensional memories by using a special type of neighborhood called the *tiling neighborhood.* In this scheme, copies of the tile cover the whole memory in such a way that no tiles overlap and no part of the memory is left uncovered except at the boundaries of the memory. You and Hayes suggested that due to some physical defects the contents of a cell may become sensitive not only to the contents of cells in its neighborhood but also to those in the same column [90].

To tackle the complexity of pattern-sensitive faults, Franklin et al. [12-14] proposed a new fault concept called *Row/Column pattern-sensitivity faults*. Instead of a cell being sensitive to the contents of the neighborhood, as suggested by Hayes [19,20], and Suk and Reddy [81], they proposed that the content of a particular cell is sensitive to the contents of the cells in its row and column. They defined a model to encompass such faults and demonstrated that tests required to detect such faults must be of the order $n^{3/2}$ [12].

There are a number of physical reasons why coupling and PSF fault models have become relevant. For example, improper timing or delay may allow partial timing overlap of two wordline signals which may give rise to column based coupling or PSF faults. In addition, several manufacturing defects may give rise to coupling faults. Moreover, in DRAMs, complementary data stored around a cell, elevated substrate voltage, or well leakage, may enhance the junction leakage of the cell which may erode its stored data value.

3. DEFECT BASED RAM FAULT MODEL DEVELOPMENT

The fault models (and hence the test algorithms) proposed in the previous section were largely based upon mathematical models of faulty RAM behavior and not on the actual manufacturing defects. Moreover, even though the impact of several manufacturing defects can be explained through coupling and PSF fault models; there were no efforts to validate these models with silicon results. As a result certain algorithms are inadequate in representing the actual failure mechanisms in RAMs (like MATS, MSCAN) and others are probably overkill in complexity (like GALPAT). Therefore, the need for efficient and accurate test algorithms was increasingly felt.

3.1 Defect based SRAM Fault Models and Test Algorithms

Dekker et al. [7] applied defect-oriented analysis, IFA, to evolve a SRAM fault model and test algorithms. These procedures were reasonably short, yet powerful enough to catch defects. The main objective of their work was to show the feasibility of a fault model and test algorithm development based on actual device defects. The defects are modeled as local disturbances in the layout of an SRAM array and translated into defects in the corresponding schematic. The electrical behavior of each of these defects is analyzed and classified, resulting in a fault model at the SRAM cell level. Efficient test algorithms were developed using the fault model. The fault model, as well as the test algorithm, was validated by testing SRAM devices and by performing failure analysis. For the development of a fault model, an SRAM is divided into three blocks:

1. Memory array
2. Address decoder
3. R/W logic

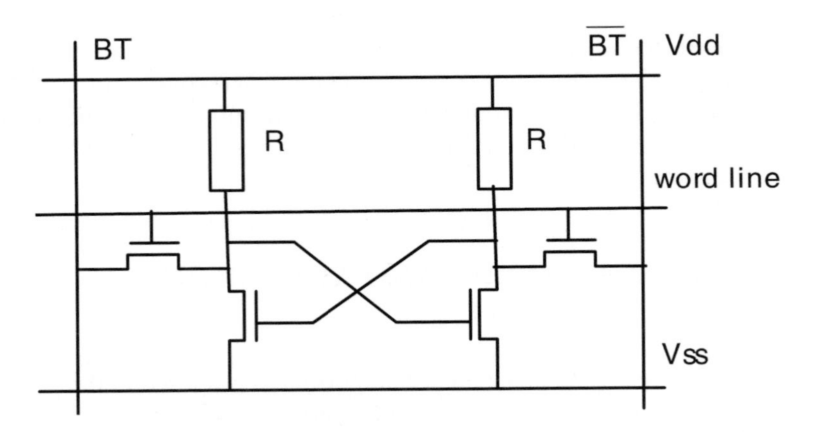

Figure 5-1. Schematic of the SRAM Cell analyzed by Dekker et al.

These building blocks are analyzed separately. However, Dekker et al. did not perform explicit IFA analysis on address decoder and R/W logic following the hypothesis of Nair et al. [56] suggesting all faults in these blocks can be mapped onto equivalent faults in the memory array. It is interesting to note that it took researchers almost two decades to find out that the hypothesis is not correct for certain classes of address decoder defects. We will discuss these defects at length in Section 5.4.

The layout of an 8kx8 double poly CMOS SRAM was used as a vehicle to perform the fault model study. The schematic of the SRAM cell is shown in Figure 5-1. This cell contains four transistors and two pull-up resistors. Each resistor is 100 GΩ and is composed of highly resistive polysilicon. In their study spot defects resulted in following defects in the layout:

Broken wires

Shorts between wires

Missing contacts

Extra contacts

Newly created transistors

A detailed two-step analysis was carried out for approximately 60 defects. In the first step, defects were placed onto the schematic (Figure 5-2). In the second step, the defects were classified according to faulty behaviors. Following this analysis, the SRAM fault model had six fault classes:

(i) An SRAM cell is stuck-at-0 or stuck-at-1

(ii) An SRAM cell is stuck-open

(iii) An SRAM cell has a transition fault

(iv) An SRAM cell is state coupled to another cell

(v) There is a multiple access fault from one SRAM cell to another SRAM cell at another address

(vi) An SRAM cell suffers from a data retention fault in one of its states. The retention time depends on the leak current to the substrate and the capacitance of the floating node.

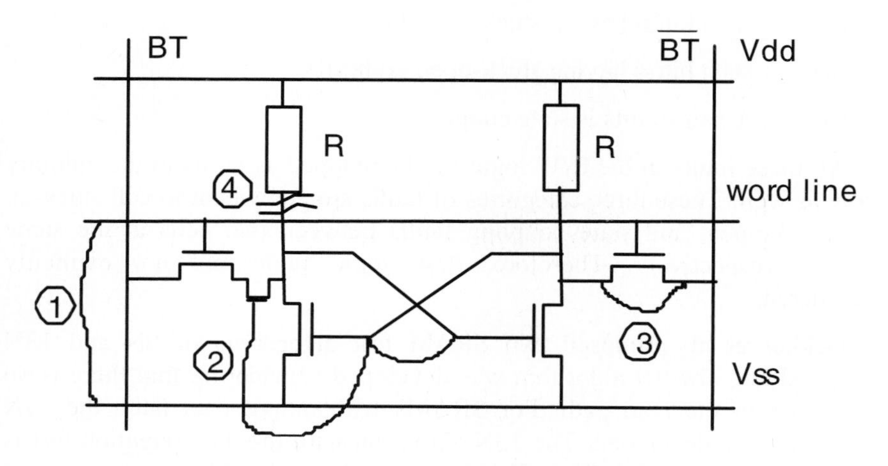

Figure 5-2. Examples of defects for different fault classes.

Figure 5-2 illustrates some of these classes of defects in the schematic. For address decoder faults a general fault model was proposed by Nair et al. [56]. Under the condition that the faulty decoder stays combinational, the decoder behaves in one of the following manners:

A decoder selects more than one address

A decoder selects no cell

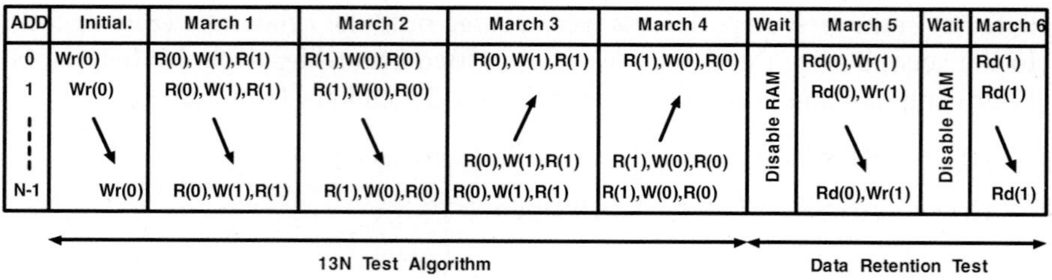

ADD	Initial.	March 1	March 2	March 3	March 4	Wait	March 5	Wait	March 6
0	Wr(0)	R(0),W(1),R(1)	R(1),W(0),R(0)	R(0),W(1),R(1)	R(1),W(0),R(0)	Disable RAM	Rd(0),Wr(1)	Disable RAM	Rd(1)
1	Wr(0)	R(0),W(1),R(1)	R(1),W(0),R(0)	↗	↗		Rd(0),Wr(1)		Rd(1)
⋮	↘	↘	↘	R(0),W(1),R(1)	R(1),W(0),R(0)		↘		↘
N-1	Wr(0)	R(0),W(1),R(1)	R(1),W(0),R(0)	R(0),W(1),R(1)	R(1),W(0),R(0)		Rd(0),Wr(1)		Rd(1)

13N Test Algorithm ⟷ Data Retention Test

Figure 5-3. 13N SRAM Test algorithm proposed by Dekker et al.

The first situation is equivalent to a multiple access fault and the second situation is equivalent to stuck-open fault in the memory array. Similarly, R/W logic communicates data from I/O ports to the memory array and vice-versa. Faults in busses, sense amplifiers and write buffers result in following fault classes:

(i) Data bit(s) having stuck-at fault(s)

(ii) Data bit(s) having stuck-open fault(s)

(iii) A pair of bits is state coupled

All these faults in the R/W logic can be mapped as faults in the memory array as well. These three categories of faults are equivalent to cell stuck-at, cell stuck-open, and state coupling faults between two cells at the same address, respectively. Therefore, R/W logic faults are not explicitly considered.

Dekker et al. proposed two SRAM test algorithms of 9N and 13N complexities. The 9N algorithm was developed considering that there is no data latch in the read path. For SRAMs with output data latch the 13N algorithm was developed. The 13N algorithm with the data retention test is illustrated in Figure 5-3. The effectiveness of these algorithms was validated with SRAMs. Defective memories were analyzed using optical microscope and scanning electron microscope (SEM) techniques. The validation exercises had twin objectives. The first objective was to validate the fault model: Do the defects occur in real life and behave as described in the fault model? The second objective was to validate the effectiveness of the test algorithms compared to test algorithms proposed in the literature with respect to catching realistic defects.

Table 5-1. Fault clusters of observed fault classes.

Cluster	# of devices	Fault Classes
0	714	SA & total failure
1	169	Stuck-open faults
2	18	Multiple access faults
3	9	State coupling faults
4	8	?
5	5	?
6	26	Data retention faults
-	-	?
14	2	?

A set of 1192 failed devices was selected from 9 wafers belonging to 3 different batches for test data analysis. The analysis resulted in 15 clusters. The prominent ones are illustrated in Table 5-1. Most of the devices (714) appear to have SA and total chip failures and 14% of failures were stuck-open. Other distinguishable failures were data retention faults, multiple access faults, and state coupling faults. Approximately 10% of analyzed faulty behaviors could not be explained with IFA based defects and remained unexplained.

Subsequently, the effectiveness of the algorithms was compared with many other algorithms using 480 devices from the total of 1192 failed devices. For devices that suffered from total device failures, SA faults were not considered since such failures can be detected by any algorithm. IFA based 9N and 13N algorithms were found to be better than most of the other test algorithms.

3.2 Subsequent Defect-oriented SRAM Test Development

Dekker et. al demonstrated the effectiveness of IFA with device production test results. Promising industrial results on the defect-oriented testing stimulated wide ranging interest in applying IFA techniques on SRAMs from academia as well as in industry [71,42,26,91,32,23,18,8,46]. Segura and Rubio analyzed the impact of gate oxide shorts on SRAM functionality [71]. They showed that gate oxide defects may result in popular memory specific faults such as coupling faults. However, not all of these defects may be detected by traditional march tests, and they recommended I_{DDQ} testing.

Defect-oriented Cache Analysis: In a detailed industrial study Mak et al. described an IFA analysis over cache RAM [42]. Cache testing constitutes a significantly large portion of overall microprocessor test cost. In addition, owing to high layout density, Cache is most sensitive to manufacturing defects, and therefore are yield limiters. Therefore, it made a lot of sense to utilize IFA techniques to identify yield limiting layout features and rationalize the test costs. They used the Carafe IFA tool for this analysis [31].

Table 5-2. Summary of IFA versus Low Yield Analysis Methods [42].

LYA conclusions	IFA conclusions
Redundant via 1 reduces open failures in M2 cell.	Via 1 open is four times more likely in M3 cell.
Metal 1 layout reduces Bit/Bit# to V_{SS} shorts.	Bit/Bit# to V_{SS} shorts in Metal 1 reduced by 15%. However, total difference in Bit/Bit# to V_{SS} short should be insignificant.

As we know, the collection of defect statistics for a given fabrication line is extremely important to the success of the IFA methodology. Significant resources were spent on the data collection. They utilized optical in line monitors, in line failure analysis techniques, etc. to generate accurate defect statistics. In line optical methods are limited by throughput which reduces significantly as the resolution of the equipment is increased. Therefore, only some of the critical layers were monitored optically. This method was complemented by traditional methods of inline failure analysis. The results of these two methods were compared to find out their relative effectiveness as well as to fine tune techniques to catch all potential defects. In this manner defect densities for the manufacturing line were determined. The layout, together with defect densities, was analyzed to generate the weighted critical areas (WCAs). The WCA was calculated for each fault type so that different layout options could be explored in order to maximize the yield, and/or optimize the test cost.

In order to examine the effectiveness of the technique, a test chip in a 0.25 μm process was analyzed. The test chip had two different cell layouts (M2, and M3) which were analyzed using the Carafe tool. It was found that the M2 cell had the total WCA of 80% of that of the M3 cell which should result in higher yield. These results were found to be very well correlated with the independent yield analysis results from the manufacturing using the low yield analysis (LYA) techniques. Table 5-2 provides a summary of these two analysis methods. As it is apparent from the table, the IFA analysis allowed researchers to quantifying the impact of various design, layout and test modifications on product yield and quality.

SRAM Test Algorithm Comparison: An interested reader may find several memory algorithms. Most of these algorithms claim very good functional fault coverage. However, the relationship between memory function fault types and types of defects that cause failure is often not well understood. Therefore, the important question is how different functional test algorithms rank in detecting realistic defects? Using the IFA technique, Kim and Chen [26] tried to characterize the effectiveness of different SRAM algorithms for their respective realistic defect coverage. They concluded that difference among the defect coverage of 11 memory test algorithms other than the checker board and sliding diagonal tests is insignificant. However, it must be mentioned that different densities for different defects were not considered. In addition, the simulated results were not substantiated with silicon measurements.

SRAM Test Algorithm in Deep Sub-micron: As technology is scaled to 130 nm CMOS and beyond, the aluminium interconnects are replaced by copper. Copper offers improved conductivity and electro-migration. However, resistive open defects such as resistive vias are increasing with copper metallization. Such defects affect primarily the delay and stability of logical operation. On this premise recently Majhi et al. [46] investigated the impact of such defects on SRAMs. In particular their objective was to examine the effectiveness of different stress conditions for catching elusive high resistive defects.

They tested approximately 11k SRAM test instances in 180 nm technology at nominal and different stress conditions. These stress conditions were identified as (i) Very Low Voltage (VLV) testing, Vmax, and at-speed testing. Out of total failed devices, 36 were deemed as interesting since these devices passed the standard test but failed the same test under at least one of the stress conditions.

Figure 5-4 illustrates the effectiveness of these stress conditions on detection of defective SRAMs. All stress techniques detect unique failures; however, the VLV testing is most effective. These results were crosschecked with the simulation and authors concluded that VLV testing targets mainly resistive bridges while V_{max} targets resistive open defects, and finally at-speed tests target timing related failures caused by resistive shorts or opens. It is extremely important to see the above mentioned study in the proper context. Let us assume that all except 36 of 11k SRAMs are good (unlikely scenario). If none of these stress tests were performed, then the outgoing DPM will be 3272. Needless to say that such an escape rate is unacceptable.

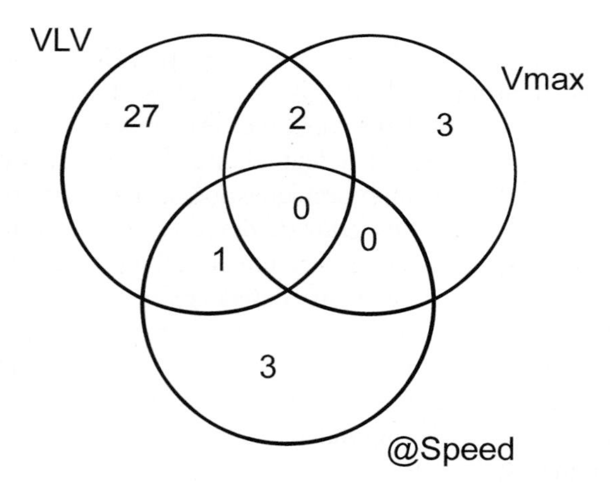

Figure 5-4. Venn diagram of failing devices at different stress conditions.

3.3 Defect based DRAM Fault Models and Test Algorithms

Defect-oriented testing strategies were also employed for rationalizing DRAM test procedures and costs. Oberle and Muhmenthaler [57] used defects as a basis for a DRAM test pattern fault simulator. Unlike Dekker, they derived realistic fault information from DRAM failure analysis. They developed test algorithms that covered those failures, and the effectiveness of their tests was verified by the fault simulator.

Dekker et al. [7] and Oberle and Muhmenthaler [57] in their respective approaches used only hard defects for fault model development. However, DRAMs are also very sensitive to subtle process variations or soft defects. Sachdev and Verstraelen [67] included soft defects in their IFA of embedded

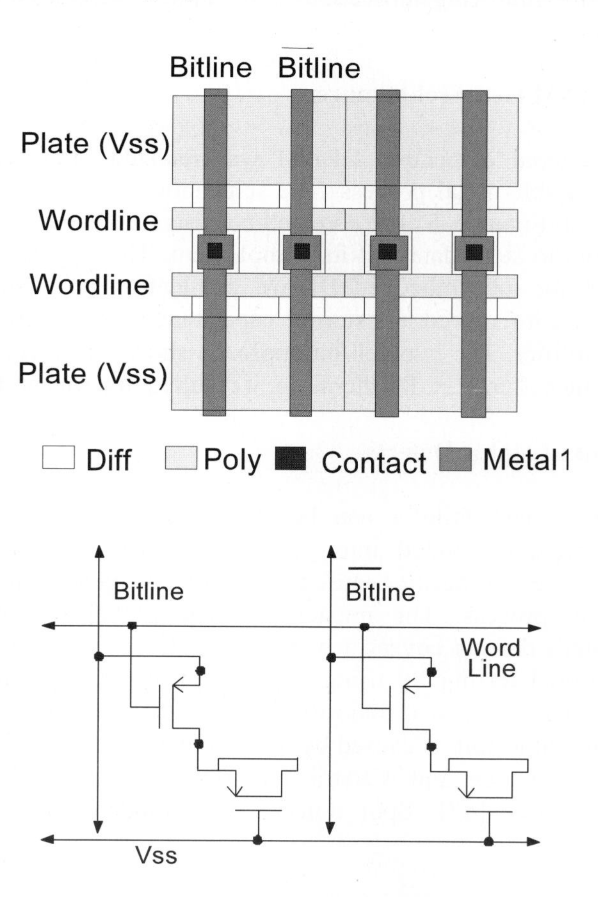

Figure 5-5. DRAM cell and the schematic.

DRAMs. The manufacturing process related defects were divided into hard and soft defect categories. In the analysis, both were separately analyzed and their respective impacts were mapped onto the circuit schematic. In this manner a better and more realistic embedded DRAM fault model was developed. For the hard defects, VLASIC [87], a catastrophic defect simulator, was utilized. Defects were sprinkled onto the layout and their impact is mapped onto the schematic (Figure 5-5 and Figure 5-6). Since VLASIC does not handle soft defects, a different analytical approach was utilized for such defects. It was assumed that soft defects will cause various 2-coupling faults. The basis of this assumption is that soft spot defects have a local influence that is likely to cause 2-coupling faults. In addition, higher order coupling faults are less likely to occur due to soft defects. Therefore, an exhaustive 2-coupling fault model is developed. Both approaches are explained in the following subsections and a fault model based upon them is developed.

3.3.1 DRAM Cell Architecture

The considered embedded DRAM was realized in a typical CMOS single-poly, double-metal process. The schematic and layout of a memory cell is shown in Figure 5-5. The core cell consists of two pass transistors and two capacitors to store data and its complement. The two-cell/bit approach was selected for the embedded DRAM application on a standard VLSI process because it required less storage capacitance for a given sensitivity of the sense amplifier. The two-cell/bit approach shows excellent bitline noise rejection in the cell matrix. Furthermore, it resulted in a robust design [80].

3.3.2 Catastrophic Defects

Integrated circuit failures can be attributed to several causes. These causes are broadly divided into global and local disturbances. Global disturbances are primarily caused by defects generated during the manufacturing process. The impact of these global (or manufacturing process related) defects covers a wider area. Hence, they can be detected before functional testing by using simple test structure measurements or supply current tests. A vast majority of faults that remains to be detected during functional testing is caused by local defects, popularly known as spot defects [43]. Sachdev and Verstraelen used only spot defects for fault modeling purposes [67]. Spot defects were modeled in the following manner:

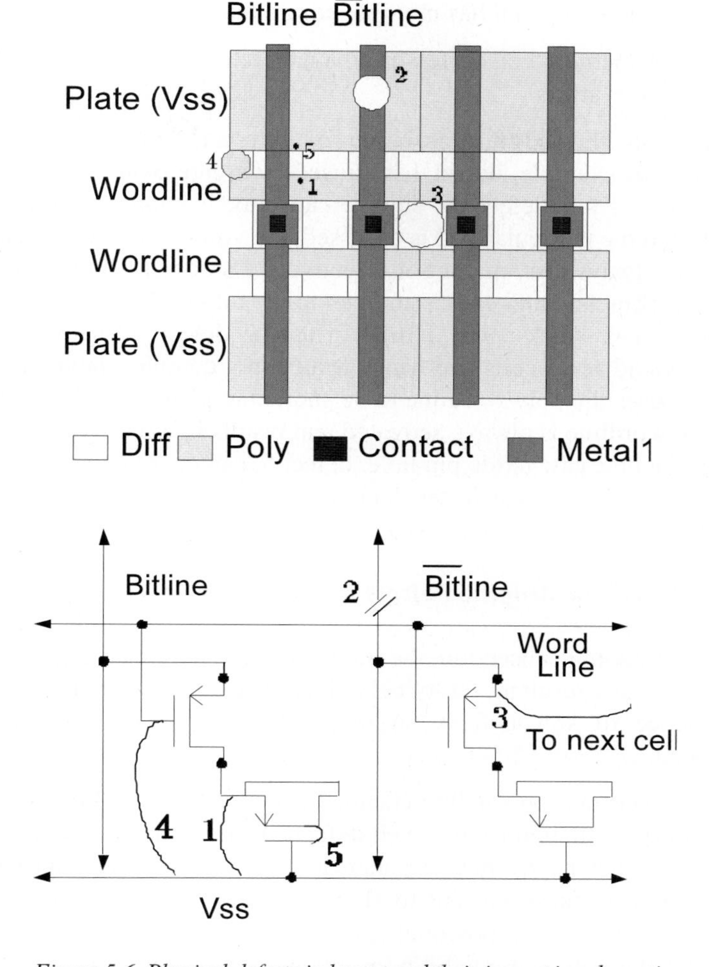

Figure 5-6. Physical defects in layout and their impact in schematic.

For hard defects, their analysis was similar to that proposed by Dekker et al. [7]. Defect analysis is performed in two steps. In the first step, defects in the layout are mapped onto the schematic as shown in Figure 5-6. In the second step, mapped defects are classified in various fault categories. In this manner, the contribution of hard defects to the fault model is determined. Hard defects resulted in following fault categories:

(i) A memory cell is stuck-at-0 or stuck-at-1

(ii) A memory cell is stuck-open

(iii) A memory cell is coupled to another cell

(iv) A memory cell has multiple access faults

(v) A memory cell suffers from a data retention fault in one (or both) of its states

Figure 5-6 shows the translation of physical defects into a circuit schematic. For example, defect 1 is caused by a gate oxide pin hole. Under the influence of this defect the storage capacitor is shorted to the wordline (poly). When the particular cell is accessed, the voltage on the corresponding wordline is always low, making it effectively a stuck-at-0 fault. Defect 2 is an absence of metal1 and causes an open in the bitline. Due to this defect the cell behaves as stuck-open. Extra diffusion, defect 3, causes coupling between two adjacent cells and hence results in a coupling fault. Extra poly, defect 4, causes the poly wordline to be shorted to plate poly (V_{SS}). With this defect, the wordline is always activated and results in a multiple access fault. A high resistance gate oxide pin hole, defect 5, can cause data on the storage capacitor to leak at a rate faster than stipulated and hence results in a data retention fault.

3.3.3 Non-Catastrophic Defects

In the foregoing subsection, the impact of hard defects was demonstrated on the circuit and resultant faulty behaviors are explained. In this subsection, the influence of soft defects on the development of a fault model is investigated.

Most defects are too small to change the connectivity or logic function of a circuit [83]. On the one hand, such defects degrade the circuit performance and, on the other hand, they can increase the mutual capacitance between adjacent cells or cause current to flow from one cell to another. In other words, they may cause potential coupling faults. Furthermore, Bruls [5] highlighted the potential reliability problems resulting from such defects. As mentioned earlier, owing to their nature of operation, DRAMs, are much more susceptible to such defects than SRAMs or logic circuits. Therefore, coupling faults should be carefully investigated for DRAMs. Abadir and Reghbati [1] defined coupling between two cells as:

A pair of memory cells, *i* and *j*, are said to be **coupled** if a transition from *x* to *y* in one cell of the pair, say cell *i*, changes the state of the other cell *j*, from 0 to 1 or from 1 to 0.

Other investigators, e.g., Nair et al. [56] and Suk and Reddy [82], defined coupling faults in RAMs in a similar fashion. As explained in previous section, Savir et al. [70] also gave a comprehensive definition of coupling faults based upon transitions in the coupling cell. According to these definitions, it is the transition in the coupling cell, say i, which initiates the coupling fault. Dekker et al. [7] defined the concept of state coupling for SRAMs, signifying the importance of the state of the coupling cell rather than its transition. He defined state coupling as:

A memory cell, say cell *i*, is said to be ***state coupled*** to another memory cell, say *j*, if cell *i* is fixed at a certain value $x(x \varepsilon \{0,1\})$ only if cell *j* is in one defined state $y(y \varepsilon \{0,1\})$. State coupling is a non symmetrical relation.

Clearly, in their definition it is the state of the coupling cell that introduces and maintains the coupling fault in the coupled cell. The important difference between the two definitions is explained as follows: According to the former definition, the coupling is introduced into the coupled cell (*j*) at the time of transition in the coupling cell (*i*). Thus, any subsequent write operation on cell *j* will overwrite the coupling fault and it will take another similar transition in cell *i* to introduce the coupling fault into cell *j* again. However, according to the latter definition, as long as the coupling cell, say *i*, is in a particular state, the coupled cell, say *j*, is also in a particular state and a write operation on cell *j* should not be able to modify its content. In other words, a particular state of the coupling cell, *i*, causes a SAF in the coupled cell, *j*.

Before attempting to map either of these definitions onto DRAMs, an important difference between DRAMs and SRAMs should be brought out. In a DRAM, a cell is driven only when it is accessed and it remains undriven when not accessed. While performing a write or a refresh, a cell is driven by the bitline driver. A read operation, destroys the cell content, hence, the value is written back into the cell. Thus, a read, write or refresh operation on a cell causes it to be driven. Therefore, at the time of access, the coupling cell, say *i*, is driven and the coupled cell, say *j*, is not driven. This situation is different from that of a SRAM. In a SRAM at the time of the coupling fault both the coupling cell as well as the coupled cell are driven. This special nature of DRAM operation has a twofold impact on nature of coupling faults:

(i) The coupling cell *i* has the stronger capability of introducing the coupling in cell *j* only when it is in driven state. If the cell *i* is not driven, it will have a marginal ability of causing the coupling fault in cell *j*.

(ii) The coupled cell *j* is very vulnerable to coupling when it is not driven and coupling cell *i* is being driven (accessed).

Table 5-3. Possible 2-coupling faults between cell i and j.

Coupling fault	Nature of coupling
1	$(i=0) \rightarrow (j=0)$
2	$(i=0) \rightarrow (j=1)$
3	$(i=1) \rightarrow (j=0)$
4	$(i=1) \rightarrow (j=1)$
5	$(j=0) \rightarrow (i=0)$
6	$(j=0) \rightarrow (i=1)$
7	$(j=1) \rightarrow (i=0)$
8	$(j=1) \rightarrow (i=1)$
9	$(i=x) \rightarrow (j=x)$
10	$(i=x) \rightarrow (j=x')$
11	$(j=x) \rightarrow (i=x)$
12	$(j=x) \rightarrow (i=x')$
13	$(i=0) \equiv (j=0)$
14	$(i=0) \equiv (j=1)$
15	$(i=1) \equiv (j=0)$
16	$(i=1) \equiv (j=1)$
17	$i \equiv j$
18	$i \equiv j'$

From this analysis it appears that the definition of coupling based on transitions is enough for DRAMs. But a closer inspection reveals that this is not the case. In DRAMs, a refresh on cell *i* would re-initiate the coupling in cell *j*, but it would not change the contents of cell *i*. Consequently, a transition based definition of coupling does not represent actual coupling in DRAMs. Looking at Dekker's definition of the state coupling, let us assume that there exists a coupling between cells *i* and *j* such that logic 1 in *i* forces logic 0 in *j*. A write(1) on *i* will initiate logic 0 in *j*. Now, write(1) is also performed on *j*. Thus, the coupling is lost. It is assumed that cell *j* will maintain logic 1 despite the fact that cell i is still logic 1. It is based upon the fact that at that moment cell i is not driven and hence is not capable of introducing the coupling into cell j again. However, a refresh on cell i can now reintroduce the coupling into cell j. Moreover, as explained above, a read operation on a cell in DRAMs is destructive by nature. This means that the read value is immediately restored (or written) in the respective cell. Consequently, a read operation is also capable of introducing the coupling. Thus, neither definition is suitable for DRAMs. The coupling in DRAMs is dynamic because it is initiated only when the coupling cell, say i, is in the driven state and coupled cell, say j, is not in the driven state. Furthermore, when cell i is not in the driven state it is not capable of initiating the coupling and at best it can charge-share with cell j. In this mechanism the contents of both cells are modified and the fault should be detected. Thus, for DRAMs Sachdev and Verstraelen [67] define dynamic coupling as follows:

Two DRAM cells, say i and j, are assumed to be ***dynamically coupled,*** if the *driven state* of cell i with value $x(x \varepsilon \{0,1\})$ causes cell j to be in state $y(y \varepsilon \{0,1\})$.

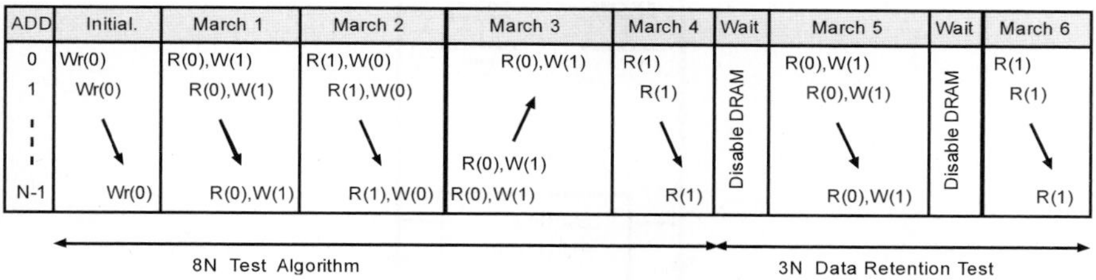

ADD	Initial.	March 1	March 2	March 3	March 4	Wait	March 5	Wait	March 6
0	Wr(0)	R(0),W(1)	R(1),W(0)	R(0),W(1)	R(1)	Disable DRAM	R(0),W(1)	Disable DRAM	R(1)
1	Wr(0)	R(0),W(1)	R(1),W(0)	↗	R(1)		R(0),W(1)		R(1)
⋮	↘	↘	↘	R(0),W(1)	↘		↘		↘
N-1	Wr(0)	R(0),W(1)	R(1),W(0)	R(0),W(1)	R(1)		R(0),W(1)		R(1)

8N Test Algorithm ←→ 3N Data Retention Test

Figure 5-7. The 8N DRAM test algorithm with data retention sequence.

Here no assumption is made about the symmetrical or asymmetrical nature of the coupling. For two arbitrary cells i and j, all possible couplings are shown in Table 5-3. It is assumed that $i_{address} < j_{address}$ while modeling these coupling faults. The other condition, $i_{address} > j_{address}$, has been taken into account by $j \rightarrow i$ coupling faults. For example, coupling fault 1 occurs when logic 0 in cell i forces a logic 0 in cell j. The coupling fault 9 illustrates that logic value x (0 or 1) in cell i forces the same value in cell j. Thus, it appears that coupling faults 1 or 4 are a subset of coupling fault 9. We shall later see that not to be the case. The detection of coupling fault 9 does not guarantee detection of coupling faults 1 and 4. Therefore, these faults should be separately considered. Coupling fault 13 involves a bidirectional coupling. However, it is different from a bridging fault because it is defined only for logic 0. Coupling fault 17 is a bridging fault and coupling fault 18 is a bidirectional inverting bridge. It is important to note that the definition of coupling is extended to include symmetrical behavior as well.

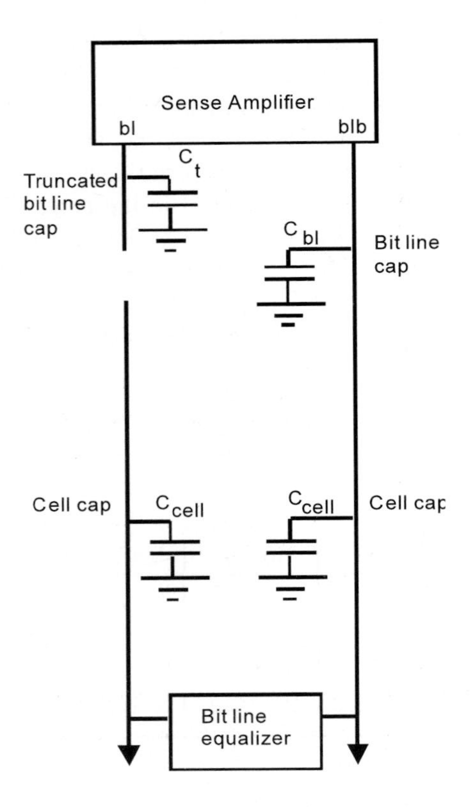

Figure 5-8. The stuck open fault.

3.3.4 DRAM Test Algorithms

In this section, we first propose a test algorithm for bit oriented DRAMs with combinational R/W logic. We, then extend the algorithm to cover sequential R/W logic and word-oriented DRAMs. The complexity of the basic algorithm is 8N, where N is the number of memory addresses. A data retention test is added to cover data retention faults. Figure 5-7 shows the flow of the algorithm. It has an initialization step and a set of four marches. In the data retention test, the DRAM is disabled for data retention time and then accessed. The coverage of the fault model by 8N algorithm is explained below.

Stuck-at faults: It is rather easy to demonstrate the 100% stuck-at fault coverage of the proposed algorithm. The stuck-at-0 fault in any arbitrary cell *i* will be detected in march 2. Similarly, stuck-at-1 in cell *i* will be detected in march 1.

Stuck-open faults: The detection of stuck-open faults is explained with the help of Figure 5-8, showing a simplified diagram of the data path from storage cells to the local sense amplifier. C_1 ... C_n are the storage cell capacitances. C_{bl} is the bitline (or bitbar line) capacitance and C_t is the truncated bitline capacitance owing to the stuck-open fault. The bit and bitbar lines are terminated on the local sense amplifier. The bitline equalizer circuit is connected to the other ends of the bit and bitbar lines. The complete read operation is divided into individual steps. In the first step, the bit and the bitbar line voltages are equalized to an approximate voltage of $V_{DD}/2$. Subsequently, the selected wordline goes low enabling the storage capacitor and line capacitance to charge-share. The bitline capacitance is approximately 10 times that of cell storage capacitance. Thus, after the charge-sharing, the voltage swing on bitlines is of the order of 400-500 mV (for a DRAM operating with 3.3V process). Bit and bitbar lines swing in opposite directions because they charge-share with complementary data. The sense amplifier raises the differential voltage to V_{DD} (or V_{SS}) level.

In the first loop the memory is initialized with logic 0. All stuck-open bits on the faulty bitline fail to initialize. However, truncated bitline capacitance, C_t, gets and maintains a logic 0. Subsequently, in the second loop read(0) and write(1) operations are carried out. Depending upon the location of the open defect, there are several possibilities. Three typical cases are discussed below:

(a) The stuck-open disconnects all bits from the sense amplifier: Ct at the beginning of the second loop has logic 0. In the beginning of the first march, read(0) is performed. The bitbar line after equalization and charge-sharing swings toward logic 1. However, C_t does not get equalized and does not charge-share with the storage capacitor and hence maintains logic 0. The sense amplifier raises voltage levels on the bit and bitbar lines and outputs a logic 0, so the fault is not detected. After the read, a write(1) is carried out, and so C_t obtains logic 1. Moving forward in the same loop, after some time again the same bit and bitbar lines are accessed (let us say location C_2). Now C_t has logic 1. At this moment the truncated bitline has a voltage level of V_{DD} and bit has the voltage $V_{DD}/2 + 0.4V$. This causes the sense amplifier to converge on logic 1 instead of logic 0 and hence the fault will be detected.

(b) The stuck-open disconnects only the bitline equalizer circuit from the sense amplifier: In this case, the bitline does not become equalized.

However, it charge-shares with the respective cell storage capacitances. It can be shown in the same manner as above that such a defect will be detected.

(c) *The stuck-open disconnects a particular cell from the bitline:* Let us assume that the metal1 to diffusion contact between the bitline and a particular cell, say *i*, is missing. This defect would cause cell *i* to be stuck open and hence the cell will not charge-share with the bitline. Thus, at the time of read on this cell, the corresponding bitline has only the equalized voltage. The sense amplifier performance will deteriorate and it will not be able to converge to either logic 1 or 0 in a given time. It should be detected in the output levels. An intermediate voltage level would cause the output driver to have a DC path through it and should be detected by an increased supply current as well.

Similar analysis can be carried out for the stuck-open faults on the wordline. It can be shown that a stuck-open fault on a wordline behaves like case (c) of the above analysis.

Multiple Access Faults: A multiple access fault occurs when more than the addressed cell are accessed during some cell operation. The decoder multiple access faults can be modeled as coupling faults in the matrix and need not be considered explicitly [56].

Coupling Faults: The algorithm covers all the modeled coupling faults. The performance of the 8N algorithm for modeled 2-coupling faults is shown in Table 5-4. The three columns of the table depict the coupling faults when faults are sensitized and when they are detected, respectively. For example, coupling fault 1, $(i=0) \rightarrow (j=0)$, is introduced by the initialization step of march 1, as well as by that of march 2. However, it is not detected by march 1, since in this march the coupled data (logic 0) is same as the original data in the coupled cell. This coupling is detected by march 2. Similarly, coupling fault 10 is sensitized several times but is detected only by march 4. The march 4 is added to detect this coupling fault $(i=x) \rightarrow (j=x')$. Owing to the nature of the coupling, this coupling fault can only be detected by a single element march in forward direction (address order $0 \rightarrow N-1$). The complement of this coupling fault, coupling fault 12, is introduced by initialization step and is detected by the first march.

Data Retention Faults: The data retention for logic 1 can be covered by disabling the DRAM for the stipulated time and applying march 5. The data retention for logic 1 is tested by once again disabling the DRAM for the stipulated time and subsequently applying march 6. Alternatively, the DRAM can be disabled after march 1 and 2, respectively, to test for data retention faults.

Table 5-4. Performance of the 8N DRAM test algorithms for 2-coupling faults.

Coupling fault	Introduced in	Status
1	Init, march 1,2	Detected, march 2
2	Init, march 1	Detected, march 1
3	march 1,2	Detected, march 2
4	march 1	Detected, march 1
5	Init, march 1,2	Detected, march 2
6	Init	Detected, march 1
7	march 1	Detected, march 2
8	march 2,3	Detected, march 3
9	Init, march 1	Detected, march 1
10	Init, march 1,2,3,4	Detected, march 4
11	Init, march 1,2,3	Detected, march 3
12	Init	Detected, march 1
13	Init, march 1,2	Detected, march 2
14	Init, march 1	Detected, march 1
15	Init	Detected, march 1
16	march 1	Detected, march 1
17	Init, march 1	Detected, march 1
18	Init	Detected, march 1

3.3.5 Extensions

Dekker et al. [7] highlighted the problem of stuck-open detectability for sequential R/W logic. They suggested that adding one extra read operation in a 2-element march can ensure the transparency of the R/W logic for stuck-open faults. For such applications, march 3 of the algorithm is modified to include an extra read operation. The resultant algorithm, shown in Figure 5-9, has the complexity of 9N. For word oriented DRAMs different data backgrounds are needed to cover intra-word faults [7]. If m bits per word are used, a minimum of k data backgrounds will be needed, where

$$k = \lceil \log m \rceil + 1$$

$$\lceil x \rceil = \min\{n \varepsilon Z | n \geq x\} \quad (5.1)$$

When m is a power of 2, the formula simplifies to:

$$k = \log m + 1 \quad (5.2)$$

However, different data backgrounds should be used only if bits constituting a word are adjacent to each other in the layout. On the other hand if different bits are not adjacent to each other, then the possibility of intra-word faults does not arise. Therefore, one data background is sufficient.

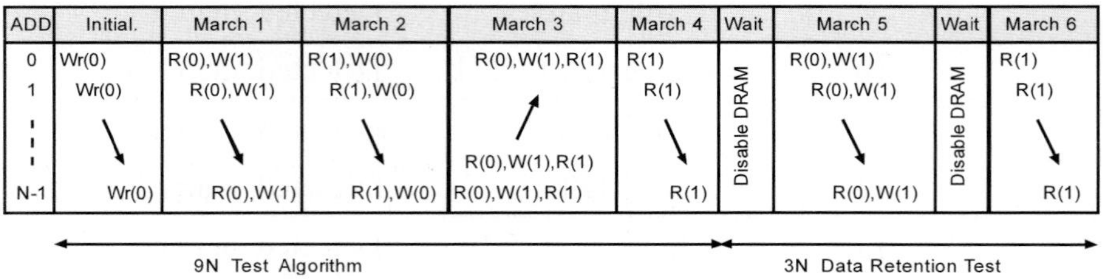

ADD	Initial.	March 1	March 2	March 3	March 4	Wait	March 5	Wait	March 6
0	Wr(0)	R(0),W(1)	R(1),W(0)	R(0),W(1),R(1)	R(1)	Disable DRAM	R(0),W(1)	Disable DRAM	R(1)
1	Wr(0)	R(0),W(1)	R(1),W(0)	↗	R(1)		R(0),W(1)		R(1)
⋮	↘	↘	↘	R(0),W(1),R(1)	↘		↘		↘
N-1	Wr(0)	R(0),W(1)	R(1),W(0)	R(0),W(1),R(1)	R(1)		R(0),W(1)		R(1)

9N Test Algorithm — 3N Data Retention Test

Figure 5-9. The 9N DRAM test algorithm with data retention test sequence.

3.3.6 Results and Model Validation

The results obtained in previous sections are validated by applying the proposed algorithm on Philips 4k×8 embedded DRAM modules. The purpose of validation is to demonstrate the effectiveness of the algorithm and to verify the foregoing analysis. A large number of devices from 34 wafers was tested with the proposed 9N algorithm. Out of the total number of tested devices, 579 were failed by the algorithm. Figure 5-10 shows the effectiveness of each march element in catching failed devices. For example, if a device failed in the first and third marches then the respective march failure numbers are increased by one. As expected, the most complex march element, march 3, caught the largest number of failures. In most cases, a large number of bits was failed which was also failed by the first or second march. Therefore, the performances of first and second marches in catching faults were comparable to that of march 3.

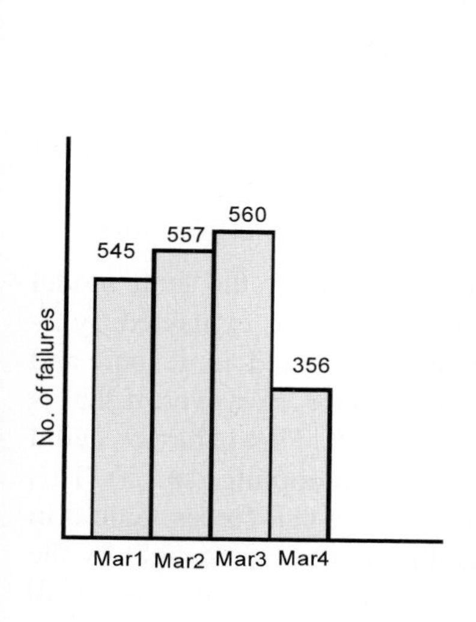

No.	Devices	Signature	Explanation
1	318	FFFF	Wordline stuck-at, read write failures, opens
2	201	FFFP	Wordline stuck-at, read write failures
3	21	PFFF	Bitline stuck-at 0, bit stuck-at 0
4	18	FPFP	Bitline stuck-at 1, bit stuck-at 1
5	11	PFPF	Coupling fault 3
6	3	PFPP	Coupling fault 1
7	3	FPFF	Combination of stuck-at 1 and coupling 10
8	3	FFPF	Combination of coupling faults 9 & 10
9	1	FFPP	Coupling fault 10

Figure 5-10. Number of failures per march element, and different failure categories and explanations.

The table in Figure 5-10 gives a concise analysis of failures. The first and second columns of the table show the type of failure and number of failed devices, respectively. In the third column, pass/fail response of the failed devices in terms of marches is shown. For example, FFFF means that the device is failed in all four marches. Similarly, FPFP means that the device is failed in the first and third marches and is passed in the second and fourth marches. The fourth column lists an explanation for each failure. As shown in this table, 318 devices failed in all marches. Bitmaps of the failed devices typically show wordline and/or bitline failures that could have been caused by bridging and open defects. Other possible causes include, read/write failures owing to the bridging of bit and bitbar lines, etc. Another set of 201 devices failed the first three marches, however, they passed the fourth march. In such cases, a typical bitmap showed wordline and/or bitline failures.

Twenty one devices were found to have a bit stuck-at-0 behavior. Therefore, except for the first march, these devices failed in the rest of the marches. Similarly, eighteen devices were found to be stuck-at-1. These devices passed second and fourth marches but failed in other two. These failures were caused by bridging or open defects in cells. An important difference between these two set of failures and failures mentioned previously was the number of failed bits. In the latter case, individual bits were failed, while in the former, majority or a large number of bits failed.

A small number of failures could not be explained by the fault model based upon catastrophic defects. However, these could be explained by 2-coupling faults model. For example, eleven devices failed in second and fourth marches. This appears to be a stuck-at-0 behavior. However, if that is so, then devices should fail in the third march as well. This behavior could be explained by $(i=1) \rightarrow (j=0)$ coupling fault model (coupling fault 3). This coupling fault is initiated when logic 1 in the coupling cell forces logic 0 in the coupled cell. Since $i_{address} < j_{address}$ this coupling fault is not caught by the third march. As another example, a small set of three devices passed all marches except the second march. This is not a stuck-at-0 behavior. A stuck-at-0 behavior should also be detected by the third and fourth marches. However, this failure can be explained as a coupling fault 1. Three devices failed in all marches except in the second march. Their bitmaps revealed that in the first and third marches, a set of bits were failed and in the fourth march different set of bits were failed. Thus, these failures are thought to be a combination of the stuck-at-1 failures detected by first and third march elements and the coupling fault 10 detected by the fourth march element. Similarly, three more devices failed in all marches except the third march. However, in first two marches, one set of bits failed and in the fourth march a different set of bits failed. This behavior can be explained by a combination of coupling fault 9 and coupling fault 10. However, these coupling faults influenced different bits on the die. The coupling fault 9 caused a set of bits to fail in the first two marches but could not be detected by the rest of marches. The coupling fault 10 influenced another set of bits and could only be detected by march 4. Finally, one device failed only in the first two marches and its behavior coincided with that of coupling fault 9. In this fashion, all failures could be explained by the proposed fault model.

3.4 TCAM Fault Models and Test Algorithms

Content addressable memories (CAMs) are like random access memories (RAMs) with additional search capabilities. The user provides a RAM with an address, and data is either written to or read from that location. A CAM can also do this, but has a built-in reverse-lookup capability. In a CAM, the user can provide search data, sometimes called the search key, and the CAM will report which addresses contain that data.

There are two types of CAMs: Binary and Ternary. Binary CAMs store and search only '0's and '1's, so they are limited to exact-match SEARCH operations. Ternary CAMs (TCAMs) can store and search an additional "mask" state, denoted 'X', which corresponds to the Boolean "don't care". Therefore, TCAMs can also perform partial matching, which is extremely useful in applications such as packet forwarding and classification in network routers. Despite these attractive features, the complex integration of memory and logic in TCAMs makes their testing very time consuming.

3.4.1 TCAM Architecture

A typical TCAM chip consists of three major parts: (i) TCAM arrays for ternary data storage, (ii) peripheral circuitry for READ, WRITE, and SEARCH operations, and (iii) test and repair circuitry for functional verification and yield improvement. The peripheral circuits include decoders, bit line sense amplifiers (BLSAs), search line (SL) drivers, match line sense amplifiers (MLSAs), and priority encoders (PEs). The test and repair circuitry includes on-chip test structures and redundancy. Each row in a TCAM array stores a word. Within a word, a bit is located by its column number. All the TCAM cells in a row share a word line (WL) and a match line (ML). Similarly, all the TCAM cells in a column share bit lines (BLs) and SLs. TCAMs can have multiple matching addresses, so PEs are used to determine the highest priority match to output.

TCAM Cell

Each TCAM cell consists of two RAM cells and a comparison logic circuit. This allows the storage and searching of the three states encoded into two stored bits. Figure 5-11 illustrates dynamic and static TCAM cells. The 6T dynamic cell (Figure 5-11(a)) is smaller, but it requires a specialized embedded DRAM process. The conventional 16T static cell (Figure 5-11(b)) is more attractive due to its compatibility with the standard logic process. In both cases, the comparison logic (N1 though N4) connects to the SLs (N1 and N3) and to the storage nodes (N2 and N4). During the SEARCH operation, if the cell is a match, the ML cannot discharge, and if the cell is a mismatch, the ML will discharge either through N1 and N2, or N3 and N4. For example, if '0' was stored (encoded as '0 1') and '1' was searched (encoded as '1 0') there would be a discharge path for the ML through

N3-N4 in the dynamic cell, and N1-N2 in the static cell. The resulting value on the ML is sensed by its MLSA (not shown).

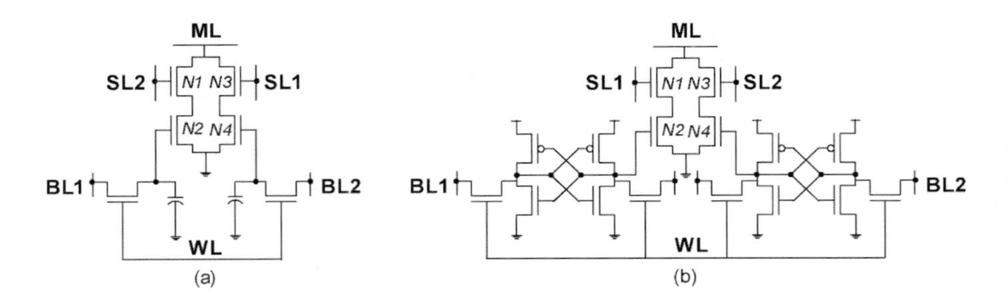

Figure 5-11. 6T dynamic TCAM cell (a), conventional 16T static TCAM cell (b).

Priority Encoder

TCAMs need wide-input PEs to resolve multiple matches across the entire array. Conventionally, the lowest-address word has the highest priority, and the application software stores data into the appropriate memory address. Generally, PEs consist of two stages: (i) multiple match resolver (MMR), and (ii) match address encoder (MAE).

An MMR is an n-bit input, n-bit output datapath circuit. An output bit is a '1' ("match") if (i) the corresponding input bit is a '1', and (ii) all the higher priority input bits are '0's. The function of an MMR can be described by the Boolean expressions in equation (5.3). That is, of the multiple high input signals, only the highest priority signal will remain high at the output and the others will be turned off.

$$Out_0 = ML_0$$

$$Out_1 = ML_1.ML_0'$$

$$\ldots$$

$$Out_n = ML_n.ML_{n-1}'.....ML_1'.ML_0' \qquad (5.3)$$

A wide-input MMR is implemented as a tree structure. There are multiple levels of smaller 8- or 16-bit MMRs with their level in the tree indicated by a prefix (e.g. L1-MMRs connect to the MLSAs, L2-MMRs connect to the L1-MMRs, etc.). Each small MMR indicates to the next higher level if it has a matching address. If the higher level MMR detects multiple lower level MMRs with matches, it will disable those MMRs containing only lower-priority matches. When the operation is complete, only the L1-MMR with the highest-priority match will remain active and pass its match to the MAE. The MAE is a ROM that encodes the MMR's single high output as the address of the corresponding matching word.

3.4.2 TCAM Testing

TCAM test issues have not been addressed adequately. Most of the previous work on CAM testing is focused on binary CAMs [75,35,36]. Algorithms developed for binary CAMs cannot be directly applied to TCAMs because its masking capabilities and the differences in their comparison logic circuits.

TCAM Cell Fault Analysis

Wright et al performed a transistor-level fault analysis on a TCAM cell to develop a defect-oriented test algorithm [88]. RAM testing is a mature area of research, so existing algorithms can provide adequate fault-coverage for the RAM cells, and most defects in RAM cells result in stuck-at faults [7], so their fault analysis assumes that the defects in RAM cells cause stuck-at faults (SA1 and SA0) in the storage nodes.

TCAM cells are symmetric, so the fault analysis only needed to be performed on one half of the cell since the results are equally valid for the other half. The fault analysis results in five possible transistor-level faults: (i) source/drain contact defect, (ii) gate contact defect, (iii) gate to drain oxide failure, (iv) gate to source oxide failure, and (v) sub-threshold conduction. Table 5-5 lists these faults applied to one half of a TCAM cell (defects #1 through #12) and their detection methods. It also lists other possible inter-transistor faults (defects #13 through #19) and their detection methods. Table 5-5 assumes that the 6T dynamic TCAM cell in Figure 5-11(a) is used. When the static cell is used, the detection methods can be altered accordingly. An equivalent Table 5-5 for Figure 5-11(b) will replace transistors N1 and N2 by N3 and N4 respectively, and vice versa. The last operation in each method in the column "Detection Method" refers to the result under correct operating conditions. The column labeled "Induced Fault" refers to the type of functional fault that a test algorithm would detect as a result of the specific defect and detection method. For example, defect #3 makes N2 appear to be stuck-open (SOP) since the source or drain contact has a defect. Similarly, defect #16 allows conduction through N3 and N2, making N4 appears to be stuck-on (SON) from a functional perspective.

Table 5-5. Possible TCAM cell faults.

#	Defect	Detection Method	Induced Fault
1	Storage node SA0	Write "1"; SL2 = "1"; SL1 = "0"; (c) Search for Mismatch	N2 SOP
2	Storage node SA1	Write "0"; SL2 = "1"; SL1 = "0"; Search for Match	N2 SON
3	N2 S/D Contact	Write "1"; SL2 = "1"; SL1 = "0"; Search for Mismatch	N2 SOP
4	N2 Gate Contact	Write "1"; SL2 = "1"; SL1 = "0"; Wait; Write "0"; Search for Match	N2 SON*
5	N2 G/D Oxide Failure	Write "0"; SL2 = "1"; SL1 = "0"; Search for Match	N2 SON
6	N2 G/S Oxide Failure	Write "1"; SL2 = "1"; SL1 = "0"; Search for Mismatch	N2 SOP
7	N2 Sub-V_T Conduction	Write "0"; SL2 = "1"; SL1 = "0"; Search for Match	N2 SON
8	N1 S/D Contact	Write "1"; SL2 = "1"; SL1 = "0"; Search for Mismatch	N1 SOP
9	N1 Gate Contact	Write "1"; SL2 = "1"; SL1 = "0"; Wait; SL2 = "0"; Search for Match	N1 SON*
10	N1 G/D Oxide Failure	SL2 = SL1 = "0"; Search for Match	N1 SON
11	N1 G/S Oxide Failure	Write "1"; SL2 = "1"; SL1 = "0"; Search for Mismatch	N1 SOP
12	N1 Sub-V_T Conduction	Write "1"; SL2 = "0"; SL1 = "1"; Search for Match	N1 SON
13	N1-Gate to N2-Gate Short	Write "0"; SL2 = "1"; SL1 = "0"; Search for Match	N1 SON
14	N1-Source to N4-Drain Short	Write "1"; SL2 = "0"; SL1 = "1"; Search for Match	N1/N4 SON
15	N1 Gate to N4 Drain Short	Write "1"; SL2 = "0", SL1 = "1"; Search for Match	N4 SON
16	N2 Gate to N4 Drain Short	Write "mask" ('0 0'); SL2 = "0", SL1 = "1"; Search for Match	N4 SON
17	N1 Gate to N3 Gate Short	Write "1"; SL2 = "0", SL1 = "1"; Search for Match	N1 SON
18	N1 Gate to N4 Gate Short	Write "0"; SL2 = "0", SL1 = "1"; Search for Mismatch	N4 SOP
19	N2 Gate to N4 Gate Short	Write "1"; (b) SL2 = "0", SL1 = "1"; Search for Match	N4 SON

The detection methods for defects #4 and #9 have a "wait" operation whose duration determines the resistance range of defects covered by the detection methods. For example, a longer "wait" can detect a larger resistance range of defects. Such a precisely controlled "wait" operation is not always practical to implement. Therefore, high-level algorithms were developed assuming that weak defects ultimately result in SON or SOP faults as shown in the last column of Table 5-5.

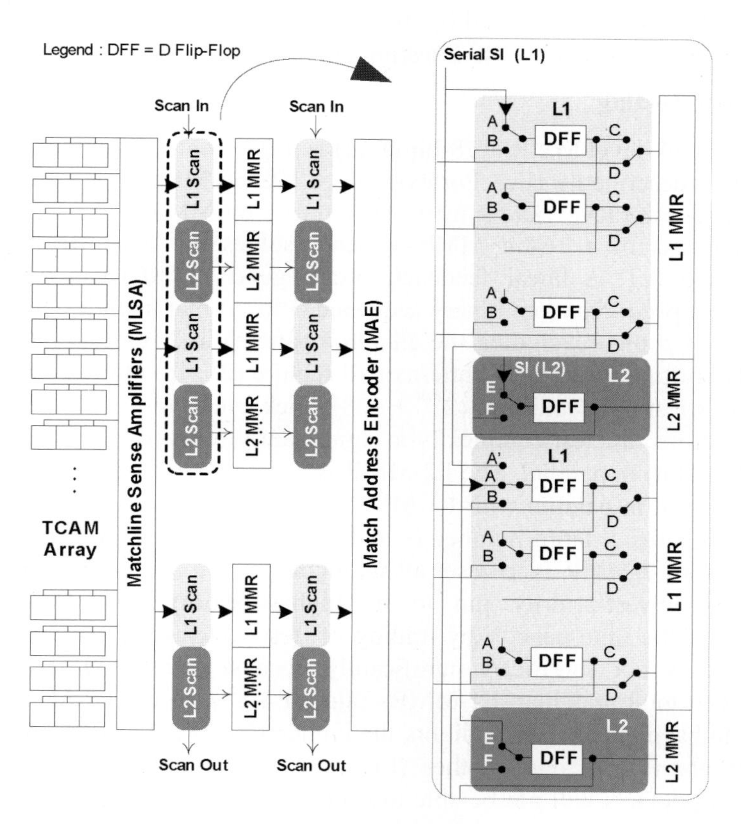

Figure 5-12. Scan chains for MMR testing.

Test Algorithms

Each individual TCAM cell can discharge its ML, which is shard by all the cells in a word. Since each discharge path must be tested, conventional TCAM test schemes have the complexity of $O(nl)$, where n is the number of words and l is the number of bits per word. For an 18Mb TCAM, this test complexity corresponds to O(18M), which makes TCAM-testing a time consuming process. The other TCAM components, such as the PE, also need to be tested in addition to the array. Since the SEARCH operation proceeds from TCAM array to MMR to MAE, these components should be tested in reverse order.

MAE Testing

The MAE is tested by encoding every possible address and checking that the output is correct. The test starts by resetting the MAE scan chain to all '0's. A '1' is shifted followed by '0's, and the MAE outputs are checked for the correct address. Thus, MAE testing only requires n shifts.

MMR Testing

The block-level MMR (128-bit input) is implemented as a tree of smaller MMRs as described earlier. For example, a 128-bit MMR is implemented in two levels. First level (L1) is made of sixteen 8-bit MMRs. The second level (L2) consists of a 16-bit MMR which resolves the inter-MMR priority conflicts of L1. A linear feedback shift register (LFSR) can be used to generate a pseudorandom binary sequence (PRBS) that includes all possible patterns of p bits (excluding the all-zeros pattern) in 2^p clock cycles. Thus, serially shifting of all p-bit patterns will require (2^p + p) clock cycles. Since full block testing would take (2^{128} + 128) clock cycles to test all the possible combinations, the test complexity is significantly reduced by testing the L1-MMRs (8-bit) in parallel. The L2-MMR is isolated during the L1 testing and can be tested in parallel with L1-MMRs. Exhaustive testing of a 16-bit L2-MMR requires a large number (~ 64k) of clock cycles. A faster way of testing the L2-MMR is to reset all its inputs to '0's, and then shifting '1's from its lowest-priority pin to its highest-priority pin. This method eliminates the time penalty by trading off test coverage. However, the test coverage is not sacrificed significantly because the L2-MMR is much smaller in total area than L1-MMRs (almost 1/8th), so the L2-MMR is less likely to have a defect. In addition, the inputs of the L2-MMR are physically further apart from one another (Figure 5-12), and the most commonly occurring defects will not be able to bridge two inputs of the L2-MMR that are far apart. Thus, complex test patterns (with non-consecutive active inputs) are not needed to test the L2-MMR, and a simple functional test is sufficient. If all the 8-bit MMRs are fault-free, they are re-connected in the tree-structure for block-level testing. Initially, the scan chain is reset to '0', and a string of '1's is shifted. In summary, each node of the tree is tested in isolation, and then the tree is tested as a whole.

Recently, a PE test algorithm has been reported that uses the CAM array to test stuck-at faults in the PE [38]. Since it assumes a fault-free CAM array, it cannot be used together with CAM test algorithms that require a PE, such as ours. It also assumes that the n-bit PE is designed in one level, so it does not take advantage of the PE's hierarchical structure. It can be used in conjunction with the presented scheme (e.g. in L1-MMR testing) by inserting DFT structures to stimulate the PE as shown in Figure 5-12. For 8-bit L1-MMRs, it does not make much difference in the total test complexity, but it can be beneficial for 16-bit or larger L1-MMRs.

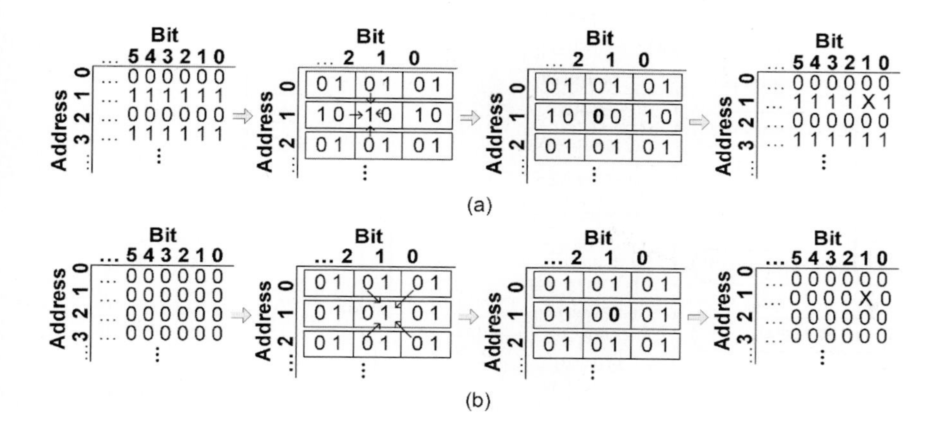

Figure 5-13. Data patterns to stimulate (a) horizontal/vertical and (b) diagonal inter-cell faults.

TCAM Array Testing

Intra-cell defects likely to result in SON or SOP faults. A high-level algorithm was developed to detect these faults with column level diagnostics. The proposed algorithm also detects horizontal, vertical and diagonal inter-cell coupling faults. Figure 5-13(a) and (b) show data patterns that stimulate horizontal/vertical and diagonal coupling faults respectively. Figure 5-13 also shows the bits in their stored ternary format ('0' ≡ '0 1' and '1' ≡ '1 0'). As shown in Figure 5-13, an inter-cell fault can change a TCAM cell's value to the "mask ('0 0')" state. The remaining inter-cell faults can be stimulated by inverting these patterns. A coupling fault can also change a TCAM cell to an invalid '1 1' state that causes transistors N2 and N4 in Figure 5-11 to conduct, and the affected word will always mismatch. However, this becomes a '0 0' fault under the inverse data conditions.

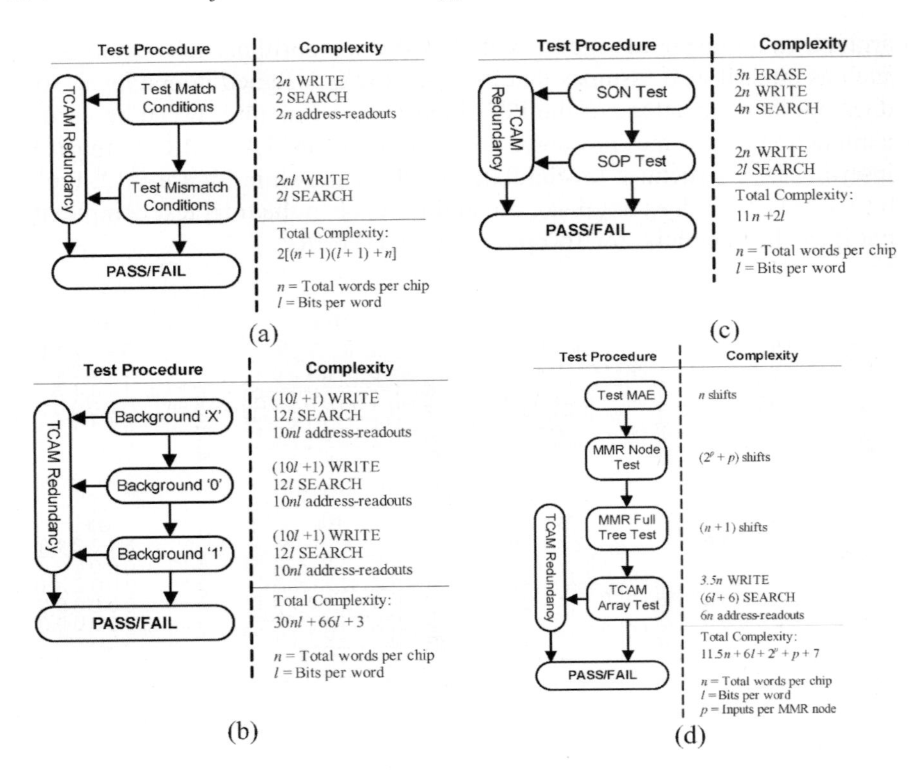

Figure 5-14. Test procedures and complexities of (a) simple, (b) Lee [39], (c) Li/Lin [40], and (d) Wright's algorithm [88].

The presented algorithm assumes at most one SON fault per word. If some words mismatch for both local and global masking, there are SON faults in both SL and BL transistors of these words. It is difficult to determine the exact column locations of such multiple faults, so they can be replaced with redundant rows.

3.4.3 TCAM Test Algorithms

A simple TCAM test algorithm individually tests each bit's ability to match and mismatch for both '1's and '0's, which can be accomplished in two steps: (i) test the ability for an address to match, (ii) test each bit's ability to mismatch. Match ability is tested by writing '000…000' to every address, and searching for '000…000' to verify that every address matches. This process is repeated using inverted values. Mismatch ability is tested by writing '000…001' to all the addresses, and then searching for '000…000' to ensure that no address matches. The SEARCH operation is repeated after shifting the pattern by one bit ('000...010') and writing it to all the addresses. This process is also repeated using inverted values. Figure 5-14(a) shows the simple test procedure and its complexity. The total complexity assumes equal time-penalties for WRITE, SEARCH, shift, and address-readout operations.

Some new TCAM array test algorithms have recently been proposed by Lee [38] and Li/Lin [39]. Lee's algorithm is developed for a single cell and is subsequently expanded to a TCAM array [38]. This bottom-up approach is not optimized to exploit the parallel search capabilities of the TCAM. In addition, it does not provide column-level resolution for SON faults and does not verify global masking in the SLs [38]. It assumes word-parallel write access, which may not be realistic in a large TCAM. It proceeds in three identical steps as shown in Figure 5-14(b). Assuming a word-parallel write access, each step requires $(10l + 1)$ WRITE + $12l$ SEARCH operations + $10nl$ address-readouts. The huge number of address readouts is caused by multiple "match" conditions in most SEARCH operations.

Li/Lin's algorithm detects a subset of faults covered by the Wright's algorithm. In addition, their algorithm lacks test procedures for inter-cell fault detection and column-level resolution for SON faults [40]. Moreover, it does not verify if the 'X' value can be properly stored and searched. Figure 5-14(c) illustrates this test procedure along with its complexity. The ERASE operation requires an additional feature called the "valid" bit, which determines if a word will participate in SEARCH operations.

Figure 5-14(d) shows the complete test flow and complexity of the algorithm proposed by Wright et al. [88]. It assumes the availability of scan chains with reset. For a typical TCAM ($l = 144$), this results in 89 operations per SON fault, which is negligible as compared to the total test complexity. Thus, it is not included in complexity calculations in Figure 5-14(d). The presented algorithm achieves column-level resolution of SON faults, which is particularly useful if both row and column redundancy are available.

4. ADDRESS DECODER DEFECTS

As pointed out earlier, the conventional wisdom suggests that RAM decoder defects can be mapped as RAM array faults that are detected during RAM array tests. Hence, no special test is needed for address decoders. However, we would like to re-examine this assumption as it is based on the analysis carried out in the 1970s [56,86]. Since then, semiconductor technology has changed significantly. Recently we came across some open defects in RAM address decoders that were not detected by linear test algorithms (e.g., march test) and resulted in field failures. This observation prompted us to look into occurrence of open defects in RAM address decoders.

Open defects or transistor stuck-open faults are known to cause sequential behavior in CMOS circuits and require 2-pattern test sequences for detection. The transistor and logic stuck-open testability has received considerable attention and a number of DfT solutions for stuck-open testing have been proposed (Chapter 3). However, application of these solutions to RAM decoders is unlikely due to performance/area constraints. Furthermore, defects in address decoders are not directly observable. One has to excite them in such a way that they are detected via the read operation of the RAM. Finally, owing to the constraints of addressing sequence, the detection of open defects by march tests is not ensured. All these reasons put together, on one hand, render existing stuck-open DfT solutions for RAM decoders impractical, and on the other hand, make testing of such faults a new challenge.

A missing contact/via is a dominant source of open defects in CMOS technology. In the case of a DRAM process, the depth of contact is much higher compared to a logic process which increases the sensitivity for open defects. According to the SIA technology roadmap for semiconductors, for a typical DRAM process the contact/via height/width aspect ratio is 4.5:1. The same for a typical logic process is 2.5:1 [74]. For future DRAM generations, the contact/via aspect ratio is expected to become 10.5:1, whereas for logic the same would become 6.2:1. The projected increase in the aspect ratio is a compromise to alleviate the large increase in per unit interconnect resistance and to prevent crosstalk [74]. Effectively, it means that for future CMOS devices in general, and DRAMs in particular, it will be much harder to make good, low resistance contacts. Furthermore, DRAM designs require the tightest metal pitch available and higher packing density, leaving no room for multiple contacts at most contact locations. Hence, open defects in RAMs require a careful investigation.

Open defects in a RAM matrix have been studied before and are known to appear as cell(s) row/column read failures or cell(s) SAFs that may be detected by march tests. However, a class of open defects in address decoders is not detected by march tests. In this section, we focus on open defects in RAM address decoders and propose test and testability strategies for their detection.

4.1 Early Work on Address Decoder Faults

A vast majority of the research on RAM testing was focused on efficient test algorithms for a variety of fault models. These fault models range from simple SAFs to complex pattern sensitive faults (PSFs) in the RAM array. However, little attention was paid to the faults in the address decoders or other RAM building blocks. Address decoder faults were assumed to be tested implicitly. An address decoder is a combinational circuit that selects a unique RAM cell for each given RAM address. Assuming that the faulty address decoder does not become sequential in operation, Thatte and Abraham [86] suggested that a faulty address decoder should behave in one of the following manners:

1. The decoder does not access the addressed cell. In addition, it may access non-addressed cell(s).
2. The decoder accesses multiple cells, including the addressed cell.

In the case of multiple accesses, the fault is viewed as RAM matrix coupling fault between different cells. In the case of no access, the cell is viewed as either SA0 or SA1. In simple terms, decoder faults manifest themselves as RAM matrix faults that are tested by the conventional algorithms.

4.2 Technological Differences

The above study was conducted for an NMOS decoder. The open defects in NMOS address decoders cause a logic SA behavior. As the technology made transition from NMOS to CMOS, the validity of the assumption was never re- evaluated. In CMOS technology, only a subset of open defects causes a logic SA behavior. The rest of the open defects cause sequential behavior in logic gates. Some of the defects causing sequential behavior in address decoders may escape detection by conventional tests.

The difference between an NMOS and a CMOS address decoder can be explained with the help of Figure 5-15. The figure shows a typical address decoder and logic implementations in NMOS and CMOS technologies. A logic gate in NMOS technology is implemented by a depletion mode NMOS load transistor and switching enhancement mode transistors. On the contrary, a logic gate in fully static CMOS technology is implemented by equal numbers of enhancement mode PMOS and NMOS transistors.

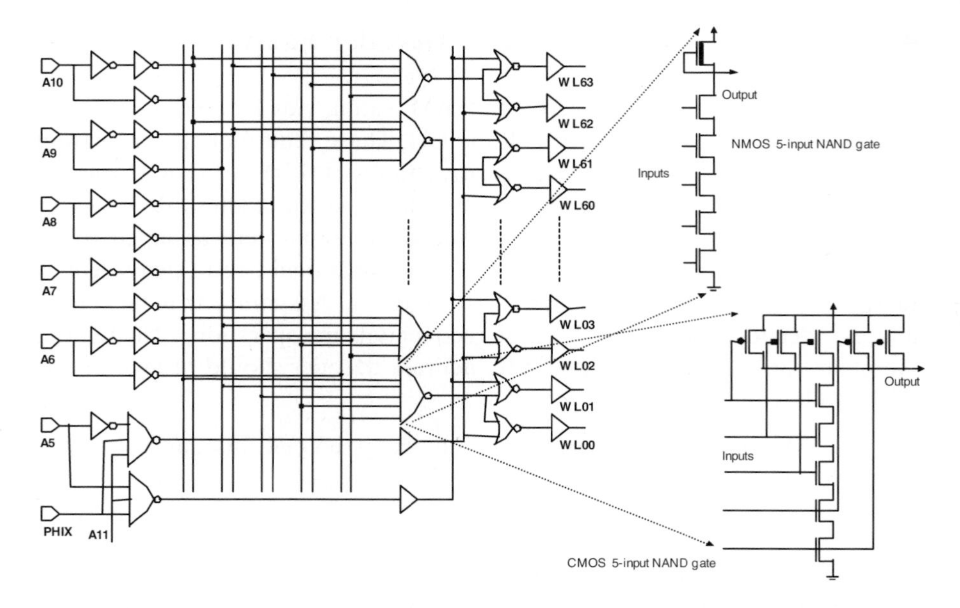

Figure 5-15. A typical RAM address decoder with implementations in NMOS and CMOS technologies.

An address decoder selects a specific wordline depending on the given input address. This requires the output of a logic gate in the address decoder to be active only for a unique input address and remain inactive for all other addresses. For example, for the 5-input NAND gates in Figure 5-16 the output is active (logic 0) only if all inputs of the gate are high and the output is inactive (logic 1) for the rest of the cases. In the case of NMOS technology, the depletion mode load transistor pulls up the output to the inactive state when inputs are not causing the gate to be in the active state. Now, an open defect in a switching transistor of the NMOS logic gate will cause the gate to stay in the inactive state when it was suppose to be in the active state. In other words, such a defect will cause the addressed decoder not to access the addressed cell. On the other hand, if there is an open defect in the load transistor, the logic gate will stay in the active state causing a multiple access fault.

In the case of the CMOS address decoder, the active state is arrived at in the same manner. However, the inactive state is reached by several parallel paths (depending upon the fan-in) selected by the input addresses. In later sections we shall discuss the faulty behavior that open defects in these parallel inactive paths can cause.

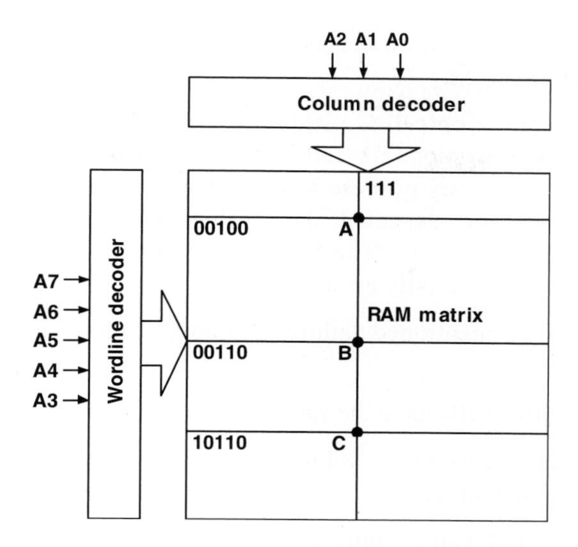

Figure 5-16. *Graphical representation of the failure.*

4.3 Failure and Analysis

Undetected faults of address decoders are explained in Figure 5-16. The figure depicts a part of an embedded SRAM block diagram showing the matrix, wordline and column decoders. An actual failure mechanism in an embedded SRAM is illustrated with the help of three cells, A, B, and C, respectively. The addresses, (A7--A0), of these cells are 00100 111; 00110 111; and 10110 111, respectively. The considered SRAM has 256 addresses (8-bits) and the word is also 8-bit wide. Different bits are not close to each other, so there is no possibility of an intra-word coupling fault. Address bits A7, A6, A5, A4, and A3 decode the word lines and the rest of the bits select the column (or bit) line. Cell C is the cell that fails conditionally. Following are the symptoms of the failure as observed:

Write address C (10110 111) with logic 1.

Write address A (00100 111) with logic 0.

Read address C: result is logic 1, which is correct.

Observation: RAM behaves normally, because address input A4 has changed.

Write address B (00110 111) with logic 0.

Read address C: result is logic 0, which is wrong!!

Observation: the failure occurs, because no address input among A3, A4 or A5 has changed.

The read operation on cell C yields a wrong data value only if between the write and read operations for cell C, some address bits (A5, A4 and A3) are kept unchanged. If any of these bits are changed, then the read operation for cell C yields the expected data value. Furthermore, the fault is completely data independent. The failure does not write data into another cell and appears as a read only error.

From the above mentioned failure symptoms following deductions are made:

1. All three cells have the same column address (111).
2. Cell C yields a read failure when address bits A5, A4 and A3 are kept unchanged.
3. The fault causes only a read failure in cell C and does not influence other cells in any manner.
4. The fault is not detected by the 6N SRAM test algorithm.

The first deduction suggests that when cell B is enabled after a cell C access, somehow cell C is also enabled (or it is not disabled). Cell C is controlled by the corresponding wordline. Consider a situation when wordlines B as well as C are enabled. If the complementary data (cell C) is written in cell B, the same is written in cell C as well. Hence, a subsequent read operation on cell C results in a read failure. The second deduction makes it clear that it is not the case with all cells of the same column. Cell C is sensitive only when address bits A5, A4 and A3 are not changed. The third deduction strengthens the first one stating that only a read on cell C is affected by the defect mechanism.

There are two possible explanations that match the above symptoms and deductions: (i) the wordline of cell C is also enabled when the wordline of cell B is enabled, and (ii) the wordline of cell C is not disabled when cell B is enabled. The first possibility is unlikely. However, it could be caused by (a) a decoder design error or (b) a low resistance bridging fault between wordlines of cells B and C. The decoder design error is ruled out since in that case a large number of devices would then fail under the test conditions. The low resistance bridging fault explanation also seems unlikely since the corresponding fault should be bidirectional. Moreover, such a defect is a typical case of a decoder fault mapped onto the matrix coupling fault that should be detected by the 6N test algorithm. Therefore, the fault that the wordline of cell C is not disabled when cell B is selected is a likely cause.

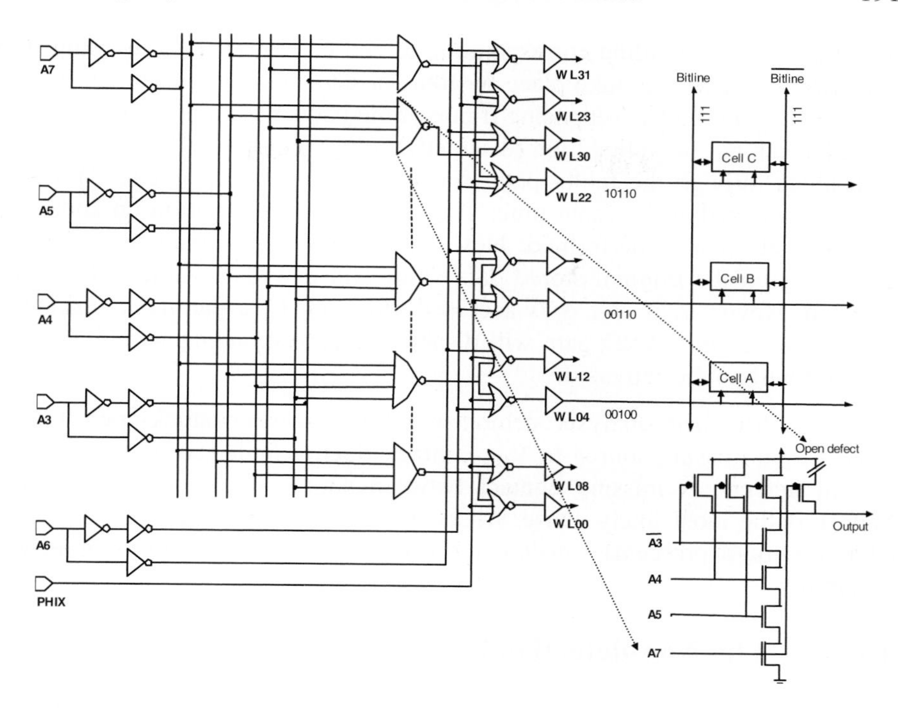

Figure 5-17. An undetected open defect in the address decoder.

The argument that cell C is not disabled can be explained with the help of Figure 5-17 that illustrates a part of the wordline address decoding logic and the corresponding bitlines. Figure 5-17 does not show the wordline drivers and input buffers. The wordline decoder has a 5-bit address. The address bits are buffered and their true and complement values are generated. The address decoding is achieved with the help of 4-input NAND gates. Subsequently, 3-input NOR gates decode the outputs of NAND gates with address bit A6. A periodic, timing signal, PHIX, forms the third input to these NOR gates. The outputs of NOR gates are buffered to drive the wordlines.

Let us assume for a moment that the NAND gate in the wordline decoder that decodes cell C has an open defect such that a p-channel transistor having A7 as its input is disconnected from VDD. Now, let us once again try to repeat the experiment carried out earlier. It is easy to notice that all the steps and observations can be re-created with the defect in cell C. In a decoder constructed with NAND gates, a simultaneous logic high on all inputs cause n-channel transistors to be in conduction mode and hence enables the particular wordline. For example, 1111 on A7, A5, A4, A3' and 0 on A6 select the wordline corresponding to cell C. However, a disabling of that particular path can take place by four (or depending upon the fan-in of the NAND gate) paths. Supposing if one of the paths has an open defect, the wordline can not be pulled high (disabled) through that path. If the wordline is disabled through the faulty path (for example, by selecting cell B), two cells are selected at the same time. Therefore, a write operation on another selected cell is also performed. Now, if a read is performed for cell C, depending on the original stored data value and new data value, a fault is detected. However, if cell A is accessed after cell C, a parallel p-channel path in the faulty NAND gate will disable the corresponding wordline and the fault will not be activated.

The subsequent analysis demonstrated a transistor stuck-open fault caused by a missing source to V_{DD} contact. Such a fault can be caused by several defects. A missing contact between source (drain) diffusion and Metal1 is the most likely cause. An open defect in the metallization layer (step coverage problem) causing a transistor stuck-open fault is another possibility.

4.4 Why Non-detection by March Tests?

The question is why the fault is not detected by a march test and how should we detect such failures. The 6N test algorithm is shown in Figure 5-18. It is a popular and time tested algorithm used within Philips for SRAM testing [50]. First, the RAM is initialized with logic 0. Subsequently, March1 reads the initialized value and writes logic 1 in each RAM cell in ascending address order. The following binary address after wordline C (address 10110) is 10111, which modifies the A3 bit. Hence, wordline C is disabled like a fault- free case (Figure 5-17). In other words, the fault is neither activated nor detected in March1. Similarly, the fault is not detected in March2. The March2 is in descending address order. After the wordline C is activated, the next wordline address (descending order, 10101), which modifies A4 and A3 bits. As a result wordline C is disabled once again and the fault is not detected. This type of fault can only be detected by a march test (or a linear algorithm), if the next wordline address causes the fault to be activated in at least one march direction, and keep it activated till it is detected by a read operation on the cell. Now, depending upon the original and over-written data values, the defect can be detected. However, this condition can not be met for all such open defects in NAND gates. Therefore, most of such defects are not detected by march tests.

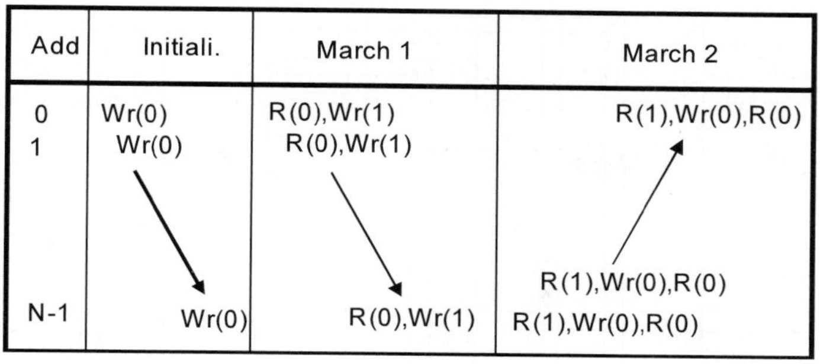

Add	Initiali.	March 1	March 2
0	Wr(0)	R(0),Wr(1)	R(1),Wr(0),R(0)
1	Wr(0)	R(0),Wr(1)	↑
	↓	↓	R(1),Wr(0),R(0)
N-1	Wr(0)	R(0),Wr(1)	R(1),Wr(0),R(0)

Figure 5-18. The 6N SRAM march algorithm.

A march test may have any address order, as long as all addresses are accessed. For reasons of simplicity, mostly ascending (descending) address sequences are selected. However, without the loss of generality, it can be argued that no addressing sequence will detect all open defects in the address decoder. Furthermore, no linear test algorithm will detect such defects since due to these defects, the address decoder is changed into a faulty sequential circuit that, in general, requires a two-pattern test. The basic assumption about address decoders that under faulty conditions they should remain combinational is violated. This is a generic problem with decoders implemented with static CMOS logic gates. When decoders are implemented with dynamic logic (or NMOS) such faulty conditions may not arise.

4.5 Address Decoder Open Defects

A linear, march test algorithm will not detect some open defects in address decoders. Other RAM test algorithms are also not likely to detect these defects owing to the fact that a two-pattern test sequence (T1, T2) for all potential defects is not ensured by them. For example, complex algorithms for neighborhood pattern sensitive faults will not be able to ensure decoder open fault detection. Arguably, the GALPAT (GALloping PATtern) algorithm [16] of complexity $O(n^2)$ will detect these defects. However, application of GALPAT even for moderately sized RAMs is not possible due to its excessive test time. On the contrary, we shall see later in this chapter that there has been a significant effort made to employ parallel test techniques to reduce RAM matrix test costs. In parallel test techniques, address decoder faults are less vigorously tested. As a result, RAM address decoder testing becomes a quality and economics issue.

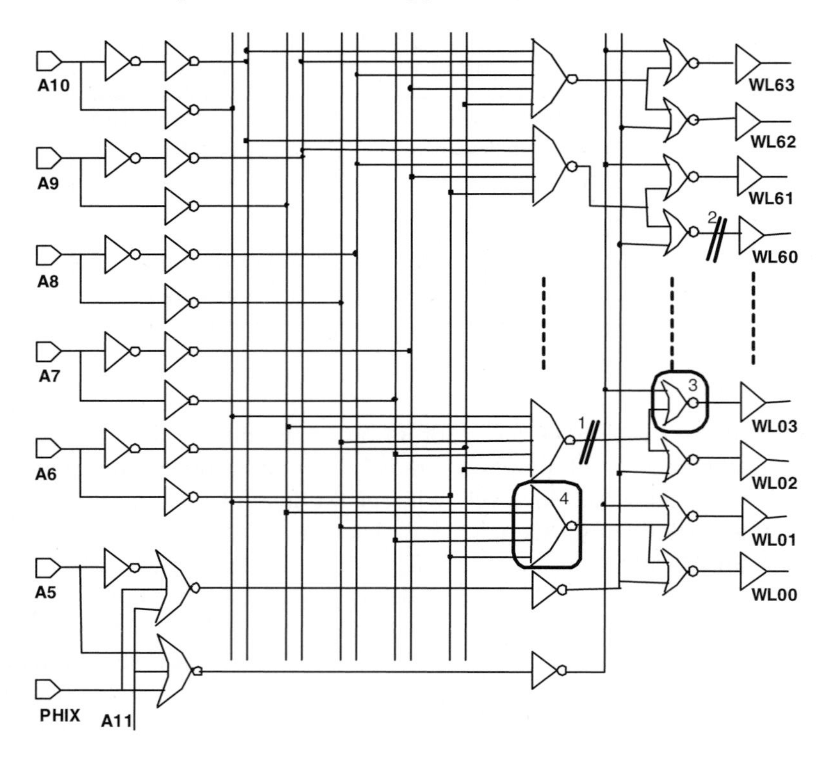

Figure 5-19. A typical wordline address decoder.

Only a subset of all open defects in an address decoder is not detected by march tests. Hence, it is logical to analyze which open defects are not likely to be detected by march tests and then devise a test only for those open defects. For this purpose, we take a row decoder of an embedded RAM (Figure 5-19). This circuit decodes a 6-bit address (A10 -- A5) into 64 wordlines. Address bit A11 determines the selected quadrant and PHIX is a periodic timing signal that controls the timing of the wordline (X) address. In this decoder, instead of 4-input NAND gates, 32 5-input NAND gates are utilized and the 6th bit (A5) is further decoded by 64 2-input NOR gates.

In general, open defects in an address decoder may occur either in between logic gates (inter-gate) or inside a logic gate (intra-gate). Defects 1 and 2 in Figure 5-19 are representatives of the inter-gate class. The inter-gate open defects cause a break in an interconnect line. Owing to this class of defects, at least one RAM cell cannot be addressed. Hence, a cell may appear to have a stuck-at fault. In other words, inter-gate open defects do not cause sequential behavior and they are detected by march tests such as the 6N algorithm. However, intra-gate open defects (defects 3 and 4) are difficult to detect because they may influence only a single transistor. Hence, they may result in a sequential behavior. If an intra-gate open defect disconnects all paths between the output and V_{DD} (V_{SS}), it effectively causes an output SA0 (SA1) fault that is detected by the march test. However, if an open defect disconnects only one of the paths between the output and V_{DD} (V_{SS}), it causes a sequential behavior.

Let us assume that defect 3 causes an n-channel transistor in a 2-input NOR gate to be disconnected from V_{SS}. Since there is only one other n-channel transistor in parallel with the defective transistor, the defect will be detected by either the ascending march (March1) or the descending march (March2) of the address space depending on which of the transistors is faulty. The condition for this detection is that the inputs of the faulty gate should be changed in a Gray code manner. Therefore, in this case address bit A11 and decoded A10--A5 bits should change in the Gray code manner. The situation becomes complex as the number of inputs in a gate rises to three or more. With the reasoning of the previous section, it can be concluded that detection of all open defects in a 3 (or more) input logic gate is not guaranteed by the march test. It can be observed in this example, that at least 3 open defects in each of the 5- input NAND gates will not be detected (the other 2 open defects will be detected by descending or ascending march elements). There are 32 such NAND gates in the decoder giving rise to at least 96 undetected defects.

4.6 Supplementary Test Algorithm

Once all likely escapes are known, a test solution may be devised. A small algorithmic loop is appended to the 6N algorithm to detect address decoder stuck-open faults not detected by the 6N algorithm. However, this algorithmic loop is specific to an address decoder and is independent of the 6N algorithm. Hence, it can be added to any other test algorithm.

Let us assume that M is the number of input bits of the wordline decoder and the number of wordlines equals 2^M. To test the row decoding logic we can select any arbitrary column address for read and write operations. In the following algorithm we set the column address to 0. As explained before, the least significant bit (in Figure 5-19, bit A5) is a don't care and remains 0 during the test. To test for the hard-to-detect opens, the NAND gates in the decoding logic should be tested in a sequential manner. For each NAND gate a logic 0 is written in the selected cell (say D) by the corresponding wordline (remember that bit A5 is set to 0). Subsequently, the wordline address is changed such that only one address bit is changed (let us say A6). This will allow the particular NAND gate to be disabled through one selected p-channel transistor. Now, logic 1 is written in the new address location (say E). If the selected p- channel transistor had an open defect, the cell D is still enabled and the write operation on cell E can also over-write the content in cell D. A subsequent read operation on cell D will detect a read failure and hence the open defect. This is repeated for all address bits to NAND gates and for all NAND gates. For example, for the 5-input NAND gate in Figure 5-19 with defect 4 (shaded) following test sequence can be applied:

1a. Keep Y decoder address constant,

1b. keep A5=0 and A11 (if available)=0

2a. Let A10A9A8A7A6 = 00000, Write(1);

2b. A10A9A8A7A6 = 00001, Write(0);

2c. A10A9A8A7A6 = 00000, Read(1);

2d. A10A9A8A7A6 = 00010, Write(0);

2e. A10A9A8A7A6 = 00000, Read(1);

2f. A10A9A8A7A6 = 00100, Write(0);

2g. A10A9A8A7A6 = 00000, Read(1);

2h. A10A9A8A7A6 = 01000, Write(0);

2i. A10A9A8A7A6 = 00000, Read(1);

2j. A10A9A8A7A6 = 10000, Write(0);

2k. A10A9A8A7A6 = 00000, Read(1);

In general, an algorithm for a given address decoder can be evolved that will supplement any RAM test algorithm. For the address decoder in Figure 5-20, such an algorithm is shown below:

In the algorithm description the address values in the read and write operations correspond to the binary code at the input bits of the wordline decoder (A10, A9, A8, A7, A6, A5). The algorithm becomes:

Column_address = 0
  For i = 0 to $2^{(M-1)}$ Do
  Base_address = 2 * i
  Write "0" to Base_address
  For j = 0 to M Do
  Write_address = Base_address XOR_{binary} 2^j
  Write "1" to Write_address

Read "0" from Base_address
End For
End For

As can be determined from the algorithm, the inner loop will be executed (M-1) times and for each i it consists of one write and one read operation. The main loop will be executed $2^{(M-1)}$ times and takes one extra write operation. This makes a total complexity of the algorithm as $(2M-1)x2^{(M-1)}$ read or write operations, where M is the number of input bits in the wordline decoder. To compare the complexity of the algorithm given above with the 6N algorithm we will consider a RAM having 6 bits devoted to the column decoding and another 6 bits to wordline decoding. The 6N algorithm will take 6×2^{12}= 24,576 read and write operations. The algorithm given above will only take 11×2^{5}= 352 read and write operations. So the additional test complexity is less than 2% of the 6N test.

It can be argued that similar open defects in the column decoder can also cause hard to detect faults. Column decoders can be analyzed to devise a suitable test algorithm.

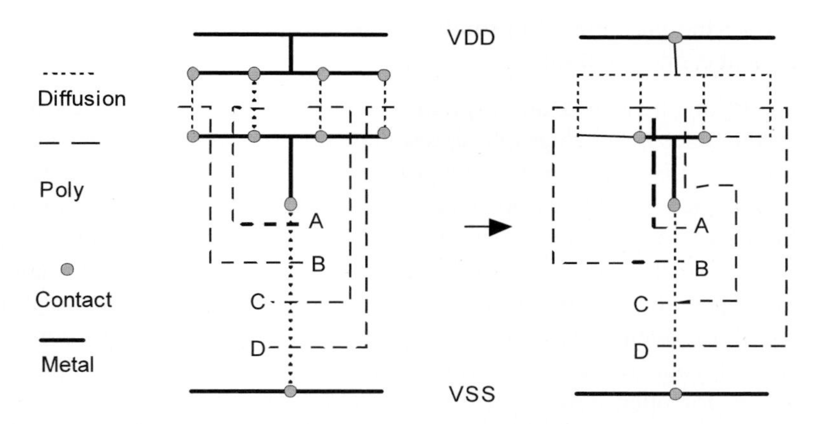

Figure 5-20. Layout transformation of a four input NAND gate for mitigating hard to detect open defects.

4.7 Testability Techniques for Decoder Open Defects

In the previous section we discussed the test escape problem arising due to open defects in RAM address decoders and evolved a test procedure to prevent test escapes. In this section, we focus on the layout level testability measures to simplify the detection of such hard to detect open defects and on building fault tolerance through logic modification. Such measures are best implemented while designing new address decoders. If these simple yet effective measures are implemented, the requirement for additional test is either completely eliminated or drastically reduced. However, the existing decoders without such measures will require the extra test procedure as proposed previously.

4.7.1 Layout Measures

The layout improvement is probably the simplest and most effective method to reduce the occurrence of such open defects. The layout of the circuit affects the testability to a great extent. Simple layout modifications may reduce the possibility of hard-to-detect faults, hence, reducing the burden on test generation. For example, Placement of multiple contacts at hard-to-detect defect locations (parallel transistors) in the decoder will make it robust against open defects. These layout techniques are well documented in the literature [29,34]. There is a need to implement such techniques in future RAM decoder designs because (i) the decoder circuitry is implicitly tested by testing only the matrix, (ii) the RAMs are often tested by linear algorithms that restrict the excitation of the decoder in a particular fashion so as to cover the fault model in small number of operations, and (iii) there are quality and economic issues.

The layout measures are explained with the help of Figure 5-20. The figure illustrates a switch graph representation of a 4-input NAND gate. This transformation is similar to the one proposed by Levitt and Abraham [34]. In the unmodified layout an open at a contact can occur at any branch of parallel transistors or metal lines. A simple test may not detect such an open defect, and all the parallel branches must be tested separately. We assume that the open defect probability due to a poor contact is relatively high compared to open defect probability due to a break in diffusion. The modified layout results in a robust design as well as in simpler test generation for open defects. It was reported [34] that though the area and delay of the transformed gate may increase marginally, the number of hard to detect faults is reduced drastically.

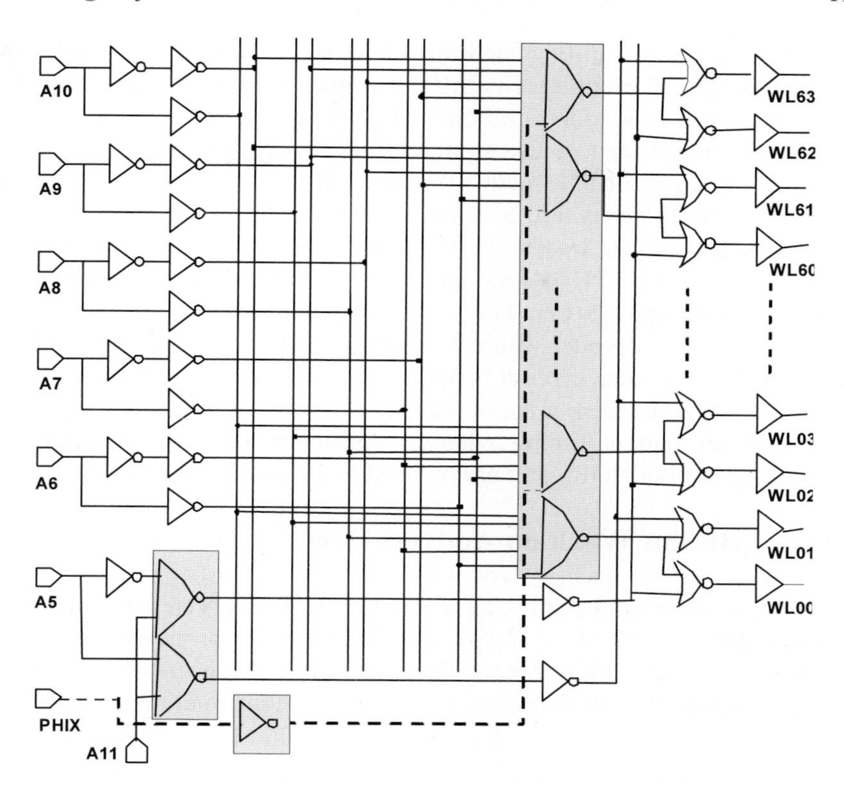

Figure 5-21. A Fault tolerant row decoder against hard to detect open defects.

4.7.2 Logical Measures: Building Fault Tolerance

The layout level techniques, in principle, can reduce the probability of occurrence of open defects in sensitive decoder locations but cannot eliminate them completely. Therefore, in this subsection we propose methods for implementing fault tolerance in the key decoder locations such that in spite of the defect, the decoder and the RAM can function correctly. The fault tolerance can be used together with the layout level transformations to enhance the robustness of the decoder. Figure 5-21 illustrates the concept of logical measures. This figure is the same as Figure 5-19 except for the highlighted areas, an inverter and an added net shown by a broken bold line. From the earlier reasoning, we conclude that opens affecting only single p-channel transistors in 5-input NAND gates are hard-to-detect. The p-channel networks in NAND gates provide the disabling paths to the wordlines (since a particular wordline is selected if and only if the output of the corresponding NAND gate is logic 0). Therefore, an extra p-channel transistor can be added in each of the 5-input NAND gates such that it provides an alternative path for wordline disabling such that before application of a new address all NAND gates are disabled. In other words, no wordline is selected. A corresponding n-channel transistor is also added to avoid logic conflicts, effectively making it a 6-input NAND gate. The modified NAND gates are shaded in the figure. The inputs of these transistors are driven by the PHIX signal which activates the wordline address. Effectively, PHIX now gates address bits A10 – A6 instead of A5 and A11 (see Figure 5-19). The extra inverter is needed to invert timing signal PHIX. In a decoder where 5-input NOR gates are utilized instead of NAND gates, the extra inverter is not needed. As far as the logic function of the decoder is concerned, it is not changed. The design of the decoder can be optimized for correct timing without sacrificing the gains. Furthermore, highlighted 3-input NOR gates are reduced to 2-input NOR gates.

4.8 Recent Work on Address Decoder Defects

In the recent past, a significant amount of work has been done on address decoder defects [2,3,15,27,58,85]. Otterstedt et. al. proposed an address generation scheme using Linear Feedback Shift Registers (LFSRs) [58]. Thaller proposed a test sequence based on the right rotation of the march tests to cover all address decoder open faults [85]. Similarly, other researchers included resistive and coupling defects in address decoders.

Perhaps, the most authoritative recent treatment on the subject is given by Azimane and Majhi [3]. They argued that address decoders are increasingly more susceptible to intra-gate resistive opens as copper is introduced in the fabrication process. In addition, they described why algorithmic solutions for resistive defects are unlikely to give high defect coverage. Resistive defects give rise to higher delay. Therefore, they suggested the manipulation of the duty cycle of the internal clock to the address decoder as a design for testability measure.

5. PARAMETRIC TESTING OF SRAMS

The memory content in contemporary Systems on Chip (SoC) is increasing significantly. It is not uncommon to find SoCs with 80-90% of transistors in memories. The vast majority of these memories are implemented using SRAMs which offer an excellent solution in power, operational speed, robustness and ease of implementation domains.

Therefore, SRAM have become the de facto standard for embedded memory implementation.

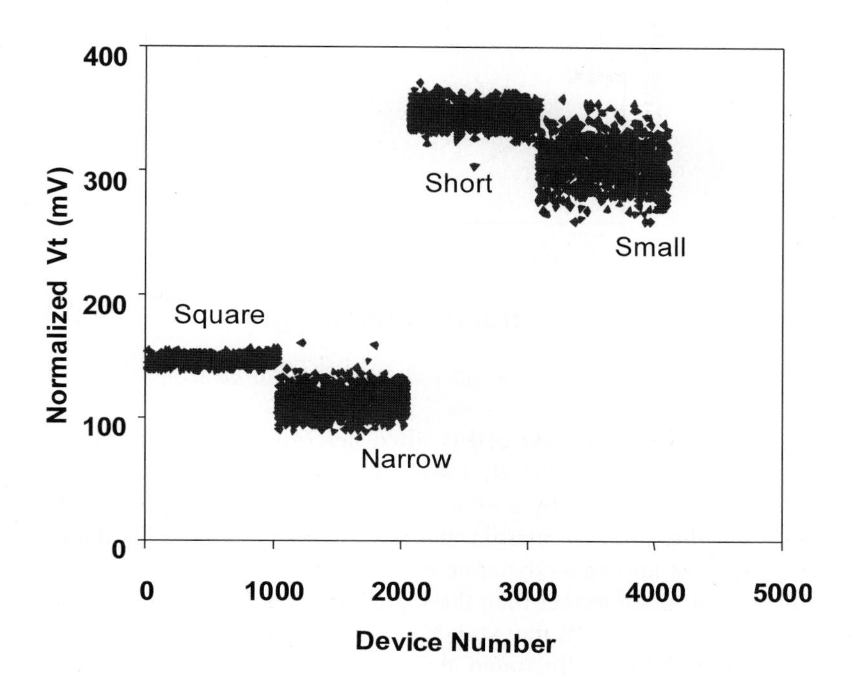

Figure 5-22. Normalized V_t for different sized transistors [courtesy Intel Corporation].

At the same time, SRAMs like any other logic or analog blocks are affected by technology scaling. In fact, they are affected drastically owing to their peculiar implementation and operational constraints. Transistor parameters become increasingly more difficult to control due to short and narrow channel effects. For example, vital transistor parameter such as the V_t becomes a function of transistor width and channel length if scaled in nano-metric dimensions, as illustrated in Figure 5-22. Smaller transistors exhibit larger variations in the threshold voltage compared to others. SRAM transistor density is extremely high; hence SRAM cell transistors are very small. Therefore, SRAM cells also exhibit variations in drive capability, cell stability, etc. In addition, a high packing density makes transistors sensitive to catastrophic as well as non-catastrophic manufacturing defects.

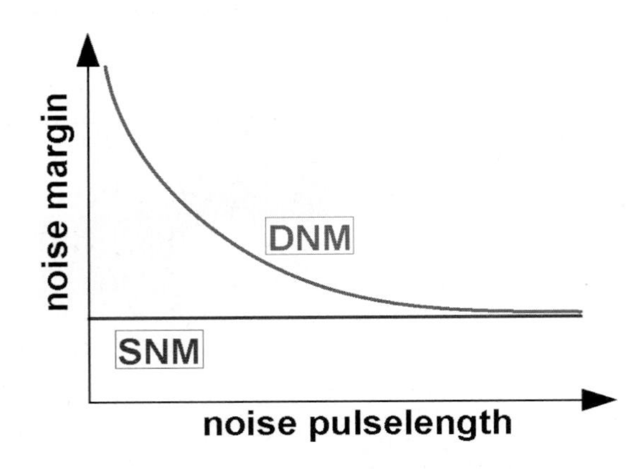

Figure 5-23. Noise margin with respect to noise duration.

The stability of an SRAM cell is often determined in terms of its static noise margin (SNM). In general, a noise margin is the maximum electrical noise that can be tolerated by a logic gate while maintaining its correct logic output. The value of noise margin increases exponentially as its duration in reduced, and it is known as dynamic noise margin. On the other hand, if the noise is present much longer than the logic gate delay, it is deemed as static. If a noise pulse is applied, the situation is quasi-static and the noise margin approaches the SNM, as illustrated in Figure 5-23.

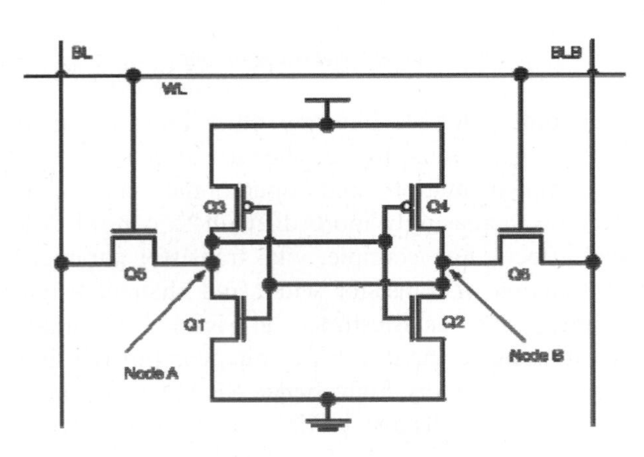

Figure 5-24. The 6T SRAM cell.

5.1 SRAM Cell and SNM

Figure 5-24 depicts the six transistor (6T) SRAM cell. Transistors Q1 and Q2 are known as drivers, Q3 and Q4 are PMOS load transistors, and finally Q5 and Q6 are the access transistors. If the wordline, WL, is high, the access transistors connect internal nodes to bit lines.

The read operation of the cell is initiated by pre-charging the bit lines to V_{DD}, and subsequently activating the WL to V_{DD}. Assuming that the node A is at logic 0, the VBL is discharged through Q1 and Q5, while the V_{BLB} remains at the pre-charged level. Transistors Q1 and Q5 form a voltage divider whose output is connected to the Q2-Q4 inverter. Q1 and Q5 are sized so that the Q2-Q4 inverter does not flip causing the read upset. The maximum voltage at node A during read is given as [65]:

$$\Delta V = \frac{V_{DSATn} + CR(V_{DD} - V_{Tn}) - \sqrt{V^2{}_{DSATn}(1+CR) + CR^2(V_{DD} - V_{Tn})^2}}{CR} \qquad (5.4)$$

Where CR is the cell ratio defined as:

$$CR = \frac{W_1 / L_1}{W_5 / L_5} \qquad (5.5)$$

If the CR is reasonably large (CR> 1.3) the cell is stable during the read operation.

During the write operations, one of the bit lines is driven low from the pre-charged value. Assuming that node B is at logic 1, the V_{BLB} is driven to logic 0. Transistors Q6 and Q4 are sized such that the voltage at node B is brought below the switching threshold of the cell. This voltage is given as [65]:

$$V_B = V_{DD} - V_{Tn} - \sqrt{(V_{DD} - V_{Tn})^2 - 2PR\{(V_{DD} - V_{Tp})\}V_{DSATp} - V^2{}_{DSATp}/2} \qquad (5.6)$$

Where the pull up ratio of the cell is defined as:

$$CR = \frac{W_4 / L_4}{W_6 / L_6} \qquad (5.7)$$

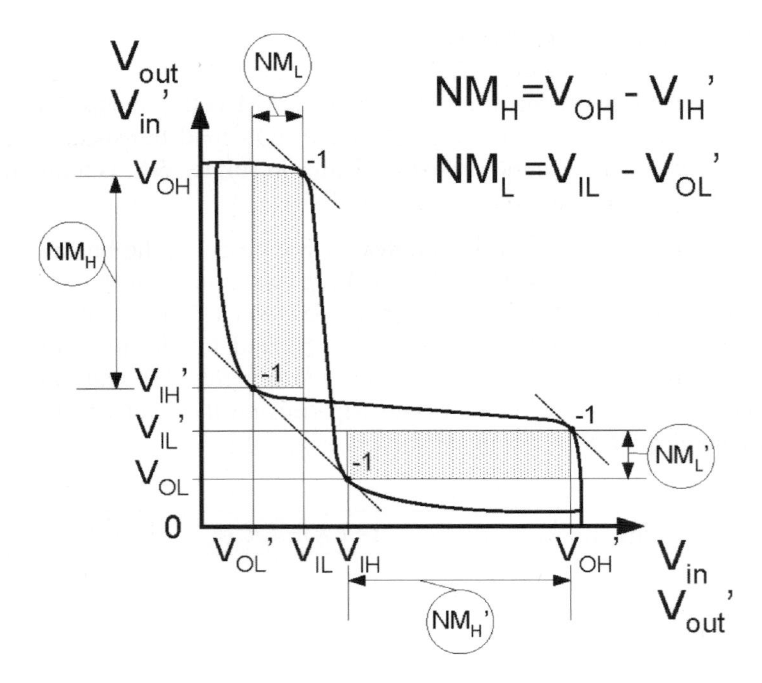

Figure 5-25. Graphical representation of SNM.

There are few definitions of SNM that can be found in the literature. However, there are two that are worth discussing here. The first one is the graphical approach where V_{OH} and V_{OL} are represented as stable points with $dV_{out}/dV_{in} = -1$ of both inverters in the cell [41].

The SNM is the sum of the two sides (NM_L and NM_H) of the largest rectangle that can be accommodated in between the voltage transfer characteristics of inverters. Using this methodology, the SNM of the cell can be computed easily. However, this technique represents the legacy from the noise margin of the inverter chain rather than the noise margin of an SRAM cell. For example, it is possible that NM_L is small which makes the cell sensitive to upset while the SNM be significantly large.

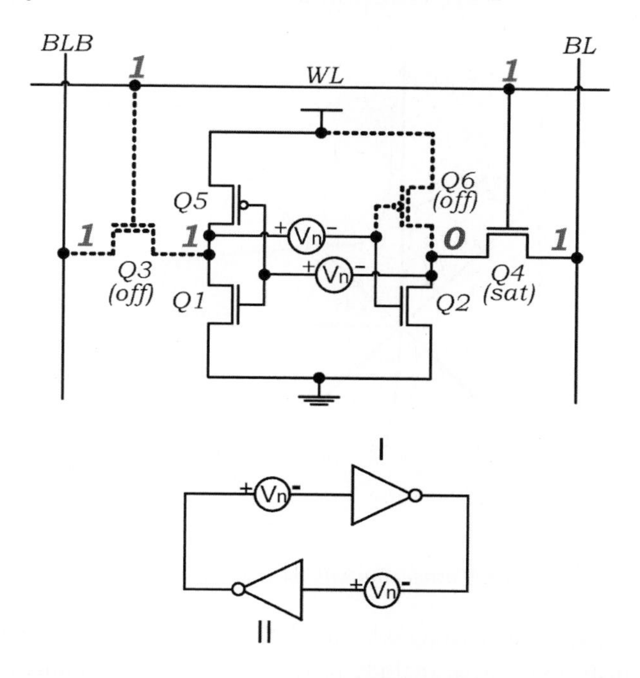

Figure 5-26. SRAM cell under read access condition, and noise sources.

In the second approach, the SNM is defined as the side of a maximal square that can be drawn between the inverter characteristics [21]. The approach can further be described with the help of Figure 5-26. Two equal and opposite noise sources are introduced in series with the inverters. The polarity of the noise sources is opposite of the stored value, therefore, at a certain voltage level both noise sources manage to upset the cell. These two noise sources provide the worst case SNM [41]. On the other hand, the best case SNM would have been provided if only one noise source had been applied, or both noise of the same polarity had been applied. The cell illustrated in this figure is under the read access condition which provides the worst case situation for SNM.

Figure 5-27 illustrates cell transfer characteristics of the read access cell in the x-y coordinate system. The u-v coordinate system is obtained by rotating the x-y coordinate system 45° counter-clockwise. This arrangement allows the determination of maximum diagonal which is parallel to the v axis. The side of the rectangle represents the SNM. The dashed curve in the figure is obtained by subtracting the mirrored characteristics from the normal characteristics. In general, both squares are not equal due to process variations. Therefore, the worst case SNM is determined by the smaller of the two squares.

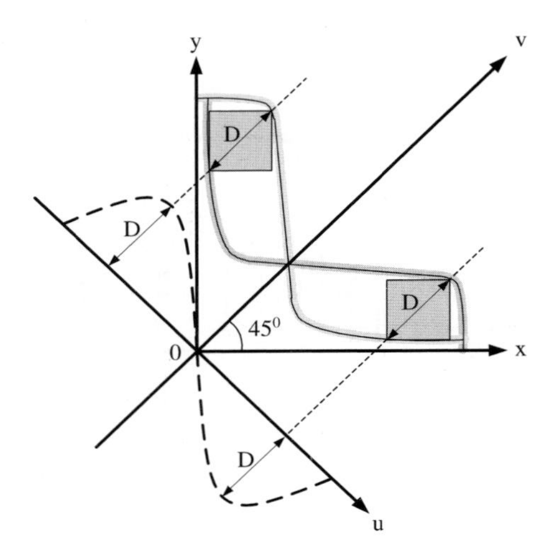

Figure 5-27. SNM estimation with the diagonal of the biggest square.

There are several reasons why the SNM of a real cell may be different from the ideal one. These include process variation, manufacturing defects, and environment issues such as temperature, power supply voltage, etc. These issues are dealt in detail somewhere else, however, in next sub-sections we will provide glimpses.

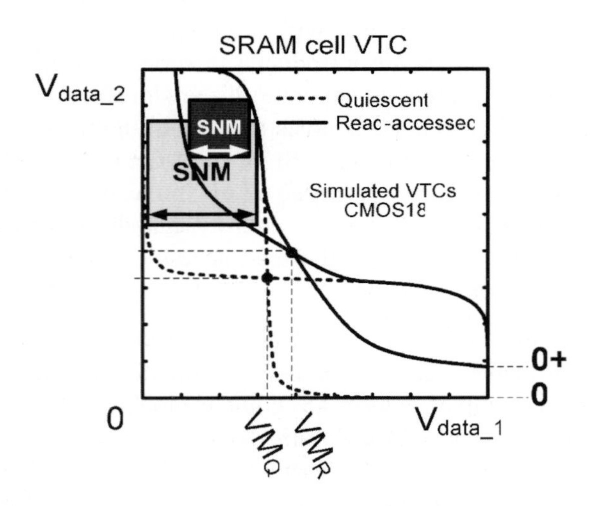

Figure 5-28. Simulated SRAM cell transfer characteristics under quiescent and read access conditions.

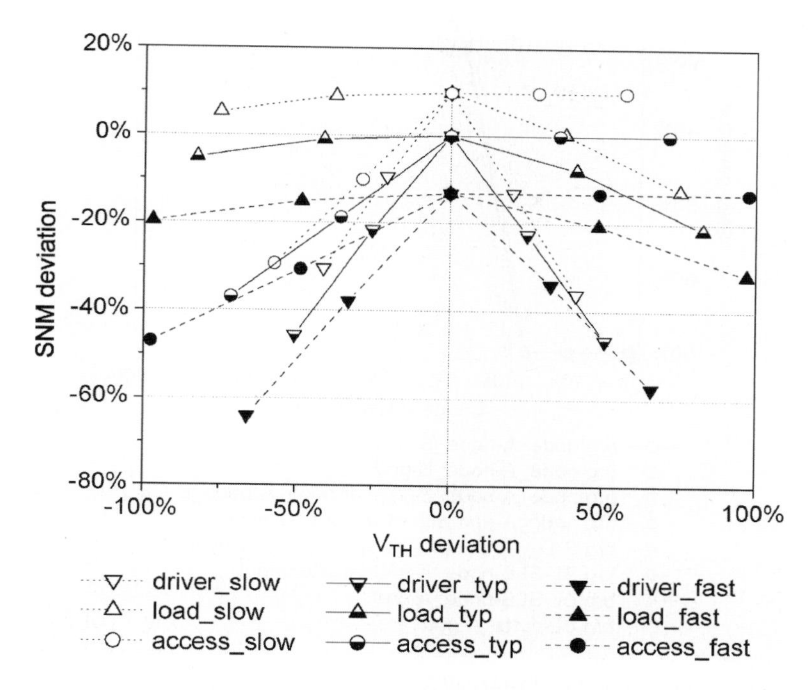

Figure 5-29. Normalized SRAM cell SNM with respect to a single transistor Vt variation for typical, fast, and slow processes.

5.2 Process Variation and SNM

Before we discuss the impact of process variation on the SRAM cell, it is pertinent to discuss the cell under read access condition. Assuming, node A in Figure 5-24 is at logic 0, when the access transistors are turned on, the voltage at node A rises and is determined by the relative strengths of the access and the driver transistor. As a consequence, the SNM of the cell is degraded significantly, as shown in Figure 5-28.

As mentioned before, process variations pose serious problems on the stability of SRAMs. For example, V_t variation of over ten percent from the typical value is not uncommon. Researchers have correlated SRAM yield to the SNM spread. It has been reported that μ - 6σ of SNM is required to exceed $0.04xV_{DD}$ to reach 90% yield on 1 MB SRAMs [79]. Transistor mismatch increases the transistor spread which reduces the μ - 6σ value as well as stability of cells.

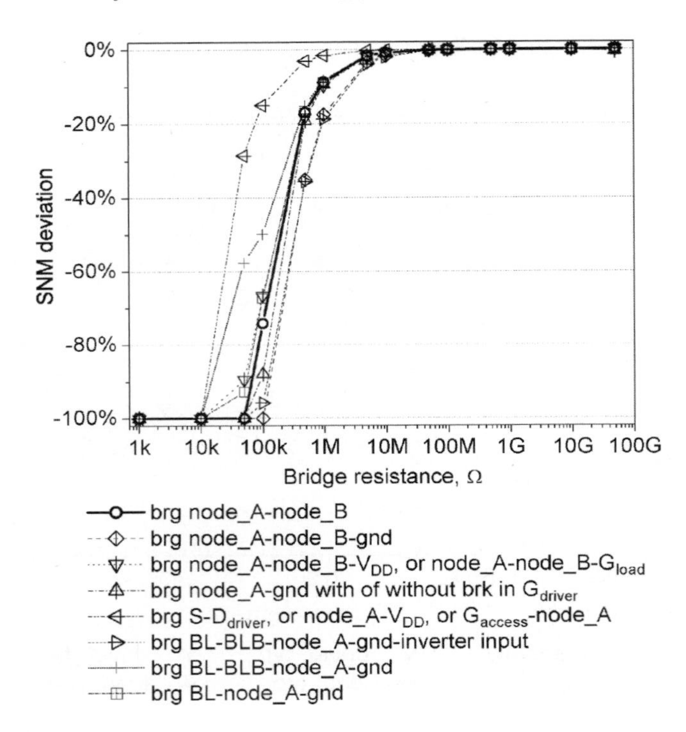

Figure 5-30. Normalized SRAM cell SNM with respect various bridging defects.

The SNM dependence on V_t variation is illustrated in Figure 5-29. The V_t of a particular transistor is changed while the other five cell transistors were kept at either nominal, or slow, or fast process condition. As expected, a change in transistor threshold resulted in transistor mismatch which degraded the cell SNM. The drive V_t variation has the largest impact on the cell SNM. Similarly, if the V_t of the access transistor is reduced, it degraded the SNM significantly. However, if the V_t of the access transistor is increased, it did not result is any significant change in the SNM. It can be explained as follows: A stronger access transistors degrade the cell logic 0 level to a larger extent which results in a larger degradation of the SNM. On the other hand, weaker access transistors reduce the logic 0 degradation under the read access condition; hence this improves the SNM marginally.

In the case of the load transistor, its V_t variation has minimal impact on SNM. The load transistor is designed to be weak compared to the driver and access transistors. Therefore, it has the least impact. It is worth noticing that the cell SNM improves with slow process while a fast process reduces the SNM. A fast process results in higher transistor leakage that degrades the SNM while converse is true in the case of slow process.

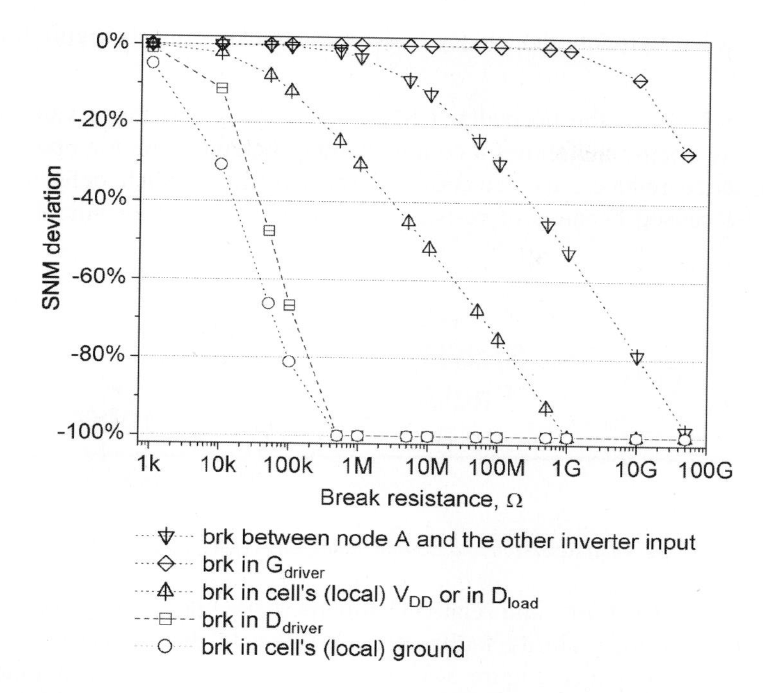

Figure 5-31. Normalized SRAM cell SNM with respect various open defects.

The situation is much more complicated in real life where V_ts of individual transistors may change in unpredictable manner. Pavlov et al. analyzed various situations and an interested reader is referred to their work [60-62].

5.3 Manufacturing Defects and SNM

Most defects have a catastrophic impact on the SRAM cell resulting in near zero SNM, and can be detected by march tests. However, non-catastrophic defects may result in an SRAM cell with reduced, non-zero SNM. Depending on the defect resistance, these defects may escape the detection through conventional test procedures.

Pavlov et al. utilized the Carafe IFA tool to introduce resistive defects in the layout. Carafe works by widening and shrinking the layout geometries to determine defect probabilities. Once a realistic defect list was compiled, the defects were simulated with varying resistances. The results of their analysis are depicted in Figure 5-30 and Figure 5-31. Figure 5-30 plots the normalized SNM for varying resistance value of various likely shorts. The detection of shorts becomes more difficult as their resistance is increases.

A short between two complementary nodes is very likely and is highlighted in the figure.

Figure 5-31 plots the normalized SNM for varying resistance value of several likely opens, and shows a complementary behavior. As the open the open resistance reduces, its detection becomes difficult. Such defects are likely to be caused because of resistive contacts, vias, or poor silicidation [52,53].

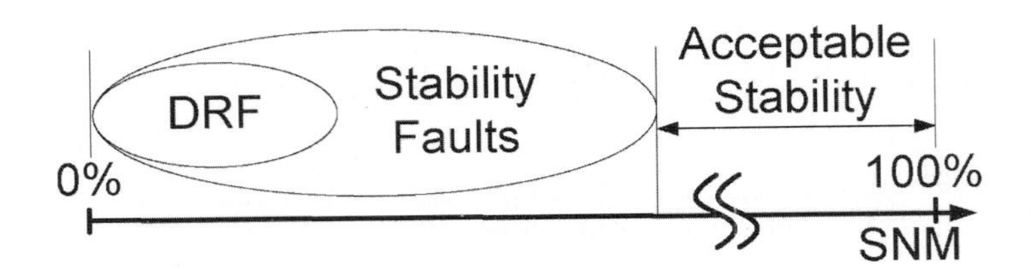

Figure 5-32. Data Retention and stability faults with respect to the SNM.

Highly resistive shorts, and relatively low resistance opens may give rise to data retention and stability faults. It is pertinent to distinguish between these two types of faults. Figure 5-32 illustrates the relation ship amongst data retention, stability faults and the SNM. If a cell is unable to hold on to its stored data, it is said to suffer from the data retention fault. Traditionally, pause test has been used to test for data retention problems. Pavlov suggested that the data retention faults are a subset of stability faults where these faults represent the extremely data instability [62].

5.4 Weak Cell Fault Model

As mentioned in previous chapters, a fault model represents the electrical impact of a physical cause. In this particular case, a fault model should be able to represent varying degrees of cell stability since stability faults are parametric in nature. Such a fault model will help us characterize defects and their detection, and devise better test methods and DfT strategies. In addition, such a fault model should be simple, and easy to use in a simulation environment.

Figure 5-33 illustrates the concept behind the weak fault model. The SNM is a measure of cell's stability and its degraded value results in stability faults which are parametric in nature. Therefore, the weak fault model should be able to mimic the degraded SNM behavior. Moreover, the range of degradation should be determined by (i) defects that are not detected by march tests, (ii) fault model complexity, and (iii) defect probabilities. A catastrophic defect is likely to degrade the SNM significantly such that it will be detected by the conventional test methods. On the other hand, a minute change in via or contact resistance will result in insignificant SNM degradation which may not be an issue. Considering these two extremes, the range of compromised SNM lies in a domain where cell SNM is significantly compromised and is unlikely to be detected by conventional means.

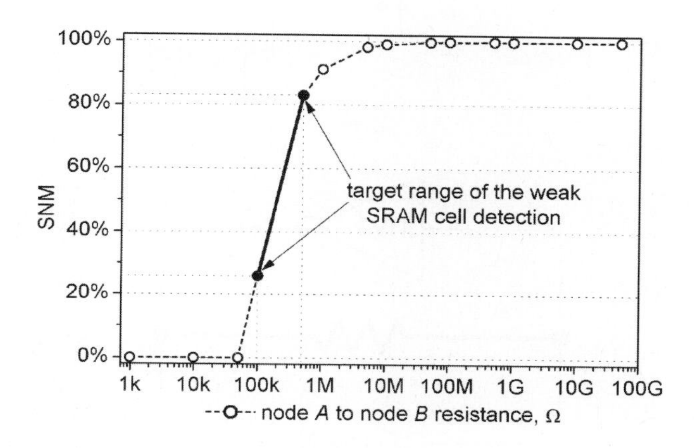

Figure 5-33. SNM range to be modeled by the weak fault model.

The weak cell fault model is shown in Figure 5-34. The added resistor between two nodes provides a negative feedback, effectively reducing the inverter gain. Reduced inverter gain results in degraded SNM. Therefore, by manipulating the resistance value the cell, the SNM can be controlled, mimicking a compromised cell. This cell now can be utilized in comparing test strategies for weak cell detection, etc.

5.5 DFT Techniques to Detect Weak Cells

There are several DFT techniques that have been proposed to detect weak SRAM cells. Weak Write Test Mode (WWTM) is one of the well known DFT techniques [52]. A weaker than nominal write voltage level is applied to the cell under test. A weak cell is overwritten while a stable cell is able to retain its value owing to its fully functional positive feedback mechanism. On the other hand, the positive feedback mechanism is compromised in a weak cell, therefore, the weak cell is flipped easily. The WWTM is realized either through a standalone circuitry or integrated into the write drivers.

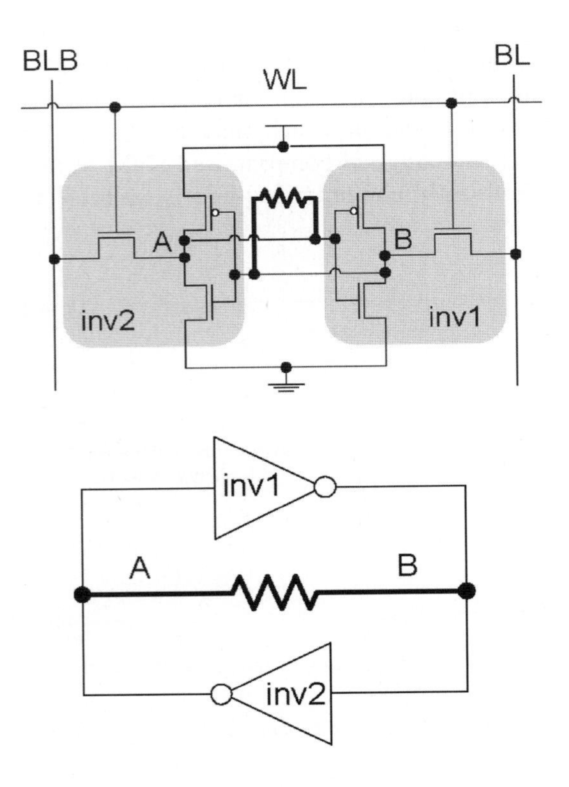

Figure 5-34. Weak cell fault model, and its equivalent circuit.

Figure 5-35 depicts the circuit topology of the WWTM. As it is apparent from the schematic, only few transistors (six to be precise) in each column are needed to provide the weak write capability. This DFT technique was implemented in a number of arrays in a Pentium processor. A number of defects were introduced using the ion-milling opens in transistor sources. Meixner and Banik found that WWTM as well as the pause (data retention test) detected open defects; however, other algorithmic march test could not detect the same. Subsequently, the WWTM was implemented in the production environment and a detailed analysis of the comparison between the pause test and the WWTM was made. It was found that the WWTM could replace the pause test with any quality impact and its implementation in production environment resulted in 20% reduction the test time. In addition, WWTM could detect 40x more failures compared to the pause test with single bit failures being the largest category with both asymmetric and symmetric failures.

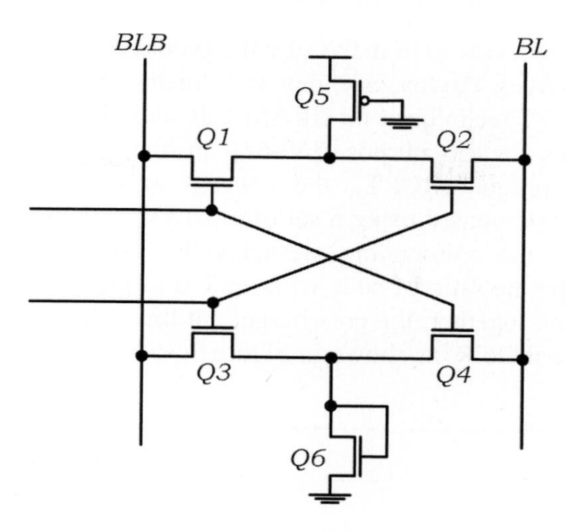

Figure 5-35. Weak write test mode for SRAM parametric testing.

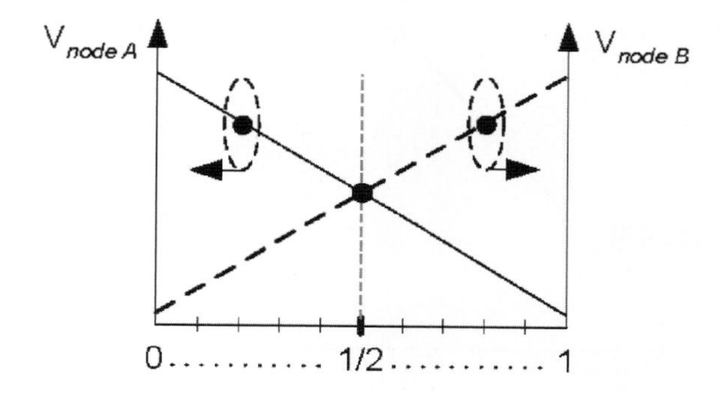

Figure 5-36. Bit and Bit line voltages as a function of R.

Process variations and transistor mismatches impose a limit on how many SRAM cells can be integrated together with a reasonable yield. For example, for a 32Mb cache SRAM, limiting the design to a single bad cell will require the design to be robust over 5σ parametric process variation. Needless to say with increasing process spread, it is becoming extremely difficult to optimize an SRAM cell over process robustness range. In addition, such a robust cell will require additional power and area which is not acceptable. Therefore, redundancy schemes and test strategies must evolve to find out not only weak cells, but also the degree of weakness. In this scenario, a single threshold approach such as WWTM though successful in the past will not be enough. Hence, multi-threshold, programmable DFT techniques must be evolved to make sure the product quality does not suffer. With these objectives, Pavlov, Sachdev and Pineda de Gyvez researched on programmable DFT techniques for SRAM cell stability. They identify three circuit techniques for this purpose [60-62]. In one of the techniques, read current ratio technique (RCRT), the concept of programmable pass/fail threshold was implemented using a set of n SRAM cells in a given column. Existing cells in the column or external cells could be utilized for this purpose. Let R be the ratio of cells with logic 0 to the total of n cells. If all cells are turned on together, the pre-charged bit line could be discharged to a voltage proportional to R, as shown in Figure 5-36.

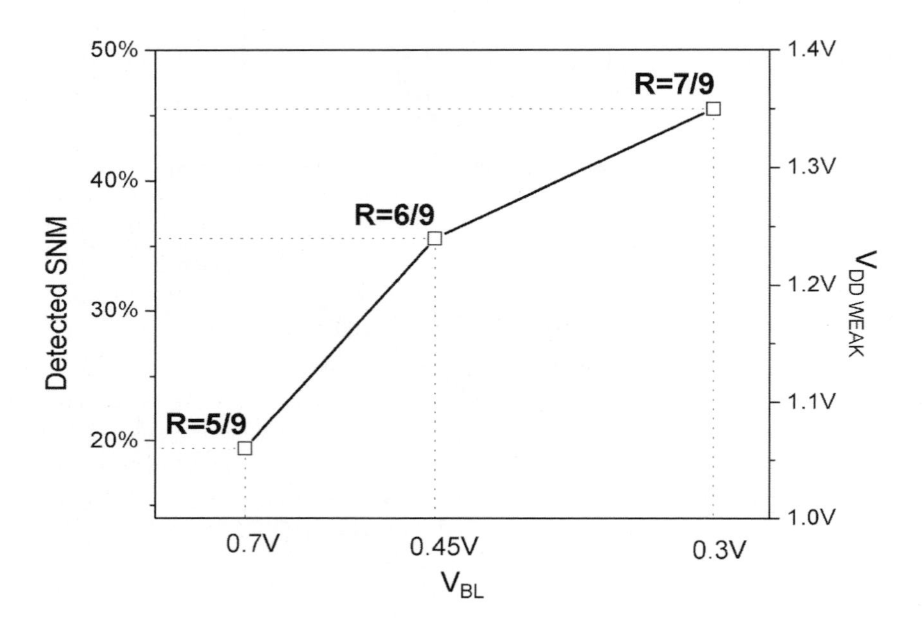

Figure 5-37. Measured detection capability of the programmable RCRT.

The circuit technique was implemented in 180 nm CMOS technology. Extensive simulations and measurements were carried out to examine effectiveness of the technique. The value of n was chosen to be 9 in the test chip. These cells were enabled together to selectively discharge the bitline in the given access time. As R increases, the corresponding bitline voltage will drop. The measured results are depicted in Figure 5-37. The cells under test were weakened by reducing their independent supply voltage (VDD_{weak}). As the value of R is increased from 5/9 to 7/9, the weak cell detection threshold increased from 18% to 46% of the nominal SNM.

6. I_{DDQ} BASED RAM TESTING

RAMs, being an array, are well suited for I_{DDQ} testing. RAMs have a well defined architecture that is suitable for I_{DDQ} based parallel testing. A relatively large number of bits can be tested in parallel without extensive circuit modifications. I_{DDQ} testing is ideally suited for parallel testing because the faulty (or fault-free) information traverses in parallel through the power buses. Therefore, no special observability conditions are needed. This property of I_{DDQ} testing has been extensively utilized to reduce the test costs of digital VLSI.

Nevertheless, a straightforward application of I_{DDQ} test technique to RAMs has a rather limited defect detection capability. Due to the very nature of RAM architecture and its operation, the I_{DDQ} test is not able to detect many of the otherwise I_{DDQ} detectable manufacturing defects. The I_{DDQ} test coverage of manufacturing process defects is enhanced extensively with minor RAM design modifications. Sachdev proposed an I_{DDQ} test mode in RAMs by modifying RAM address decoders, bitline precharging circuitry and the control unit [69]. In this test mode a majority (or all) bits can be tested in parallel. Therefore, number of required I_{DDQ} measurements is reduced drastically, making I_{DDQ}-based RAM testing practical.

In spite of several potential benefits, I_{DDQ} testing in general lost its effectiveness owing to elevated leakage current with scaling. As a consequence, no recent work is reported on I_{DDQ}-based RAM testing. However, recently SRAM hibernating circuit techniques have been reported where the supply voltage is reduced to near threshold voltage to reduced leakage and hence power [6]. Arguably, for such applications I_{DDQ} testing for SRAMs can be carried out.

7. CONCLUSIONS

RAMs enjoy a strategic position in the microelectronics industry and have been a cause of many trade battles. In terms of the volume, memories account for 20% of the semiconductor market [64]. As far as testing is concerned, RAMs suffer from quantitative issues of digital testing along with the qualitative issues of analog testing. Furthermore, RAM test cost and quality issues have become critical, jeopardizing the development of future RAM generations.

Increasing miniaturization has forced RAMs to share the same substrate with CPU or DSP cores. This merger has resulted in dramatic changes, especially in the case of DRAMs which now must be fabricated by a process developed for standard logic. This leads to new challenges in the design and testing of embedded DRAMs. Defect-oriented inductive fault analysis is carried out for an embedded DRAM module. Owing to the high circuit density, standard CMOS VLSI process implementation, and the dynamic nature of the operation, embedded DRAMs exhibit susceptibility to catastrophic as well as to non-catastrophic defects. The probability of occurrence of non-catastrophic defects is significantly high. Most of the non-catastrophic defects degrade the circuit performance and/or cause coupling faults in the memory.

The coupling faults in DRAMs need special attention. Previous definitions (transition based and state based) of coupling faults do not adequately represent their behavior in DRAMs. Such faults can be caused by catastrophic as well as non-catastrophic defects. Furthermore, they are dynamic in nature. A fault model is evolved taking into account the catastrophic and non-catastrophic defects and the dynamic nature of coupling faults. Based upon the fault model, a test algorithm of complexity 8N is developed which completely covers the above mentioned fault model. The algorithm can be easily modified if the R/W logic is sequential in nature. The resultant algorithm has the complexity of 9N. For word oriented DRAMs, different backgrounds can be utilized to cover intra-word faults. However, if the bits constituting a word are not adjacent to each other in layout, then one data background is sufficient.

The effectiveness of these algorithms has been validated with tested DRAM devices. Most device failures (89%) appear to be total chip failures or memory stuck-at faults affecting a large number of bits. Such failures could easily be explained by catastrophic bridging or open defects at various locations in the layout. This result is consistent with the results reported by Dekker et al. [7]. Typically, a large number of bits failed due to such defects and, therefore, such failures were easily detected. The wafer yield was relatively low and hence many total chip failures occurred. A significant segment of failed devices (7%) showed bit stuck-at behavior that could also be explained with catastrophic defects in the cells. These defects caused relatively small number of bits to fail. However, a small number of device failures (4%) could not be explained with catastrophic defect model and could only be explained by foregoing coupling fault model based on non-catastrophic defects.

An important class of address decoder defects are not detected by linear, march test algorithms. Other RAM test algorithms are also not likely to detect these defects owing to the fact that two-pattern test sequence (T1, T2) for all potential defects is not ensured by them. Therefore, special test vectors, or DFT techniques must be implemented to catch these defects. In last decade, several new algorithms and DFT techniques were developed to improve the defect coverage of address decoder defects.

Process variations and defects may give rise to stability faults or data retention faults as they are popularly known in SRAMs. The existing test practices of pause test or single threshold weak test mode will not be adequate for sub 130 nm CMOS SRAMs. Therefore, programmable, multi-threshold weak test methods must be implemented to weed out the marginal cells.

References

1. M.S. Abadir and H.K. Reghbati, "Functional Testing of Semiconductor Random Access Memories," ACM Computing Surveys, vol. 25, no.3, pp. 175-198, September 1983.

2. Z. Al-Ars, and A.J. van de Goor, "Static and Dynamic Behavior of Memory Cell Array Opens and Shorts in Embedded DRAMs" Proceedings of IEEE Asian Test Symposium, 2001, pp. 496-503.

3. M. Azimane, and A.K. Majhi, "New Test Methodology for Resistive Open Defects in Memory Address Decoders," Proceedings of IEEE VLSI Test Symposium, 2004, pp. 25-29.

4. M.A. Breuer and A.D. Friedman, Diagnosis and Reliable Design of Digital Systems, Rockville, MD: Computer Science Press, 1976.

5. E.M.J.G. Bruls, "Reliability Aspects of Defects Analysis," Proceedings of European Test Conference, 1993, pp. 17-26.

6. B. Calhoun, and A. Chandrakasan, "A 246kb Sub-threshold SRAM in 65 nm CMOS, Proceedings of IEEE International Solid State Circuits Conference, pp. 628-629, February 2006.

7. R. Dekker, F. Beenker and L. Thijssen, "Fault Modeling and Test Algorithm Development for Static Random Access Memories," Proceedings of International Test Conference, 1988, pp. 343-352.

8. L. Dilillo, P. Girard, S. Pravosssoudovitch and A. Virazel, S. Borri and M. Hage-Hassan, "Resistive-Open Defects in Embedded SRAM core cells: Analysis and March Test Solution, IEEE Design Automation Conference, 2005, pp. 857-862.

9. B.N. Dostie, A. Silburt and V.K. Agarwal, "Serial Interfacing for Embedded-Memory Testing," IEEE Design & Test of Computers, vol. 7, pp. 52-63, April 1990.

10. F.J. Ferguson, and J.P. Shen, ''Extraction and simulation of realistic CMOS faults using inductive fault analysis,'' Proceedings of International Test Conference, 1988, pp. 475-484.

11. A.V. Ferris-Prabhu, ''Computation of the critical area in semiconductor yield theory,'' Proceedings of the European Conference on Electronic Design Automation, 1984, pp. 171-173.

12. M. Franklin, K.K. Saluja and K. Kinoshita, "Design of a BIST RAM with Row/ Column Pattern Sensitive Fault Detection Capability," Proceedings of International Test Conference, 1989, pp. 327-336.

13. M. Franklin, K.K. Saluja and K. Kinoshita, "Row/Column Pattern Sensitive Fault Detection in RAMs via Built-in Self-Test," Proceedings of Fault Tolerant Computing Symposium, June 1989, pp. 36-43.

14. M. Franklin, K.K. Saluja, and K. Kinoshita, "A Built-In Self-Test Algorithm for Row/Column Pattern Sensitive Faults in RAMs," IEEE Journal of Solid State Circuits, vol. 25, no. 2, pp. 514-523, April 1990.

15. E. Gizdarski," Detection of Delay Faults in Memory Address Decoders," Journal of Electronic Testing: Theory and Applications, Vol. 16, No. 4, pp. 381-387, August 2004.

16. A.J. van de Goor, Testing Semiconductor Memories, Theory and Practice, John Wiley and Sons, 1991.

17. T. Guckert, P. Schani, M. Philips, M. Seeley and H. Herr, ''Design and Process Issues for Elimination of Device Failures Due to 'Drooping' Vias,'' Proceedings of International Symposium for Testing and Failure Analysis (ISTFA), 1991, pp. 443- 451.

18. S. Hamdioui, G. Gaydadjiev, and A.J. van de Goor, "The State of the art and Future Trends in Testing Embedded Memories," Proceedings of IEEE International Workshop on Memory Technology, Design and Testing, 2004, pp. 54-59.

19. J.P. Hayes, "Detection of Pattern-Sensitive Faults in Random Access Memories," IEEE Transactions on Computers, vol. C-24, no.2, pp. 150-157, February 1975.

20. J.P. Hayes, "Testing Memories for Single-Cell Pattern-Sensitive Faults," IEEE Transactions on Computers, vol. C-29, no.3, pp. 249-254, March 1980.

21. C. Hill, "Definitions of Noise Margin in Logic Systems," Mullard Technical Communication, vol. 89, pp. 239-245, February 1967.

22. L.K. Horning, J.M. Soden, R.R. Fritzemeier and C.F. Hawkins, ''Measurements of Quiescent Power Supply Current for CMOS ICs in Production Testing,'' Proceedings of International Test Conference, 1987, pp. 300-309.

23. R-F Huang, Y-F Chou and C-W Wu, "Defect Oriented Fault Analysis for SRAM," Proceedings of IEEE Asian Test Symposium, 2003, pp. 1-6.

24. J. Inoue, T. Matsumura, M. Tanno and J. Yamada, "Parallel Testing Technology for VLSI Memories," Proceedings of International Test Conference, 1987, pp. 1066-1071.

25. M. Inoue, T. Yamada and A. Fujiwara, ''A New Testing Acceleration Chip for Low-Cost Memory Test,'' IEEE Design & Test of computers, vol. 10, pp. 15-19, March 1993.

26. V-K Kim and T. Chen, "On Comparing Functional Fault Coverage and Defect Coverage for Memory Testing," IEEE Transactions on Computer Aided Design of Integrated Circuits and Systems, vol. 18, no. 11, pp. 1676-1683.

27. M. Klaus, and A.J. van de Goor, "Tests for Resistive and Capacitive Defects in Address Decoders, Proceedings of IEEE Design and Test in Europe Conference, 2001, pp. 31-36.

28. J. Knaizuk and C.R.P. Hartman, "An Optimal Algorithm for Testing Stuck-At Faults in Random Access Memories," IEEE Transactions on Computers, vol. C- 26, no. 11, pp. 1141-1144, November 1977.

29. S. Koeppe, ''Optimal layout to avoid CMOS stuck-open faults,'' Proceedings of 24th Design Automation Conference, 1987, pp. 829-835.

30. M. Kumanoya, et al., "A 90ns 1Mb DRAM with Multi-Bit Test Mode," International Solid State Circuits Conference; Digest of Technical Papers, 1985, pp. 240- 241.

31. A. Jee, and F.J. Ferguson, "Carafe: An Inductive Fault Analysis Tool for CMOS VLSI Circuits," Proceedings of IEEE VLSI Test Symposium, 1993, pp. 92-98.

32. A. Jee, J.E. Colburn, V. Swamy Irrinki and M. Puri, "Optimizing Memory Tests by Analyzing Defect Coverage," IEEE International Workshop on Memory Technology, Design, and Test, 2000, pp. 20-25.

33. K.J. Lee and M.A. Breuer, ''Design and Test Rules for CMOS Circuits to Facilitate I_{DDQ} Testing of Bridging Faults,'' IEEE Transactions on Computer-Aided Design, vol. 11, no. 5, pp. 659-669, May 1992.

34. M.E. Levitt and J.A. Abraham, ''Physical Design of Testable VLSI: Techniques and Experiments,'' IEEE Journal of Solid State Circuits, vol. 25, no. 2, pp. 474- 481, April 1990.

35. K. Lin, and C. Wu, "Testing content-addressable memories using functional fault models and march-like algorithms," IEEE Transactions on Computer Aided Design of Integrated Circuits and Systems, pp. 577-588, 2000.

36. J.-F. Li, R.-S. Tzeng, and C.-W. Wu, "Testing and diagnosis methodologies for embedded content addressable memories", Journal of Electronic Testing: Theory and Applications, vol. 19, no. 2, pp. 207-215, Apr. 2003.

37. J.-F. Li, K.-L. Cheng, C.-T. Huang, C.-W. Wu, "March-based RAM diagnosis algorithms for stuck-at and coupling faults," Proc. International Test Conference (ITC), pp. 758-767, 2001.

38. J.-F. Li, "Testing priority address encoder faults of content addressable memories," Proc. International Test Conference (ITC), paper 33.2, pp. 1-10, 2005.

39. K.-J. Lee, C. Kim, S. Kim, U.-R. Cho, and H.-G. Byun, "Modeling and Testing of Faults in TCAMs," Proc. Asian Simulation Conference (AsianSim), pp. 521-528, 2004.

40. J.-F. Li, and C.-K. Lin, "Modeling and Testing Comparison Faults for Ternary Content Addressable Memories," Proc. IEEE VLSI Test Symposium (VTS), pp. 60-65, 2005.

41. J. Lostroh, E. Seevink, and J. de Groot, "Worst-case Static Noise Margin Criteria for Logic Circuits and their Mathematical Equivalence," IEEE Journal of Solid State Circuits, vol. sc-18, pp. 803-807, December 1983.

42. TM Mak, D. Bhattacharya, C. Prunty, B. Roeder, N. Ramadan, J. Ferguson and J. Yu, "Cache RAM Inductive Fault Analysis with Fab Defect Modeling," Proceedings of IEEE International Test Conference, 1998, pp. 862-871.

43. W. Maly, "Realistic Fault Modeling for VLSI Testing," Proceedings of 24th ACM/ IEEE Design Automation Conference, 1987, pp.173-180.

44. W. Maly and M. Patyra, ''Design of ICs Applying Built-in Current Testing,'' Journal of Electronic Testing: Theory and Applications, vol. 3, pp. 397-406, November 1992.

45. Y. Matsuda, et al., "A New Parallel Array Architecture For Parallel Testing in VLSI Memories," Proceedings of International Test Conference, 1989, pp. 322- 326.

46. A. Majhi, M. Azimane, S. Eichenberger and F. Bowen, "Memory Testing Under Different Stress Conditions: An Industrial Evaluation, Proceedings of the Design, Automation and Test in Europe Conference and Exhibition, 2005, pp. 438-443.

47. P. Mazumder, ''Parallel Testing of Parametric Faults in Three Dimensional Random Access Memory,'' IEEE Journal of Solid State Circuits, vol. SC-23, pp. 933- 941, 1988.

48. P. Mazumder and K. Chakraborty, Testing and Testable Design of High-Density Random-Access Memories, Boston: Kluwer Academic Publishers, 1996.

49. H. McAdams, et al., "A 1-Mbit CMOS Dynamic RAM with Design For Test Functions," IEEE Journal of Solid State Circuits, vol. SC-21, pp. 635-641, October 1986.

50. R. Meershoek, B. Verhelst, R. McInerney and L. Thijssen, "Functional and I_{DDQ} Testing on a Static RAM," Proceedings of International Test Conference, 1990, pp. 929-937.

51. A. Meixner and W. Maly, "Fault Modeling for the Testing of Mixed Integrated Circuits," Proceedings of International Test Conference, 1991, pp. 564-572.

52. A. Meixner, J. Banik, "Weak Write Test Mode: An SRAM Cell Stability Design For Test Technique," Proceedings of IEEE International Test Conference, pp. 1043-1052, October 1997.

53. R.R. Montanes. J.P. de Gyvez, and P. Volf, Resistance Characterization for Weak Open Defects," IEEE Design & Test of Computers, vol. 19, no. 5 pp. 18-26, Sept-Oct. 2002.

54. S. Naik, F. Agricola and W. Maly, ''Failure analysis of High Density CMOS SRAMs Using Realistic Defect Modeling and I_{DDQ} Testing,'' IEEE Design & Test of Computers, vol. 10, pp. 13-23, June 1993.

55. R. Nair, "Comments on an Optimal Algorithm for Testing Stuck-at Faults in Random Access Memories," IEEE Transactions on Computers, vol. C-28, no. 3, pp. 258-261, March 1979.

56. R. Nair, S.M. Thatte and J.A. Abraham, "Efficient Algorithms for Testing Semiconductor Random Access Memories," IEEE Transactions on Computers, vol. C- 27, no. 6, pp. 572-576, June 1978.

57. H.D. Oberle and P. Muhmenthaler, "Test Pattern-Development and Evaluation for DRAMs with Fault Simulator RAMSIM," Proceedings of International Test Conference, 1991, pp. 548-555.

58. J. Otterstedt, D. Niggemeyer, and T.W. Williams, "Detection of CMOS Address Decoder Open Faults with March and Pseudo Random Memory Test," Proceedings of IEEE International Test Conference, 1998, pp. 53-62.

59. C.A. Papachristou and N.B. Sahgal, "An Improved Method for Detecting Functional Faults in Semiconductor Random Access Memories," IEEE Transactions on Computers, vol. C-34, no.2, pp. 110-116, February 1985.

60. A Pavlov, M. Sachdev, and J. Pineda de Gyvez, "An SRAM Weak Cell Fault Model and a DFT Technique with Programmable Detection Threshold," Proceedings of IEEE International Test Conference, pp. 1106-1115, October 2004.

61. A Pavlov, M. Azimane, J. Pineda de Gyvez, and M. Sachdev, "Programmable Techniques for Cell Stability Test and Debug in Embedded SRAMs," Proceedings of IEEE Custom Integrated Circuits Conference, pp. 443-446, September 2005.

62. A Pavlov, M. Azimane, J. Pineda de Gyvez, and M. Sachdev, "Wordline Pulsing Technique for Stability Fault Detection in SRAM Cells, Proceedings of IEEE International Test Conference, pp. 33.1-33.10, October 2005.

63. R. Perry, ''I_{DDQ} testing in CMOS digital ASICs,'' Journal of Electronic Testing: Theory and Applications, vol. 3, pp. 317-325, November 1992.

64. B. Prince, Semiconductor Memories, Chichester, UK: John Wiley and Sons, 1991.

65. J. Rabaey, A. Chandrakasan, and B. Nicolic, "Digital Integrated Circuits: A Design Perspective, Second Edition, Prentice Hall, 2003.

66. M.A. Rich and D.E. Gentry, "The Economics of Parallel Testing," Proceedings of International Test Conference, 1983, pp. 728-737.

67. M. Sachdev and M. Verstraelen, ''Development of a Fault Model and Test Algorithms for Embedded DRAMs,'' Proceedings of the International Test Conference, 1993, pp. 815-824.

68. M. Sachdev, ''Transforming Sequential Logic in Digital CMOS ICs for Voltage and I_{DDQ} Testing," Proceedings of European Design and Test Conference, 1994, pp. 361-365.

69. M. Sachdev, ''Reducing the CMOS RAM Test Complexity with I_{DDQ} and Voltage Testing,'' Journal of Electronic Testing: Theory and Applications (JETTA), vol. 6, no. 2, pp. 191-202, April 1995.

70. J. Savir, W.H. McAnney and S.R. Vecchio, "Testing for Coupled Cells in Random Access Memories," Proceedings of International Test Conference, 1989, pp. 439- 451.

71. J. Segura and A. Rubio, "A Detailed Analysis of CMOS SRAMs with Gate Oxide Short Defects," IEEE Journal of Solid State Circuits, vol. 32, no. 10, pp. 1543-1550, October 1997.

72. Semiconductor Industry Association (SIA), "The National Technology Roadmap for Semiconductors," pp. 94-99, 1994.

73. A.H. Shah, et al., "A 4-Mbit DRAM with Trench Transistor Cell," IEEE Journal of Solid State Circuits, vol. SC-21, pp. 618-627, October 1986.

74. J.P. Shen, W. Maly and F.J. Ferguson, "Inductive Fault Analysis of MOS Integrated Circuits," IEEE Design & Test of Computers, vol. 2, no. 6, pp. 13-26, 1985.

75. P. Sidorowicz, "Modeling and testing transistor faults in content-addressable memories," International Workshop on Memory Technology, Design, and Test, pp. 83-90, 1999.

76. J.M. Soden, C.F. Hawkins, R.K. Gulati and W. Mao, ''I_{DDQ} Testing: A Review," Journal of Electronic Testing: Theory and Applications, vol. 3, pp. 291-303, November 1992.

77. T. Sridhar, "A New Parallel Test Approach for Large Memories," Proceedings of International Test Conference, 1985, pp. 462-470.

78. F.A. Steenhof, C.G. van der Sanden and B.C. Pham, "Design Principles of a DRAM Cell Matrix for Embedded Applications," Nat.Lab. internal technical note, TN 250/90.

79. P. Stolk, H. Tuinhout et al., "CMOS Device Optimization for Mixed-Signal Technologies," Proceedings of IEEE International Electron Devices Meeting, pp. 10.2.1-10.2.4, October 2001.

80. S.T. Su and R.Z. Makki, ''Testing of Static Random Access Memories by Monitoring Dynamic Power Supply Current,'' Journal of Electronic Testing: Theory and Applications, vol. 3, pp. 265-278, August 1992.

81. D.S. Suk and S.M. Reddy, "Test Procedure for a Class of Pattern-Sensitive Faults in Random Access Memories," IEEE Transactions on Computers, vol. C-29, no.3, pp. 419-429, June 1980.

82. D.S. Suk and S.M. Reddy, "A March Test for Functional Faults in Semiconductor Random Access Memories," IEEE Transactions on Computers, vol. C-30, no.12, pp. 982-985, December 1981.

83. M. Syrzycki, "Modeling of Spot Defects in MOS Transistors," Proceedings of International Test Conference, 1987, pp. 148-157.

84. E. Takeda, et al., "VLSI Reliability Challenges: From Device Physics to Wafer Scale Systems," Proceedings of IEEE, vol. 81, no. 5, 1993, pp. 653-674.

85. K. Thaller, "A Highly Efficient Transparent Online Memory Test," Proceedings of IEEE International Test Conference, 2001, pp. 230-239.

86. S.M. Thatte and J.A. Abraham, "Testing of Semiconductor Random Access Memories," Proceedings of Fault Tolerant Computing Symposium, 1977, pp. 81-87.

87. H. Walker and S.W. Director, "VLASIC: A Catastrophic Fault Yield Simulator for Integrated Circuits," IEEE Transactions on Computer Aided Design of Integrated Circuits and Systems, vol. 5, no. 4, pp. 541-556, 1986.

88. D. Wright, and M. Sachdev, "Transistor-Level Fault Analysis and Test Algorithm Development for Ternary Dynamic Content Addressable Memories," IEEE International Test Conference, September 2003.

89. H. Yokoyama, H. Tamamoto and Y. Narita, ''A Current Testing for CMOS Static RAMs,'' Proceedings of IEEE International Workshop on Memory Technology, Design and Testing, August 1993, pp. 137-142.

90. Y. You and J.P. Hayes, "A Self Testing Dynamic RAM Chip," IEEE Journal of Solid State Circuits, vol. SC-20, no.1, pp. 428-435, February 1985.

91. K. Zarrineh, A. P. Deo and R. D. Adams, "Defect Analysis and Realistic Fault Model Extensions for Static Random Access Memories," IEEE International Workshop on Memory Technology, Design, and Test, 2000, pp. 119-124.

Chapter 6

DEFECT-ORIENTED ANALOG TESTING

Testing is becoming a substantial barrier to continued RF IC cost reductions because of the additional complexities required by new standards -including multi-band compatibility, higher linearity, lower bit-error rate, and long battery life. BER testing is in fact a preferred functional test method in RF systems. Typical test costs as a percentage of the manufacturing cost are commonly low for most digital products but for RF devices it is projected that this percentage will increase to 50%[2]*. The overall cost of an RF system consists of manufacturing, testing (wafer sort and final testing) and packaging. The traditional test flow for dc wafer testing is mainly digital. It uses cheap testers and prunes away defective devices. Typically, in this flow RF is bypassed due to the high cost of RF testers. Unfortunately, defective devices in the RF path are packaged before they are thrown away resulting in a significant loss if we consider that packaging can represent 30% of the overall cost. Current test practices are expensive, among other reasons, because of the required tester infrastructure, long test times, cumbersome test preparation, lack of appropriate defect and fault models, and lack of standardized test methods. Analog circuits due to their non-binary operation are influenced by process defects in a different manner compared to digital circuits. Seemingly an innocuous defect for digital logic may cause unacceptable degradation in analog circuit performance. This chapter surveys the advances in the field of defect-oriented analog testing and summarizes strengths and weaknesses of the method for analog circuits.*

[2] ITRS projections

1. INTRODUCTION

In the previous chapters we demonstrated the application of defect-oriented test techniques on solving digital and quasi-digital (RAM) test problems with reasonable success. In this chapter we apply the same methodology to analog circuits. However, analog test complexity is different from that of digital circuits. The emergence of mixed signal ICs further complicates the test issues. In general, analog testing poses challenges still to be surmounted by researchers. Several reasons are attributed to the inherent analog test complexity [7,9,41,56] and a number of solutions have been suggested [4,7-9,12-15,24-27,30,31,38,40,41,49,51-58]. However, in spite of these attempts and proposed solutions, almost all analog circuits are presently tested in a functional manner.

For any test strategy to succeed in terms of test quality and global applicability, it should have a sound basis. For example, poor performance of the stuck-at model based digital test schemes amply demonstrate how without a firm basis, test strategies can fail to deliver quality products [21,43]. Therefore, we set the following objectives:

- To propose an analog test methodology based on a firm foundation. The proposed test strategy is based on manufacturing process defects that provide an objective basis for analog fault model development and test generation.
- To assess the effectiveness of the methodology from two standpoints: (a) contribution of inductive fault analysis (IFA) towards testing silicon devices in the production environment and, (b) contribution of IFA towards robust analog design against process defects, quantifying the fault coverage of analog tests, and examining the practicality of analog DfT schemes.

Analog circuits, due to their non-binary circuit operation, are influenced by defects in a different manner compared to digital circuits. This poses additional challenges for modeling of defects in analog circuits. In fact, the analog fault modeling is identified as a critical factor in the success of any analog DfT scheme [45]. Furthermore, we explore the concept of structural test vectors in analog domain and examine the potential of simple test stimuli in fault detection.

2. ANALOG TEST COMPLEXITY

Considerable effort has been devoted to identify the causes of the analog test complexity [7,9,33,41,56]. These are summarized as follows:

- Unlike digital circuits, analog circuits do not have the binary distinction of pass and fail. The time and voltage continuous nature of their operation makes them further susceptible to defects. Therefore, test procedures are needed to discriminate between various faulty conditions and the non-faulty condition.
- Analog systems are often non-linear and their performance heavily depends on circuit parameters. Process variations within allowable limits can also cause unacceptable performance degradation. Deterministic methods for modeling such variations are often inefficient.
- In digital circuits, the relationship between input and output signals is logical (Boolean) in nature. Many digital DfT schemes simplify this relationship to reduce the test complexity. On the other hand, the input-output relationship in analog circuits is non-Boolean. Such behavior is complex and difficult to model.
- Digital DfT schemes based on structural division of the circuit, when applied in analog domain, are also largely unsuccessful because of their impact on the circuit performance.
- In digital domain, there exist a wide range of well defined and industrially accepted fault models. These models or abstractions form the basis for representing the faulty circuit behavior as well as test pattern generation. In analog domain the effectiveness of these models is questionable. Moreover, in the absence of an acceptable fault model, test generation has been ad-hoc and testing has been largely functional (specification oriented) in nature.
- Since different specifications are tested in different manners, analog functional testing is costly and time consuming. Moreover, often extra hardware is needed to test various specifications.
- Limited functional verification does not ensure that the circuit is defect-free and escaped defects pose quality and reliability problems.

Analog testing also suffers from automatic test equipment (ATE) related issues. For example, when noise level in test environment is not acceptable, complex shielding techniques are required. Furthermore, the integrity of test depends on interface, interconnections, probe card, etc.

3. PREVIOUS WORK

Analog fault modeling and diagnosis received much theoretical attention in the late 1970s and 1980s. Duhamel and Rault presented an excellent review of the topic [7]. These theoretical works relied on the characteristic matrix of the circuit under test for testability and diagnosability. Though those methods had a broad scope, their application to specific circuits has not been successful. The analog fault detection and classification can broadly be divided into following categories:

3.1 Estimation Method

This method can further be subdivided into an analytical (or deterministic) method and a probabilistic method. In the former, the actual values of the parameters of the device are determined analytically or based on the estimation criteria (physical or mathematical). The least square criterion approach represents this class. Typically, in this approach a factor of merit, s_i , is associated with each parameter as:

$$s_i = \sum_{j=1} \left(g_i - \gamma_j (X_i)\right)^2 \tag{6.1}$$

where g_j is the measured value of the characteristic γ_i, and x_i is a vector x1, ..., xn of parameters which have their nominal values, except for xi. The factor of merit associated with xi is taken as the minimum value of s_i. The most likely faulty parameter is the one that, given all other parameters are at their nominal value, minimizes the difference between nominal and measured characteristics.

In probabilistic methods the values are inferred from the tolerance of the parameters. For example, inverse probability method is the representative of this class. Elias [8] applied statistical simulation techniques to select parameters to be tested. On this basis, he also formulated the test limits.

3.2 Topological Method

This method is also known as simulation-after-test (SAT) method. The topology of the circuit is known and SAT method essentially reverse engineers a circuit to determine the values of the circuit component parameters. A set of voltage measurements is taken and then numerical analyses determine parameter values [4,12,26,30,31,40,51,56]. SAT methods are very efficient for soft-fault diagnosis because soft faults are based on a linearized network model. However, this method is computation intensive and for large circuits the algorithms can be inefficient.

One of the first theoretical studies of the analog circuit fault-analysis problem was initiated by Berkowitz [4]. He mathematically defined the concept of network-element-value solvability and studied the measurement conditions required to solve the problem. Trick et al. [50] and Navid and Willson Jr. [26] proposed necessary and sufficient conditions for network solvability problem. Trick et al. used only voltage and single frequency sinusoidal input to determine the parameter value for linear passive circuits. Navid and Willson Jr. suggested that for small signal analysis, non-linear active elements, like transistors and diodes, can be linearized around their operating points. They proposed an algorithm covering the solvability problem for linear passive network as well as active devices. Rapisarda and Decarlo [31] proposed the tableau approach for analog fault diagnosis instead of transfer function oriented algorithms. They argued that tableau approach with multi-frequency excitation would provide simpler diagnostic solution. Salama et al. [40] proposed that large analog circuits can be broken into smaller uncoupled networks by nodal decomposition method. These subnetworks can be tested independently or in parallel. Every subnetwork is associated with a logical variable σ, which takes the value 1 if the subnetwork is good and 0 if it is faulty. Furthermore, every test is associated with a logical test function (LTF) that is equal to the complete product of variables σ_{ji}. If the network passes the test, then

$$T_{J_t} \equiv \sigma_{j1} \cap \sigma_{j2} \ldots \cap \sigma_{jk} \tag{6.2}$$

where

$$J_t \equiv j_1, j_2, \ldots j_k \tag{6.3}$$

j_i refers to network s_i, k is the number of subnetworks involved in the test.

Hemink et al. [12] postulated that the solvability of the matrix depends on the determination accuracies of the parameter. The set of equations describing the relations between parameters and measurements can be ill-conditioned due to `almost' inseparable parameters. Further, they contended, that solving such a set of equations inevitably leads to large computation errors. They overcome this problem by an improved algorithm that finds the sets of separable high-level parameters and computes the determination accuracy of the parameters. Recently Walker et al. [56] developed a two-stage SAT fault-diagnosis techniques based on bias modulation. The first stage, which diagnoses and isolates faulty network nodes, resembles the node fault location method. The second stage, a subnetwork branch diagnosis, extracts faulty network parameters. The branch diagnosis is achieved by element modulation, a technique that varies the value of the element externally as a modulated element. The diagnostic technique requires a single test frequency and the ability to control the network bias from external source.

3.3 Taxonomical Method

This method is based upon a fault dictionary. This is also known as simulation-before-test (SBT) method [7,9,13,15,24,27,41]. The fault dictionary is a collection of potential faulty and the fault-free responses. During the actual test the measured value is compared with the stored response in the dictionary. A fault is detected if at least for a set of measurements the actual response differs from the fault-free response by predetermined criteria. The accuracy of the method depends on the accuracy of the fault dictionary [7].

The fault-free and faulty circuit responses are measured at certain key points. The number of test points depends on the diagnosis resolution and test stimuli. Schemes based on this method can be segregated according to the input stimuli and fault dictionary construction. For example, this method can be implemented with DC signals [13,24], or various time-domain signals [49] or AC signals [27]. The DC fault dictionary approach is simple but it can not detect purely capacitive or inductive defects. Such defects often give rise to parametric or soft faults that are more readily detected by the transient or AC dictionary approach. Slamani et al. [41] made a combined dictionary for DC, transient and AC input stimuli to predict the defective component. They claimed that this method could detect wide ranging defects, from tolerance deviation to catastrophic faults. Sachdev [33] made a similar fault dictionary from the catastrophic processing defect information using inductive fault analysis (IFA) [40].

4. DEFECT BASED REALISTIC FAULT DICTIONARY

Application of a defect-oriented approach [40] in solving analog test problems has gained popularity in the recent past [22,33-35,44,45]. It is proposed as an alternative to analog functional testing. However, this proposal is not without controversy. What makes this topic so controversial? The critics of IFA based analog testing are quick to remind that the test issues of analog circuits are more qualitative than quantitative. It is not uncommon to come across an analog circuit having signal to noise ratio (SNR) of 100 dB or operation frequency of few hundred MHz or input-offset voltage less than 20 mV. Secondly, analog circuits often exploit a number of circuit and device level parameters (e.g., transistor matching, layout considerations, transistor sizing, etc.) to achieve the maximum possible performance. Unfortunately, such clever techniques render the circuit vulnerable to several factors since the maximum possible circuit performance is achievable only under the optimal fabrication and operating conditions. Thirdly, in the case of analog circuits, the range of optimal conditions is substantially narrower than that of their digital counterparts. For example, in digital circuits, typically the critical path is the most sensitive for performance (parametric) degradation. While in analog circuits, the parametric requirement is much higher and widely distributed over the circuit layout. Therefore, any sub-optimal performance of one or more parameters may have significant impact on the performance. A good test program should test for all such sub-optimal performances. Finally, one may ask how comprehensive and accurate is the yield loss model based upon defects alone in the case of analog circuits? Since these are formidable concerns, according to critics, the functional (specification) testing is the only alternative to ensure the circuit performance, specifications and quality.

On the other hand, those who have faith in IFA based analog testing will argue that IFA based testing combines the circuit topology and process defect data to arrive at the realistic fault data that is specific to the circuit. This information can be exploited by test professionals to generate effective and economic tests. The same information can be used by analog circuit designers to design robust and defect tolerant circuits. Secondly, this is a structured and globally applicable test methodology that substantially reduces the test generation cost. Finally, they cite numerous examples of digital domain where IFA based tests contributed significantly to test simplification and test quality improvement [40]. We address the assessment of analog IFA from two standpoints: (i) contribution of IFA towards testing silicon devices in the production environment, and (ii) contributions of IFA towards robust analog design against process defects in quantifying the fault coverage of analog tests, and in examining the practicality of analog DfT schemes.

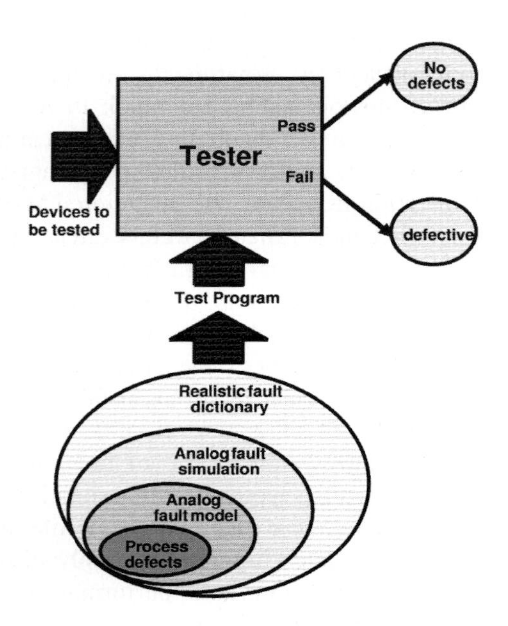

Figure 6-1. A realistic defect based testability methodology for analog circuits.

In the classical sense, the defect based fault dictionary can be categorized as a SBT approach. All forms of fault simulation are carried out before the test. Figure 6-1 illustrates basic concepts of the defect based fault dictionary. The manufacturing process defects, catastrophic as well as non-catastrophic, form the core of the methodology. Realistic defects are sprinkled over the circuit to determine the realistic fault classes. These faults are simulated with given test vectors. The fault simulation is carried out in a Spice-like simulator to achieve accurate results. Alternatively, if fault simulation at circuit level is not possible owing to the circuit complexity, a high level model of the circuit may be used. The results of the fault simulation are compiled into a fault dictionary. A fault is considered detected if the faulty response differs from the nominal response by a predetermined criterion. Next, a test program is prepared taking the fault dictionary into account. The effectiveness of the fault dictionary depends on several factors. Foremost is the defect population and relative probabilities of occurrence of various defects. It is not possible to carry out an exhaustive fault simulation with all permutations of defect (fault) impedances. Therefore, the effectiveness of the dictionary depends how representative are the faults of actual defects and how accurate is the simulator. Finally, the dictionary effectiveness also depends on pass/fail criterion. Nevertheless, the defect based fault dictionary forms the basis for a structured analog testing.

Mathematically, we can define this concept as follows: let F be the fault matrix of all faults in a given CUT and let F0, the first element of the matrix, be the fault-free element. Moreover, let S be the matrix of stimuli applied at CUT inputs and let D be the matrix of the fault-free and faulty responses (i.e., the fault dictionary). Furthermore, let us assume that in a given circuit, there are n faults, then the size of the fault matrix taking into account the fault-free element as well, is $(n+1).1$. The fault matrix, F , can be written as follows:

$$\mathbf{F} = \begin{bmatrix} F_0 \\ F_1 \\ F_2 \\ \circ \\ \circ \\ F_n \end{bmatrix} \tag{6.4}$$

For the formulation of the stimuli matrix, let us assume that CUT has m inputs. Therefore, any arbitrary test vector S_i consists of $s_{i1}, s_{i2}, \ldots, s_{i3}$. In order to simplify the analysis we assume that for any S_i all constituents put together excite the faulty CUT in a particular way. Therefore, the constituents of Si can be treated as scalars. This is not an unreasonable assumption since in analog circuits, unlike digital circuits, the circuit function depends on the continuous operation of its components and stimuli. The analysis will hold even in the absence of this assumption, however, it would require rigorous mathematics. Furthermore, in spite of this assumption, one has total freedom to select the constituents of a given (S_i) stimuli. Hence, the stimuli matrix can be formulated as:

$$\mathbf{S} = \lfloor S_1 S_2 \ldots S \;\ldots S_t \rfloor \tag{6.5}$$

Where t is total number of inputs. The fault dictionary D is a function of the fault matrix as well as the stimuli matrix.

$$\mathbf{D} = f(\mathbf{F} \times \mathbf{S}) \tag{6.6}$$

For each fault detection mechanism such as voltages on different outputs or dynamic current, formulation of different matrices will be required. Alternatively, like the stimuli matrix, different detection mechanisms can be treated as scalar fields of each dij. Elements of the matrix **D** are given as follows:

$$d_{ij} = f_{ij}(F_i \times S_j) \tag{6.7}$$

where $0 \leq i \leq n$ and $1 \leq j \leq t$

We simulate the CUT to find out all d_{ij} of the fault dictionary. It is possible to compute these elements when function f_{ij} is known. The first row of **D** gives the fault-free responses. The size of **D** is $(n+1).t$.

4.1 Implementation

The implementation issues of the fault dictionary are segregated as follows: (i) related to defects and fault modeling, and (ii) related to the analysis flow. The former is concerned with collecting of defect data for a given fab, modeling of defects for a given fault simulator, etc. The latter is concerned with establishing an analysis flow, determination of pass/fail criterion, etc.

Defects and their impact on the device performance have been studied in detail in the literature [11,18-21,42]. Broadly speaking, causes of IC functional failures can be separated into global and local process disturbances. Global disturbances are primarily caused by defects generated during the manufacturing process. The impact of these global (or manufacturing process related) defects covers a relatively wider chip area. Hence, they are detected before functional (or structural) testing by using simple test-structure measurements or supply current tests. A vast majority of faults that have to be detected during functional (or structural) testing are caused by local defects, popularly known as spot defects [18]. Since the global defects are relatively easy to detect by other measurements, we use spot defects for fault modeling purposes. In a typical single poly double metal CMOS process, commonly found spot defects are:

- Short between wires
- Open in a wire
- Pin hole; oxide, gate oxide, *pn*-junction
- Extra contact or via
- Missing contact or via

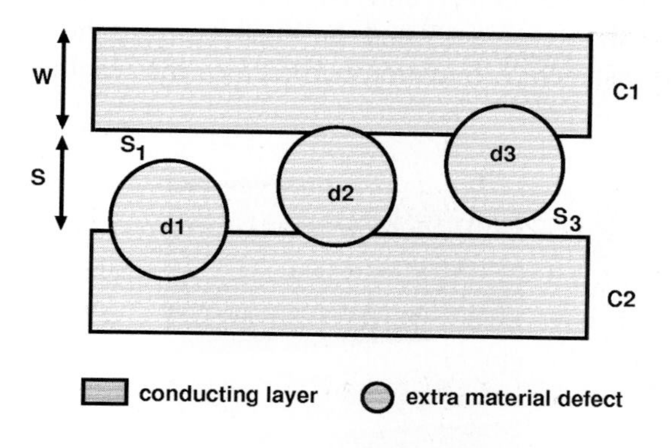

Figure 6-2. Catastrophic and non-catastrophic shorts.

Spot defects can also be categorized into two classes: (i) Defects causing a complete short or open in the circuit connectivity. These are often referred as catastrophic defects. (ii) Defects causing an incomplete short or open in the circuit connectivity. These are often called non-catastrophic or soft defects. Figure 6-2 shows catastrophic and non-catastrophic defects caused due to spot defects between two conducting layers C1 and C2. Defect d2 causes a catastrophic short (low resistance bridge) between both conductors. Therefore, the defect modifies the circuit characteristics and performance drastically. However, defects d1 and d3 do not cause complete shorts but reduce the spacing to S_1 and S_3 , respectively. Reduced spacing causes high impedance bridging defects between conductors that can be modeled as a parallel combination of a resistance R and a capacitance C. The values of R and C for d1 are given by following equations:

$$R = \frac{\rho_{SiO_2} \times S_1}{A} \tag{6.8}$$

$$C = \frac{\varepsilon_{SiO_2} \times A}{S_1} \tag{6.9}$$

In these equations, ρ_{SiO_2} is the resistivity and ε_{SiO_2} is the permittivity of the insulator between the conductors C1 and C2. The S_1 is the reduced spacing between the conductors which were otherwise a distance S apart.

The area between the defect and conductor is represented by A. The resistance of the short is directly and the capacitance of the short is inversely proportional to distance *S*. Equation 6.10 shows the resultant impedance of such a short.

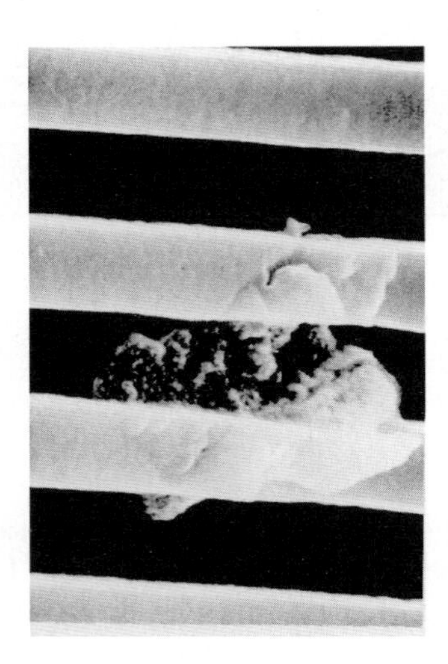

Figure 6-3. Photograph showing a high resistive short in the metallization layer.

$$Z_{short} = \frac{R}{1 + j2\pi \times fRC} \approx R \qquad (6.10)$$

As can be concluded from (6.10), the impedance of the short is a function of the spacing S and also depends inversely on frequency and phase relationship of the two conductors. At low frequencies, the model of such defects is mainly resistive. However, above certain transition frequency (f_T) it becomes primarily reactive. The transition frequency, f_T , depends on the defect geometry, spacing S, and the resistivity and the permittivity of the insulating layer. A particular soft defect may have very little impact on low frequencies but at high frequency it may be significant. Figure 6-3 shows a photograph of a non-catastrophic short in metallization layer. The extra material defect reduces the distance between two metal conductors giving rise to a high impedance bridging fault. However, for most applications and technologies the impedance of the short can be approximated as purely resistive.

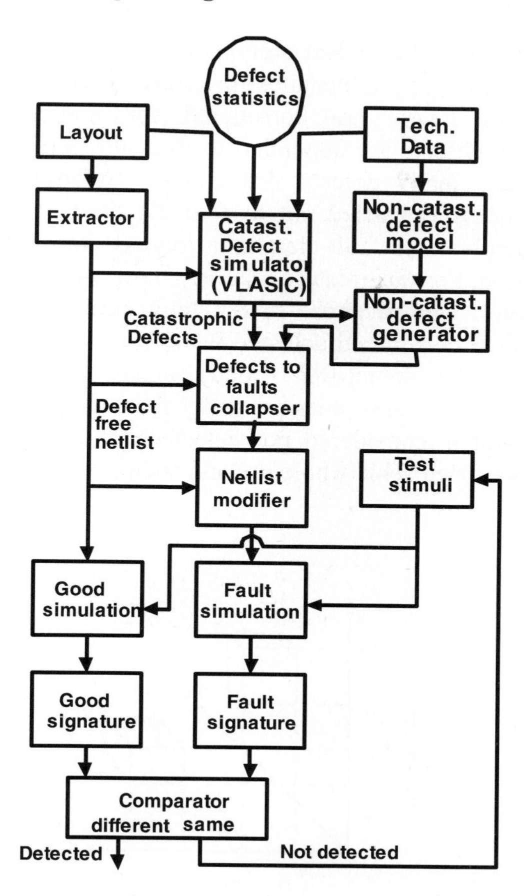

Figure 6-4. A block diagram of the realistic defect based testability methodology for analog circuits.

The block diagram of the environment is shown in Figure 6-4. The process technology data, defect statistics and the layout of the circuit under investigation are simulation inputs. The defect statistics block contains the process defect density distributions. For example, probability of shorts in metalization is significantly higher than that for open defects in diffusion. A catastrophic defect simulator, like VLASIC [57] determines the realistic fault classes specific to the circuit and layout. VLASIC mimics the sprinkling of defects onto the layout in a manner similar to a mature, well-controlled production environment. The output of the simulator is a catastrophic defect-list. Analog circuits are also susceptible to non-catastrophic or parametric defects. Such defects are often called near-misses. We assume that such defects can occur at all places where catastrophic defects are reported by the defect simulator. However, pinhole defects are inherently parametric (high impedance) in nature. Therefore, only shorts and opens in various layers are considered for non-catastrophic defect generation. These defects are appended to the catastrophic defect list. The defect-list contains many defects that can be collapsed in unique fault classes. This process is carried out to find the likely fault classes in the layout. Subsequently, each fault class is introduced into a defect-free netlist for fault simulation. For the greatest accuracy, fault simulation is based upon a circuit simulator. The response of the fault simulator is called a fault signature. A fault is considered detected if the corresponding fault signature is different from the defect-free (good) signature by a predetermined threshold. If a faulty response does not differ from the good signature by the threshold, the fault is considered not detected by the stimulus and hence another stimulus is tried. This whole process is carried out for all faults.

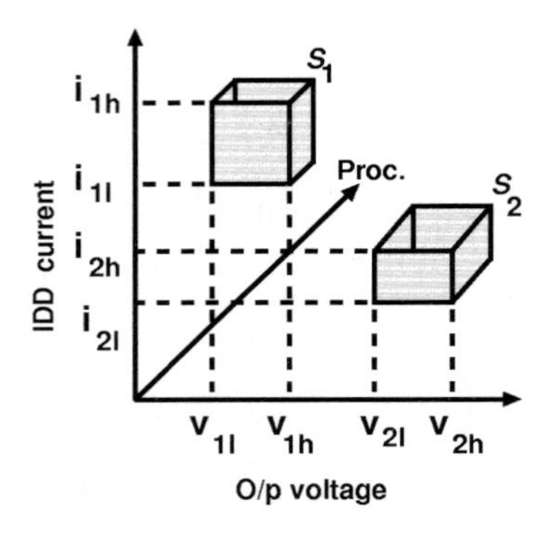

Figure 6-5. The good signature spread.

A few things are worth highlighting in the above mentioned analog fault simulation methodology. First, unlike digital circuits, analog circuits lack the binary distinction of pass and fail. In fact, the decision of pass or fail in analog circuits is not clear cut. It depends on several variables including input stimulus, output measurement parameters (output voltage, IDD current, etc.), circuit performance specifications and permitted environmental conditions (e.g., supply voltage, process, temperature, etc.). In other words, there is no absolute reference for fault detection. A reference has to be evolved for a given circuit under given conditions. This generation of a reference is a tedious exercise and it should be created for each set of input stimuli. The impact of faults is measured against these set of references. Therefore, a reference response or good signature is a multi-dimensional space and the faulty circuit must exhibit a response outside this space to be recognized as faulty, at least by one of the test stimuli.

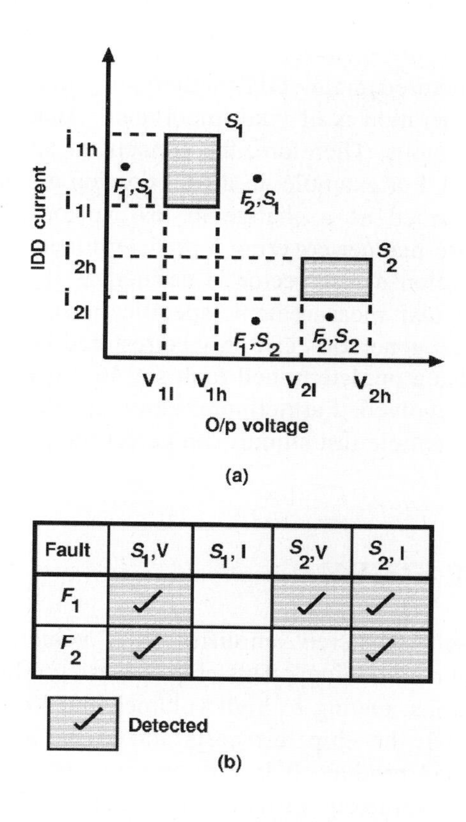

Fault	S_1,V	S_1, I	S_2,V	S_2, I
F_1	✓		✓	✓
F_2	✓			✓

Figure 6-6. (a) *The fault detection, and (b) construction of a fault dictionary.*

The graph in Figure 6-5 illustrates this concept. In this graph, two axes form the primary output measurement parameters and the third axis forms an environmental condition (e.g., fabrication process spread). A set of graphs can be plotted essentially showing a possible good signature spread. The good signature spread (shaded area) is generated for each of the given test vector. A fault is considered detected by a given test vector if the faulty output response of the circuit lies outside the good signature space. For example in Figure 6-6(a), the fault F_1 is detected by test vector S_1 with the output voltage measurement. However, it is not detected by the IDD measurement since the faulty current lies within the good current spread. On the other hand, same fault is detected by test vector S_2 with output voltage as well as IDD measurements. The information about fault detection is compiled into a fault dictionary D. Figure 6-6(b) shows the fault dictionary. Rows of the fault dictionary show different fault classes (i.e., F_1 .. Fn) and columns show stimuli (i.e., S_1 .. S_t) with voltage (V) and current (I) as subfields.

Finally, for a structured analog DfT methodology to succeed, an effective and efficient test generation is of vital importance. Analog signals are time and amplitude continuous. Therefore, the concept of analog test vectors is not very well defined. For example, in digital domain a binary change in the input stimulus is termed as a change in test vector. These vectors are generated in a precise manner covering a predetermined fault set. However, in analog domain, often a test vector is defined as a set of input stimuli required for a particular measurement (specification). The parity between digital and analog test generation can only be restored if the basis for analog test generation is also a predetermined fault set. In this manner, true analog test vectors can be evolved. Furthermore, since all likely fault classes are known, in principle, simple test stimuli can detect the presence (or absence) of a defect.

5. A CASE STUDY

We use a class AB stereo amplifier as a vehicle to examine the effectiveness of this methodology. This chip is mass produced for consumer electronics applications. Owing to high volumes and low selling cost, it is desirable to cut down the chip test costs and at the same time maintain quality of the shipped product. It is a three stage amplifier. The first and second stages are completely differential in nature. The outputs of the second stage feed the output stage which drives a load of 32 ohms. It was designed in a standard 1.0 micron single poly double metal CMOS process. The chip contains two identical amplifiers (channels A and B) and a common biasing circuit. Since both the channels of the class AB stereo amplifier are identical, only one amplifier is considered for testability analysis.

5.1 Fault Matrix Generation

VLASIC was utilized to introduce 10,000 defects into the layout. Since most of the defects are too small to cause catastrophic defects, only 493 catastrophic defects were found. These defects were further collapsed into

60 unique fault classes. Table 6-1 shows the relevant information about various fault classes due to catastrophic defects. A catastrophic short in metal layers was modeled as a resistor with nominal value of 0.2 ohm. Similarly shorts in poly and diffusion layers were modeled with a resistor of 20 and 60 ohms, respectively. Extra contact and via were modeled with a resistor of 2 ohms. Thick oxide defects were modeled as a resistor of 2k ohms. The gate poly is doped n-type and all gate oxide shorts occurred in n-channel transistors causing shorts between the gate and the source or drain of transistors. Therefore, such shorts were non-rectifying in nature and hence were modeled as a 2k resistor. The n-channel transistor is more susceptible to gate oxide shorts and most of the gate oxide shorts are likely to occur between the gate and source or drain [43].

Table 6-1 Catastrophic fault classes and their fault models.

Defect	Number	%	Model (Ohm)
Shorts	22	37	0.2,20,60
Extra contact	10	17	2
Oxide pinhole	15	25	2k
Gox. pinhole	7	11	2k
Junc. pinhole	6	10	2k
Open	0	0	--
Total	60	100	

As mentioned before, soft faults were evolved from the hard fault data. Soft faults were generated at locations of shorts and opens in interconnect, contacts and vias. Therefore, 32 soft fault classes (first 2 rows of Table 6-1) were evolved. Rodriguez-Montanes [32] reported that the majority of bridging defects are below 500 ohms. Therefore, the resistance of non-catastrophic defects was chosen as 500 ohms. The capacitance was calculated from the technology data keeping the spacing between the defect and the conductor (s) as 0.1 micron. The computed value is 0.001 pF.

All catastrophic defects in the defect-list generated by VLASIC were shorts in nature caused by extra material, oxide pin-holes or extra contacts. None of the defects caused an open circuit. However, given the defect densities for various defects in the fabrication process, it was hardly surprising. The shorts in the back-end of the process constitute the majority of the spot defects. Furthermore, the layout geometries in analog circuits are often non-minimum size and multiple contacts and via contacts are utilized to reduce the contact resistance. All this put together made occurrence of an open in the given layout less probable. However in real life, the nature of above mentioned defects can vary a great deal and hence no simulation can claim to be exhaustive. Nevertheless, these numbers are consistent with the resistivity of respective layers and the published data. Furthermore, for such an analysis, the order of defect resistance is more important than the absolute value.

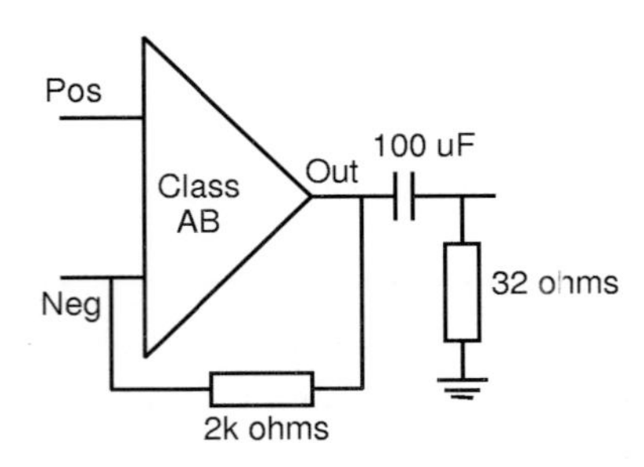

Figure 6-7. Fault simulation configuration for the Class AB Amplifier Results.

5.2 Stimuli Matrix

For this case study, we divided test signals in three categories: (i) DC stimuli, (ii) 1 kHz sinusoid stimuli and (iii) AC stimuli. Often the analog circuit function depends on the continuous operation of all sub-blocks. Therefore, it is quite likely that a catastrophic fault would change the DC operating point of the circuit and hence will be detected by a DC test stimuli. This may also hold true for some high impedance non-catastrophic faults as well. A lower frequency sinusoid was chosen since many fault classes may not be excited under DC conditions. Finally, AC stimuli were chosen because the impact of many non-catastrophic faults is frequency dependent and it is worthwhile to analyze the frequency response as well.

For the simulation of fault classes, the amplifier is put into the configuration shown in Figure 6-7. A load of 32 ohms is placed on the output with a DC blocking capacitor. A 2k ohms feedback resistor is placed between the output and the negative input. Furthermore, a full load is put at the output. Before proceeding to fault simulation, the defect-free response is compiled. A fault is considered detected if defect-free and faulty responses differ by at least 1 volt for output voltage measurement or by 0.5 mA for supply current measurement. For AC analysis a fault is considered detected if it modifies the frequency response by 3 dB. For the DC analysis the positive input is held at 2.5 volts and the negative input is swept from 0 to 5 volts. The output voltage and current drawn from VDD are sampled when negative input is 1,2, and 3 volts, respectively. Similarly, for low frequency transient analysis, a 1 kHz sinusoidal signal is applied and root mean square (rms) values of output voltage and IDD are calculated. For AC analysis different frequency signals on the negative input are applied while the positive input is held at 2.5 volts. In this configuration, the gain of the amplifier is measured.

5.3 Simulation Results

Figure 6-8(a) shows the result of fault simulation for catastrophic faults. In this figure results are independently segregated according to the mode of the analysis. On the X-axis, the mode of analysis means the type of input excitation. The detection mechanisms are represented by output voltage, supply current and gain of the amplifier. The Y-axis shows the percentage of faults detected by each mode of analysis and detection mechanisms independently. For example, DC voltage detection of a fault means that the particular fault was detected by output voltage measurement when input excitation was DC. The third column in DC analysis shows how many faults were detected either by DC voltage or by DC current. Therefore, it is the logical OR of first two columns of the same analysis. Similarly, the third column in transient analysis is the logical OR of first two columns of transient analysis.

In DC analysis, 68% of the faults were detected by output voltage. However, the current is a better measure for fault detection and all faults were detected by it. Needless to say that when both detection mechanisms, voltage and current, are considered together, all faults were detected. Though the voltage detection of faults in transient analysis is higher than that of DC analysis, the results, in general, were less encouraging. The 72% faults modified the voltage signature of the device and 83% faults modified the current drawn from V_{DD}. When both mechanisms were considered together, 90% of the faults are detected. Lower than expected performance of transient analysis compared to DC analysis can be attributed to the fact that transient defect-free current or voltage signature is sinusoidal and comparison of two sinusoids on tester (or in simulation) is more difficult than the comparison of two DC currents. Therefore, in transient analysis, fault detection is carried out manually. If a fault modified the response more than the determined threshold, it is considered detected. In AC analysis, 90% of the faults modified the frequency response of the circuit and hence are detected.

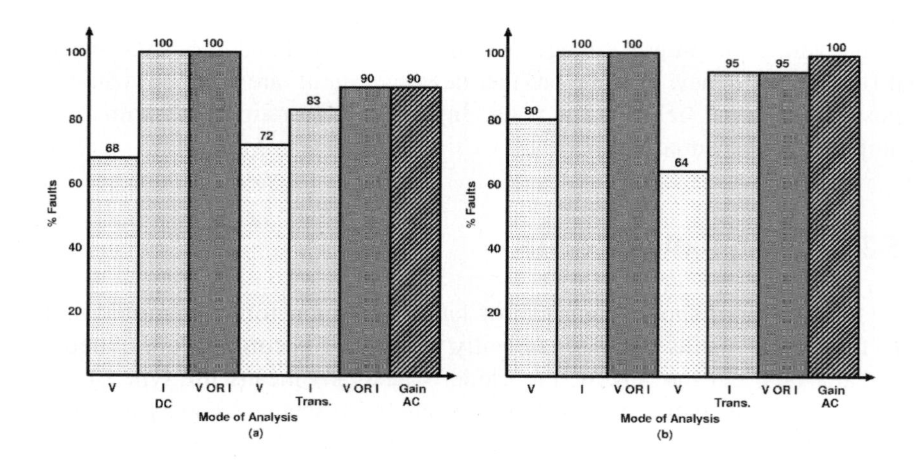

Figure 6-8. Fault simulation results for (a) catastrophic defects and (b) non-catastrophic defects].

Figure 6-8(b) shows the fault simulation results for non-catastrophic faults. The effectiveness of current in DC analysis for fault detection is once again demonstrated. However, for such defects, gain of the amplifier is also an important detection mechanism. All faults were detected by both analyses. Given the model of these non-catastrophic faults (500 ohms) it was expected.

5.4 Silicon Results

Conventionally, devices are tested by verifying a set of DC and AC specifications. The DC specifications include input offset voltage, input bias current, common mode voltage range, output voltage swing, output impedance, output current, etc. The AC specifications include total harmonic distortion (THD), signal to noise ratio (SNR), slew rate, output power, etc.

Table 6-2. Good signature spread for manufactured chips.

	DC	Trans.	AC
Voltage	0.15 (1) V	0.1 (1) V	2 (3) dB
Current	9 (0.5) mA	5 (0.5) mA	—

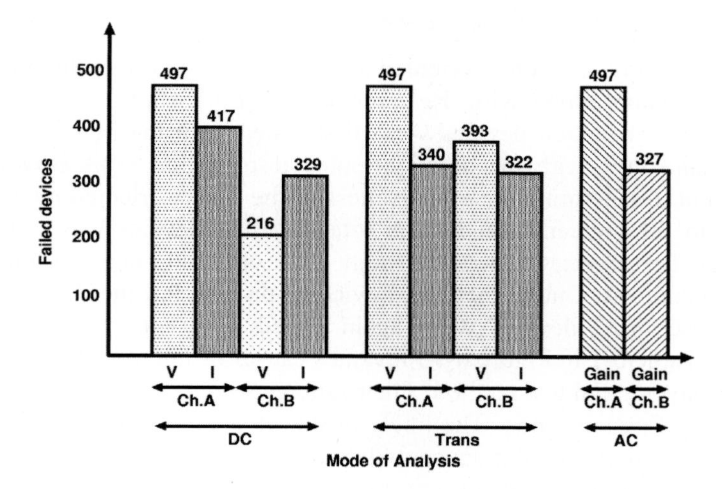

Figure 6-9. First silicon results for Class AB amplifier.

A set of 18 passed devices and 497 failed devices with the conventional test process are selected. The passed devices are selected to observe the spread of good signature compared to simulated thresholds. The comparison is shown in Table 6-2. The values shown outside brackets represent actual (measured) and inside brackets represent simulated thresholds for pass/fail. The actual voltage spread is much smaller than the simulated threshold, however, actual current spreads are at least an order of magnitude larger than the simulated threshold. One of the explanations for high current spread is that for this experiment the device is excited in a different manner compared to its normal usage. Therefore, current spread in this configuration was not controlled.

The performance of the defect-oriented tests on failed devices is shown in Figure 6-9. For channel A, DC and transient voltage as well as the gain measurements caught all faulty devices. The performance of current measurement was less satisfactory. This difference between simulated and silicon results was due to high current spread in defect-free silicon signature. The current measurement with DC was more effective than that with transients. On the other hand, for channel B, DC voltage caught fewer defects compared to DC current measurements. However, transient voltage was more effective than the transient current measurement. The gain measurement of channel B detected 327 faulty devices. In general, channel A failed more often. No failure analysis was carried out to determine the causes of the difference between channels. Probably subtle layout differences between channels are the reason for different failure rates.

Subsequently, the defect-oriented test method was put into the production test environment along with the conventional test method. A test program was evolved in which devices were first tested with the conventional test method and then with the defect-oriented test method. This exercise was carried out to determine the effectiveness of the defect-oriented method with respect to the conventional test. A total of 1894 devices was tested. The yield of the device was very high and only 11 devices failed the conventional test. Out of these 11 devices 3 did not fail the defect-oriented test. These three devices were again tested with the conventional test method. Table 6-2 in Figure 6-9 illustrates causes of failures of these devices by the conventional test method. Their failures were marginal and very close to the specification limits. The likely origin of such failures lies in inherent process variation and not in spot defects. A test methodology with spot defects as the basis can not ensure detection of such faults. Improved control of process is one possible solution to reduce such escapes.

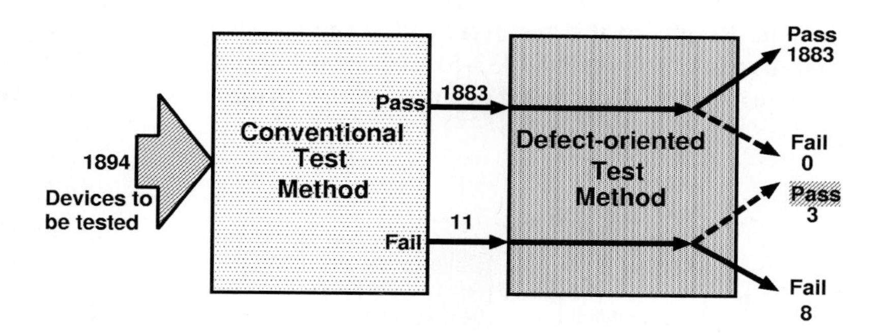

Figure 6-10. The results of the second experiment over the Class AB amplifier.

The difference in simulation and actual results is quite apparent. Several factors contribute to differences: (i) A finite sample size of defects does not cover whole spectrum of possible defects. (ii) A defect may have many possible impedances. A simulation for all observed defects and impedances is beyond the capabilities of any state-of-the-art simulator. (iii) A circuit simulator has limited capability in simulating actual silicon behavior. (iv) As mentioned before, higher silicon current spread limits the fault detection capability of current measurement method. Therefore, DC as well as transient voltage measurements appear to be more effective for fault detection in devices. (v) Finally, the defect-oriented test methodology is based upon spot defects and global or systematic defects and process variations are not taken into account. Such non-modeled faults are also possible due to differences between simulation and the actual test results.

From the second test experiment [34,35] two broad conclusions were drawn: (i) simple tests can detect catastrophic failures, however, detection of some subtle failures is uncertain, and (ii) the number of failed devices is not sufficient to draw any meaningful conclusion about the method's applicability in catching real life faults. More test data, especially on faulty devices, is needed to substantiate claims of IFA based tests. We report a relatively large experiment over the same Class AB amplifier devices with the objective to find the effectiveness of the test method on catching real life failures [36].

Figure 6-11 illustrates this experiment. A total 3270 rejected samples of Class AB amplifier were gathered from a total of 106,784 devices tested by the conventional test method. Only failed devices (3270) were considered for further testing. These devices were tested with the IFA based test method. Out of this lot, 433 devices passed the test. These passed devices from the IFA based test method were once again tested with conventional test method. Results of this test were following: (i) 51 devices passed, the test, and (ii) rest of the devices (433-51=382, or 0.4% of total tested devices) failed the test again. These failed devices (382) were subjected to a detailed analysis.

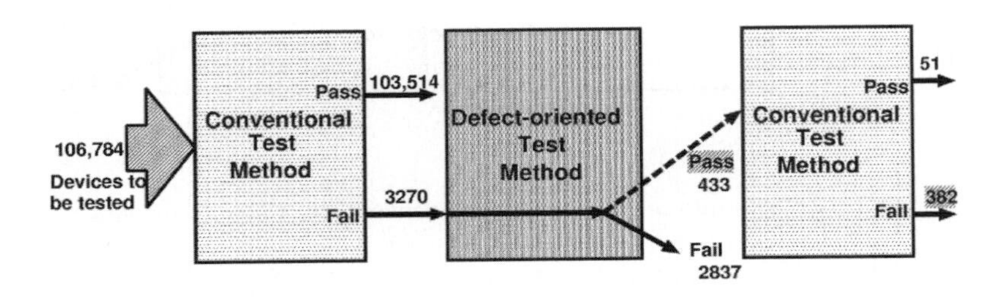

Figure 6-11. The results of the third experiment on the Class AB amplifier.

5.5 Observations and Analysis

Table 6-3 shows the result of the analysis of 382 failed devices. The input offset voltage specification contributes to the largest number of failures (182 or 47.6%) that could not be caught by the IFA based test method. The total harmonic distortion (THD) specification contributed to the second largest segment of undetected failures (123 or 32.2%). Similarly, SNR measurement failed 20 devices (5.2%). These three categories of failures contribute to the bulk (85%) of the failures that could not be detected by the IFA based test method. These failures can be attributed to un-modeled faults in the IFA process. For example, a differential amplifier has an inherent offset voltage that is the source of non-linearity in its operation. Often this offset voltage is minimized by transistor matching, layout, trimming and compensation techniques. Besides process defects, several other factors can increase the offset voltage. The increased offset voltage (within specification limits) increases non-linearity, reducing the SNR ratio of the amplifier. In the table of Figure 6-10 the device with the highest input offset voltage shows the lowest SNR and the device with the lowest input offset voltage shows the highest SNR.

In general, tighter the parametric specifications of an analog circuit, less effective IFA based test is likely to be. This is because natural process variations with higher parametric requirements will contribute to larger number of device failures. Since these are un-modeled faults, the effectiveness of IFA based test is lowered. Furthermore, the effectiveness of a process defect based yield loss model diminishes significantly with increasing parametric requirements. Therefore, a test based solely on process defects is not sufficient for ensuring the specifications of the device with high parametric specifications.

Table 6-3. Analysis of devices failed in the third silicon experiment.

Number	Percentage	Failure mechanism
2	0.5	Open/Short
41	10.7	Supply Current
182	47.6	Offset Voltage
1	0.3	Output Voltage swing DC
10	2.6	Common Voltage
0	0.0	Output Voltage AC
20	5.2	S/N ratio
123	32.2	THD at 1 kHz
0	0	X-talk
3	0.8	Ripple rejection
382	**100**	

5.6 IFA: Strengths and Weaknesses

On the basis of the above experiments, we make the following comments regarding strengths and weaknesses of IFA for analog circuits. Some of these comments are specific to the Class AB amplifier and others have general applicability.

- An IFA based test method is based upon process defects. This is in contrast with the conventional, specification based, analog test method. The IFA based test method is structured and, therefore, has a potential for quicker test generation. Though IFA based test generation requires considerable effort and resources, it is faster than the specification based test generation.

- The IFA based tests are simpler and their requirements for test-infrastructure are substantially lower compared to the specification based tests. Therefore, majority of such tests can be carried out with inexpensive testers. A vast majority of faults is detected by simple,

DC, Transient and AC measurements. For example, the Class AB amplifier devices are tested with a combination of IFA based test and limited functional test. The combined test method results in an estimated saving of 30%.

- The number of escapes (382) of the IFA based method amounts to 0.358% of tested devices (or 3,580 PPM). Clearly, it is unacceptably high. A limited specification test, as mentioned above, with IFA based test may be advantageous in quality improvement while test economy is maintained.
- The number of escapes can be reduced by a rigorous control of the fabrication process. The basis of IFA is a given set of process defects. However, this basis is not absolutely fixed because of the process dynamism. A new defect type may be introduced into the set if the process is unstable or improperly monitored. A better process control (higher Cp and Cpk) will increase the effectiveness of the IFA based test.
- Effective test generation and limit setting is of crucial importance to the success of IFA based testing. For example, when supply current measurements were implemented in IFA tests, a substantial amount of devices (41) passed the test but failed the supply current test in the conventional test method. This is because the test limits in IFA based test are determined more or less arbitrarily. The measured current on 187 good samples suggests that test limits should be more stringent. The same holds true for other detection-thresholds. Setting of 1 V or 3 dB thresholds for fault detection is not stringent enough to ensure high parametric fault requirements for the amplifier. More research is needed for test pattern generation and threshold settings.
- Design insensitivity to process variations also contributes toward the effectiveness of the IFA based test vectors. IFA based tests are ill-suited for design characterization.
- The Class AB amplifier is an audio amplifier with very tight parametric specification and relatively small number of transistors. IFA based test methods are more successful for circuits or ICs where the parametric specification are relatively relaxed and functional complexity is high. For such complex ICs, functional testing is not enough and IFA test may form the main segment of testing. On the other hand, for high performance analog ICs the IFA based simple test may form the basis of wafer sort, rejecting all potentially defective devices. The subsequent limited functional test will be applied only to potentially good devices. The combination of these two will not only improve the

economics of testing but will also result in better quality of tested devices.

- Quantifying fault coverage of a given set of test vectors for an analog circuit is an unexplored area. The IFA based test generation provides a methodology by which test vectors and design can be fault graded. Once, fault coverages of different tests are known, ordering of tests may improve test economics. Tests that do not contribute to fault detection may be discarded. Furthermore, the impact of test vectors on outgoing quality can also be quantified.
- IFA based test method is limited by the availability of CAD software tools and requires high computer resources in terms of CPU power and data storage. A substantial analysis effort is needed before an IFA based test method can be implemented. Furthermore, due to computational and CAD tool related constraints only cells and macros can be analyzed. Therefore, ideally this analysis should be carried out in the design environment on a cell by cell basis. A bigger design should be partitioned into suitable smaller segments for analysis.

6. INPUT STIMULI GENERATION

The difference between functional testing and structural testing is that the ATPG in structural testing is derived from the circuit implementation rather than from the circuit specification. Given that typically the transistor count of analog circuits is not large, structural testing can benefit from inductive fault analysis techniques (and alike), e.g. layout inspection tools to map defects to faults. In this way the ATPG is targeted to a set of realistic faults. Additionally, it is possible to derive figures of merit such as defect and fault coverage to measure the ATPG effectiveness.

Structural testing focuses on the development of dc and transient testing of analog circuits. In transient testing, the circuit under test is excited with a transient test stimulus and the circuit response is sampled at specified times to detect the presence of a fault. The transient waveform can be formed from piecewise linear segments that excite the circuit in such way that the sensitivity of the fault to the specific stimulus is magnified. These waveforms can have a periodic shape, or even arbitrary shapes or as recently proposed they can have a binary shape with distinct duty cycles [52,53]. It is also possible to structurally test the circuit by testing its dc conditions, e.g. by inspecting quiescent currents.

A dc-based analog test pattern generator can be derived by appropriately sweeping the power supply and then observing the corresponding behavior of the supply current or any other observable node. This power supply ramp method is a technique developed in the early 90s that showed potential for detecting defects [47]. In this section we explore further this technique and evaluate its use as a structural test capable of substituting or complementing conventional RF tests. In particular, we carry out an extensive study benchmarking functional and structural fault coverage. The underlying testing approach relies on a defect-oriented test analysis that takes into account the spread of the process as well as the presence of resistive (not only catastrophic) defects. We evaluated the power supply ramp technique on a standalone amplifier. In the next sections the reader will find a description of the amplifier, test methodology, fault coverage and measurement results, and conclusions about the work.

6.1 Power Supply Ramp Input Test Stimuli

The utilized technique is illustrated in Figure 6-12. It basically consists of ramping up the power supply in discrete steps such that a current signature is generated. When the power supply is discretely ramped up, all transistors in the circuit pass through several regions of operation, e.g. subthreshold (region A), linear (region B), and saturation (region C). The advantage of a transition from region to region is that defects can be detected with distinct accuracy in each of the operating regions, e.g. a bridge drains a distinct current depending upon whether the transistors are saturated or in the linear region. This method provides, thus, multiple observation points. It is simple; it is not a functional test and can easily be implemented in any tester. Observe that the typical current signature of this analog circuit follows a *tanh* function. It is expected that a defective device will present an abnormal current signature, e.g. a signature that deviates from the golden *tanh* form. Hence, a simple pass/fail test procedure can be put in place by comparing signatures. If fault diagnosis is desired, a fault dictionary database can be built and then the tested current signatures can be matched against signatures in this database [47].

The first stage of a receiver is typically an amplifier whose main function is to provide enough gain to overcome the noise of subsequent stages (such as a mixer). Besides providing this gain while adding as little noise as possible, an amplifier accommodates large signals without distortion, and frequently must also present a specific impedance, such as 50Ω or 100Ω, to the input source. This last consideration is particularly important if a passive filter precedes the amplifier, since the transfer characteristics of all filters are quite sensitive to the quality of the termination.

The above mentioned methodology will be illustrated by means of an example in the remaining sections.

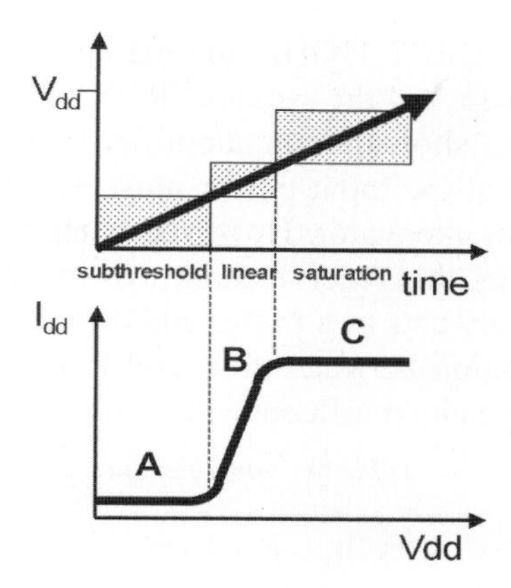

Figure 6-12. Power supply ramp with corresponding quiescent current signature: the amplifier.

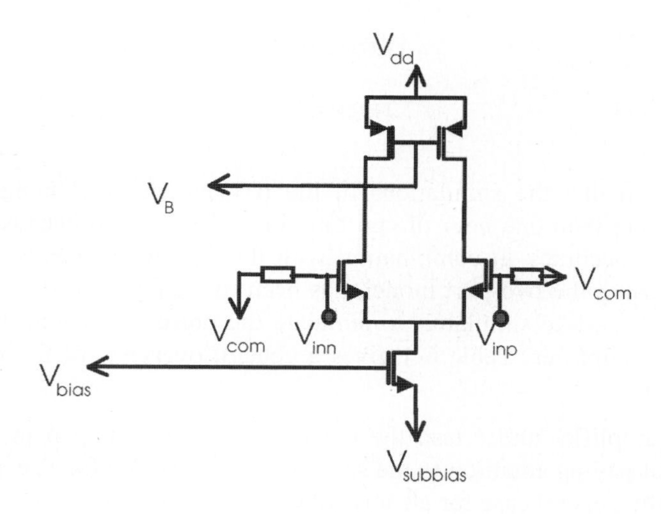

Figure 6-13. Low Noise Amplifier (LNA).

6.2 Amplifier Specs

The schematics of the amplifier are shown in Figure 6-13. The amplifier uses a power supply voltage V_{DD}=1.8V, and a common voltage V_{com}= 0.8V (common voltage at the input and output).

The input frequency f_0 is 2.45GHZ and the amplifier's input and output impedances are matched to the source R_S and load R_L impedances, respectively. Simulations show that the amplifier is linear for –70dBm < P_{RF} < 0dBm. This means that the input power must be 10^{-10}W<P_W<10^{-3}W. The amplifier seems to be an attenuator. However, when the amplifier is used in a complete receive chain, this receive chain will indeed have a gain. In this case, this amplifier merely acts as a buffer and driver towards the mixer. The amplifier provides isolation between the I and Q paths, to prevent that the two mixers in the receive chain influence each other.

Table 6-4. Amplifier specs.

Parameter	$R_S = R_L$ =50 Ω	$R_S = R_L$ =50 Ω
Voltage Gain	-29.586dB	-23.595dB
Noise figure	18.443dB	15.512dB
1dB compression	-11.666dBm	-14.414dBm
IP3	2.249dBm	-0.722dBm

We noticed that the simulations of the IP3 and the 1dB compression point take more than one hour of cpu time in an HPPA7300 because of the required high accuracy in combination with the sweep statements. An AC simulation, using the two-port model, has been run using PSTAR (a Philips proprietary Spice-like simulator) to simulate the noise factor and the noise figure of the amplifier. Table 6-4 gives a general overview of the specs of this amplifier.

For the amplifier under test, the power supply was ramped in discrete steps without paying attention to the speed of the ramp. We foresee, that this will not be the general case for all sorts of circuits, particularly for circuits in which dynamic responses are quite important. Figure 6-14 shows the amplifier's simulated current signature against the power supply voltage sweep and under statistical process variations. The statistical process variations account for inter and intra die variations. Since structural testing assumes that the circuit is properly designed and that it can sustain process variations, one could safely assume a tolerance band for the current signature.

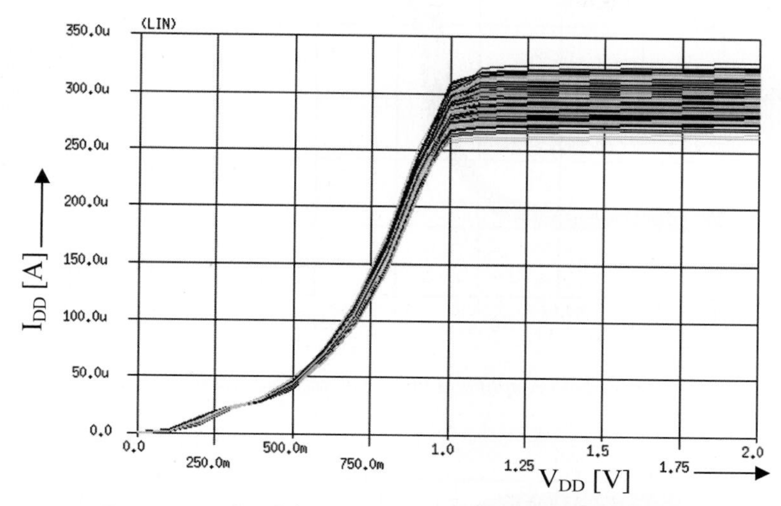

Figure 6-14. Amplifier current signatures with process spread.

In other words, any tested signature that falls within this band implies that the circuit is operating correctly. This simple assumption allows us to take into account the effect of process shifts. Assuming that we have a functional correct design, parametric faults could be identified through current signatures that are slightly out of the tolerance band. For this particular design we can see that for voltages greater than 1V there is a wide I_{DDQ} band. Figure 6-15 and Figure 6-16 show the functional behavior of the amplifier as a function of the power supply ramp. In both plots we see that the amplifier presents a typical behavior for $V_{DD} > 1V$. Observe that process shifts have a negligible effect on both the voltage gain and the noise figure. The voltage gain was simulated at 2.45GHz.

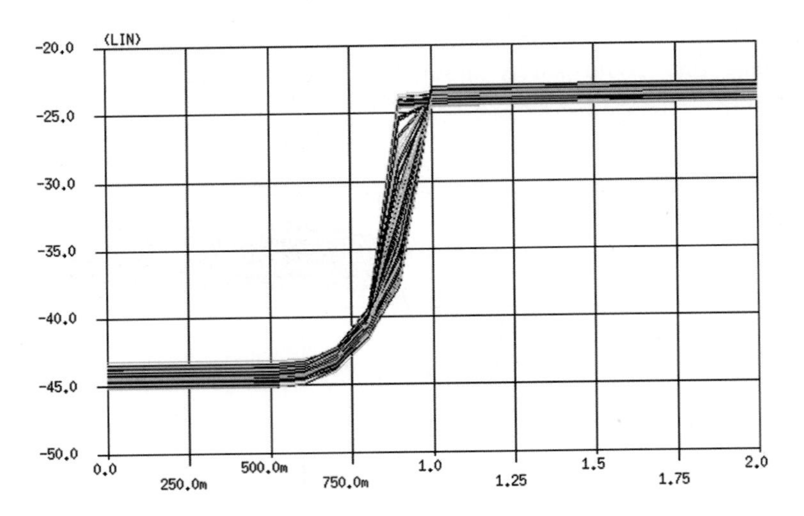

Figure 6-15 Amplifier *gain vs. power supply ramp.*

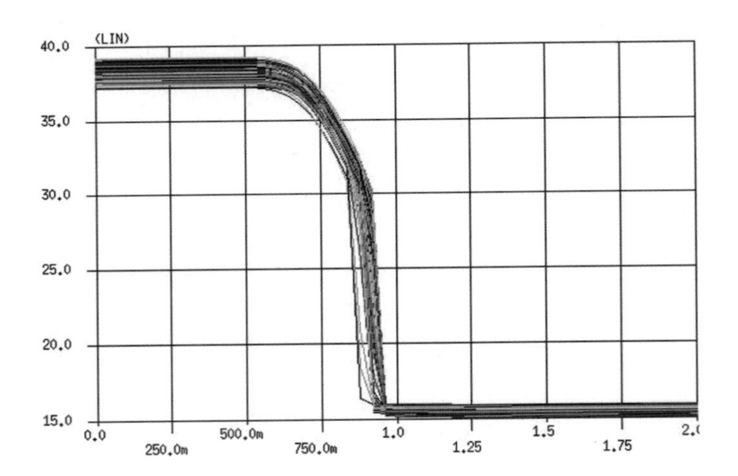

Figure 6-16 Amplifier noise figure vs. power supply ramp.

Figure 6-17 shows a correlation of quiescent current against voltage gain for various power supply values. This plot is useful for identifying I_{DD} limits such that the corresponding voltage gain is within specs. Thus, the power supply method can also be used to capture parametric faults that could be outside the limits or close to the edges using this correlation figure.

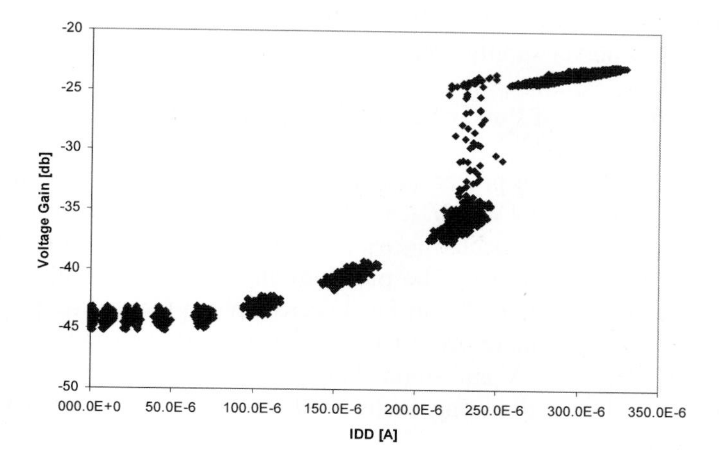

Figure 6-17. Correlation of I_{DDQ} against voltage gain for various power supply voltages.

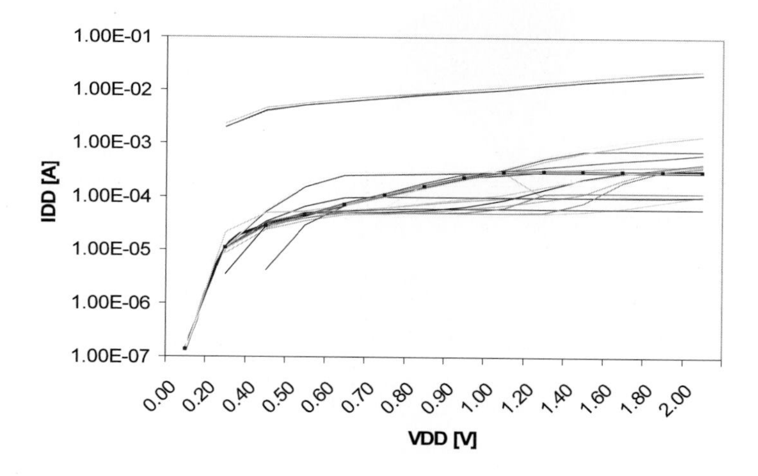

Figure 6-18. Current signatures for the first 50 bridges.

A total of 147 bridges were detected from this inductive fault analysis. Figure 6-18 shows the current signatures for the first 50 ones. Actually, this figure illustrates the potential of the method. One can see signatures that are completely different from the golden one. One can also see some signatures whose quiescent values are comparable to the golden's for $V_{DD} > 1V$, but that have entirely different values for $V_{DD} < 1V$. Conventional techniques that focus on one observation point only, say at nominal V_{DD}, would fail to detect the latter kind of faults. In summary, we have distinct fault observability for each power supply voltage.

6.2.1 Efficiency of Power Supply Ramp Method

Let us consider three power supply intervals, namely 0V < V_{DD} < 0.5V, 0.5V < V_{DD} < 1V, and 1V < V_{DD} < 2V. Let us investigate now how many faults can be uniquely detected in each region. Figure 6-19 shows the fault distribution per V_{DD} interval. The pie chart shows that from all bridges 17 were undetected and that 67 can be detected in any power supply region. Interestingly enough, there are 10 faults that can be detected *only* in the interval 0V < V_{DD} < 0.5V and so on. This proves the usefulness of sweeping the power supply for detecting faults as these faults would not have been detected at nominal V_{DD}.

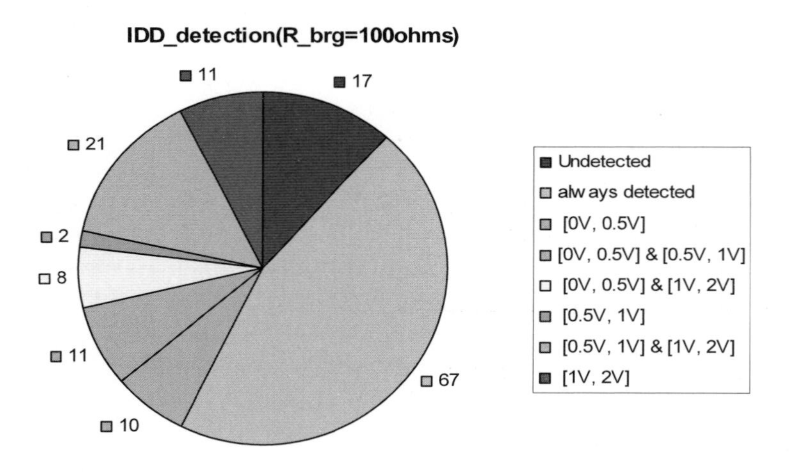

Figure 6-19. Number of detected faults per V_{DD} interval.

How easy is it to detect a given fault at distinct V_{DD} values depends on how much the faulty I_{DDQ} deviates from the golden I_{DDQ} at the preselected V_{DD} value. Figure 6-20 shows this "sensitivity" for the first 50 bridges. The horizontal axis shows the power supply voltage and the vertical axis shows the normalized sum of the absolute current difference between the faulty and golden I_{DDQ} values. This plot shows that it is easier to detect faults at lower power supplies. Of course, this sensitivity results will vary depending upon the circuit under test.

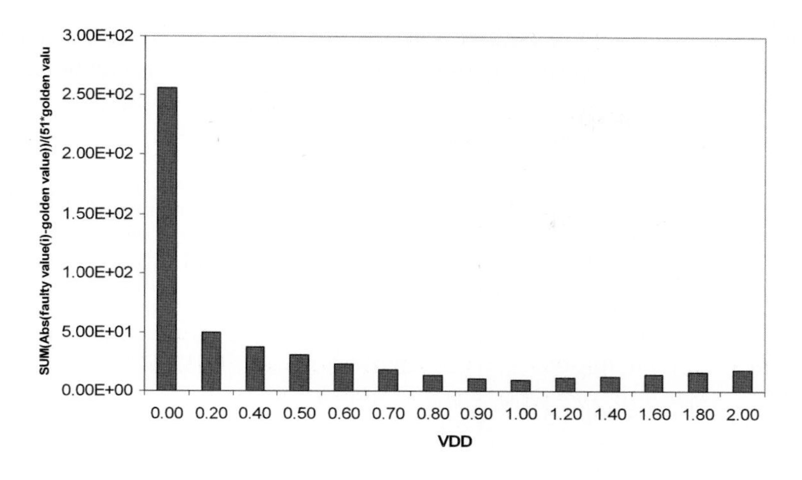

Figure 6-20. Test sensitivity of power supply ramp method.

Additional dc-voltage tests were executed as well. These tests are based on a measure of the dc voltage at the outputs of the amplifier. According to target specs, these nodes have to keep the common output voltage at 0.8V. The injection of a fault (bridge) will thus increase or decrease this value on one or both nodes. These tests are applied when the amplifier is operating as an amplifier. The values that are used for V_{DD} range from 1V to 2V. The number of the tests is equal in this case to: 2·11=22 DC tests. The electrical test limits are set at +/-50mV according to the golden value (0.8V).

6.3 Structural vs. Functional Fault Coverage

Let us now inspect the fault coverage obtained through the power supply ramp method and through evaluation of the amplifier's gain and noise figure. Fault coverage is defined as the ratio of detected faults over the whole set of faults. A weight can be given depending upon whether the faults are ranked or not. The unweighted fault coverage results of the power supply ramp method are shown in Table 6-5 for a resistive bridge of 100Ω. The current flowing at V_{DD} = 0V is because the amplifier biasing current is supplied from an independent source giving the possibility of testing the biasing circuit independent from the core supply of the amplifier. Also, the amplifier's input common mode voltage is independent of the power supply and is kept fixed while stepping the power supply. The fault coverage of the voltage gain and Noise Figure evaluated at V_{DD} = 1.8V are 75% and 83%, respectively. Interestingly enough, the fault coverage of the power supply ramp method is better than when the amplifier's gain and Noise Figure are evaluated. The Noise Figure fault coverage gives better results because each bridge is evaluated as a resistor of 100Ω which introduces additional noise to the amplifier.

Table 6-5: Fault coverage of power supply ramp method.

VDD	Low limit (A)	High limit (A)	Coverage (%)
0	5.51E-08	2.96E-07	40.136
0.2	6.63E-06	1.46E-05	46.259
0.4	2.70E-05	3.09E-05	53.061
0.5	3.90E-05	4.95E-05	52.381
0.6	6.34E-05	7.64E-05	48.980
0.7	9.39E-05	1.17E-04	47.619
0.8	1.42E-04	1.77E-04	51.020
0.9	2.08E-04	2.53E-04	55.782
1.0	2.57E-04	3.10E-04	58.503
1.2	2.60E-04	3.25E-04	68.707
1.4	2.61E-04	3.27E-04	70.748
1.6	2.62E-04	3.28E-04	68.027
1.8	2.62E-04	3.29E-04	68.707
2.0	2.63E-04	3.29E-04	72.109

Notice that for this amplifier, the fault coverage of the power supply ramp method is best when $1.2V < V_{DD} < 2V$. The combined fault coverage of the three methods is about 95%. When the additional dc-voltage test is included we obtain a 100% fault coverage. Fault coverage comparisons are shown in Table 6-5 where

- Total1 = FC(IDD, GAIN, NF) 94.6%
- Total2 = FC(IDD, GAIN, NF, DC-Voltage) 100%
- Total3 = FC(IDD, DC-Voltage) 93.9%
- Total4 = FC(NF, GAIN) 85.7%

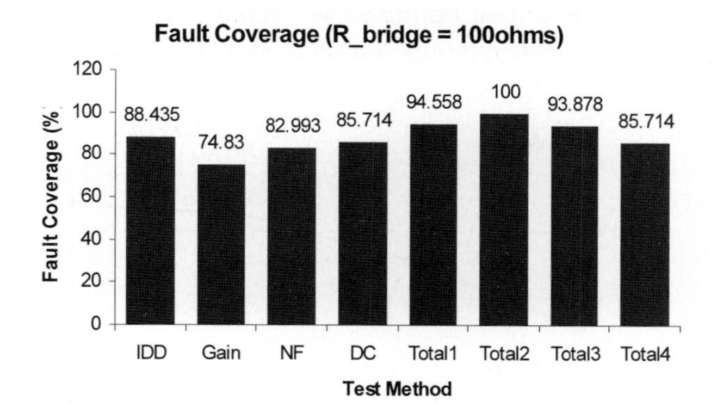

Figure 6-21. Comparison of structural and functional fault coverage for bridges.

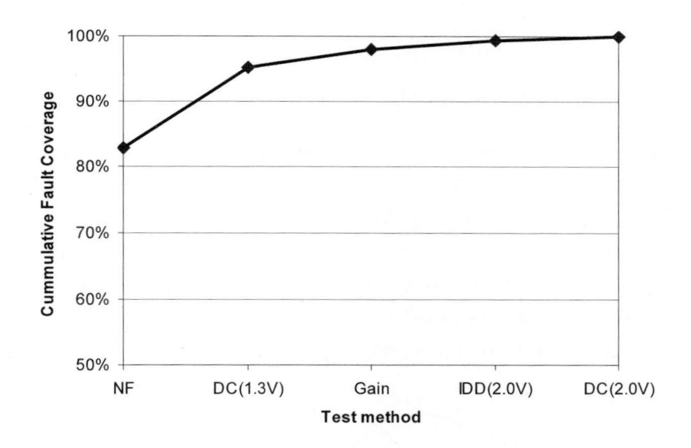

Figure 6-22. Optimized set of functional and structural tests.

Worth noting is that the fault coverage of the power supply ramp is even better than the combined fault coverage of the gain and Noise Figure. Using the DOTSS tool it is possible to carry out a test optimization, .e.g. to choose a combination of tests, say gain followed by Noise Figure followed by the power supply ramp test, etc. The advantage of using this test optimization in the power supply ramp method is that it is possible to constraint the number of current measurements to a couple of V_{DD} points. As we saw from Figure 6-21 it is possible to achieve 100% fault coverage if both functional and structural tests are executed. This can be achieved with only 5 tests as shown in Figure 6-22. Now, when no functional tests are used the fault coverage of this amplifier is 93.9%. Test optimization results show that it is possible to have this fault coverage using only 4 tests as shown in Figure 6-23.

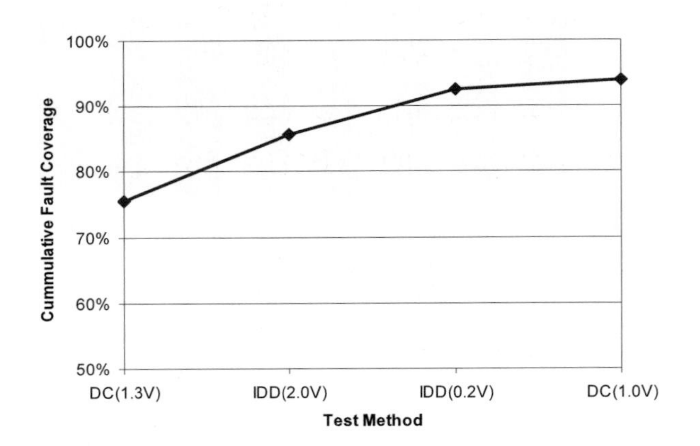

Figure 6-23. Optimized set of only structural tests.

Let us reason a bit more about these results. There are basically two lines of thinking; one is fault coverage and the other is test cost. If a low test cost is desired, then the power supply ramp method along with the dc-voltage test is sufficient since its fault coverage is comparable to the functional fault coverage, see Total3 and Total 4. On the other hand, if a better than functional fault coverage is needed, then the dc tests are good enhancements for the conventional functional tests; see Total 4 and Total 2.

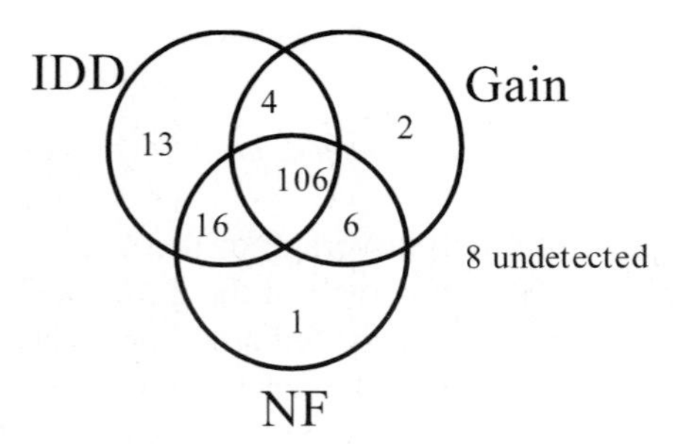

Figure 6-24. Venn diagram of test methods.

The Venn diagram in Figure 6-24 shows more details about the fault coverage of the power supply ramp, Gain, and NF test methods. We notice particularly that

- 8 faults are still undetected by all three tests.
- 13 faults are detected only by the IDD tests.
- 2 faults are detected only by the gain tests.
- 1 fault is detected only by the Noise Figure tests.

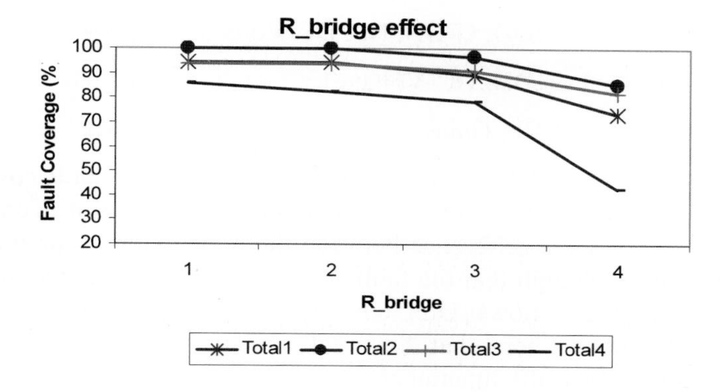

Figure 6-25. Fault coverage for "weak" bridges.

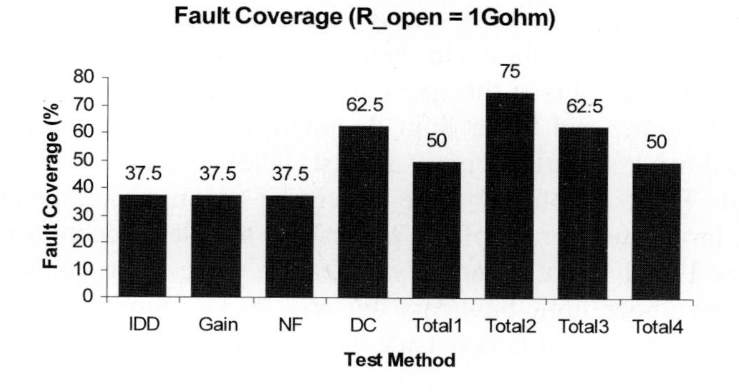

Figure 6-26. Functional vs. structural fault coverage of opens.

The previous simulations have been run on bridges with a resistance of 100Ω. The goal of this analysis is to study also the impact of weak resistive bridges. An increase in bridging resistance reduces the quiescent current, hampering, in a way the sensitivity of the power supply ramp method. The corresponding results are displayed in Figure 6-25. We notice, though, that the fault coverage of the combined dc tests outperforms the functional one for all resistance values.

Figure 6-26 shows fault coverage results for open defects where

- Total1 = FC(IDD, GAIN, NF) 50%
- Total2 = FC(IDD, GAIN, NF, DC-Voltage) 75%
- Total3 = FC(IDD, DC-Voltage) 62.5%
- Total4 = FC(NF, Gain) 50%

As with the case of bridges, we see here too that the fault coverage is improved when both structural and functional tests are used (Total2). For opens we do not see a difference between the power supply ramp and gain methods. We see though that the fault coverage of dc-voltage test method is higher. Figure 6-27 shows fault coverage results when weak opens are assumed. We notice here that for weak defects the power supply fault coverage outperforms the functional.

6.4 Experimental Results

The power supply ramp method has been experimentally verified on the amplifier. In this case, only eleven samples were tested. Their spatial location in the wafer is shown in Figure 6-28. Amplifier3 was not functional at all. Figure 6-29 shows the measured current signatures. The solid lines indicate the upper and lower limit boundaries. Both limit boundaries were determined using statistical circuit analysis where the actual wafer PCM data was used. These amplifiers were processed in an experimental lot with outcome in the fast corner of the process. A simple inspection reveals that amplifiers 1 and 9 are potentially defective since their quiescent current exceeds the upper limit boundary for $V_{DD} > 1V$. Similar for amplifier 5 whose quiescent current falls outside the lower boundary for $V_{DD} < 1V$.

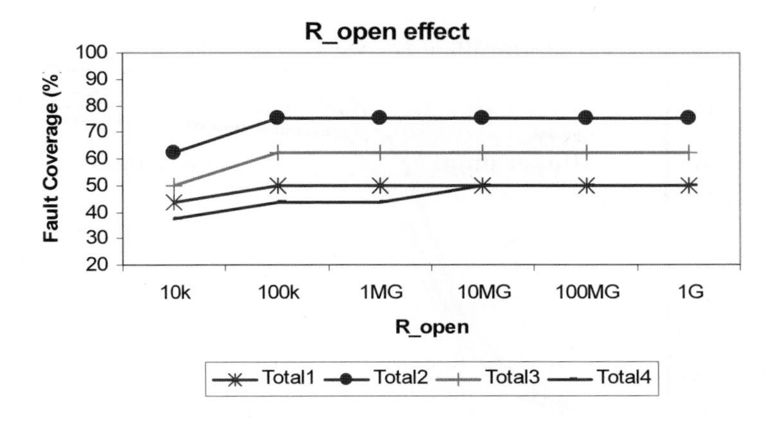

Figure 6-27. Fault coverage for weak opens.

Figure 6-30 shows measurement results of the output voltage. Labels amplifier- and amplifier+ indicate the negative and positive output voltages, respectively. The lines out of the tolerance band correspond to amplifier5 and hence we can conclude that it is defective. S-parameter measurements were done using a network analyser and then converted to voltage gain. Four measurements were performed for each frequency of interest. Figure 6-31 shows the corresponding voltage gain results. From Monte Carlo simulations we found before that the amplifier's voltage gain boundaries are –24.37dB and –22.89dB at 2.45GHz for V_{DD}= 1.8V. Table 6-6 shows the measured voltage gain at nominal V_{DD} and the measured I_{DD} at V_{DD}=1.2V. One can see that amplifiers 1 and 9 exceed IDD_{ref} which is the simulated upper I_{DD} limit of 325uA. Notice also that the voltage gain is slightly above the upper limit of –22.89dB. These amplifiers are defective suspects. In this case we can say that we have a matching suspect between functional and structural tests.

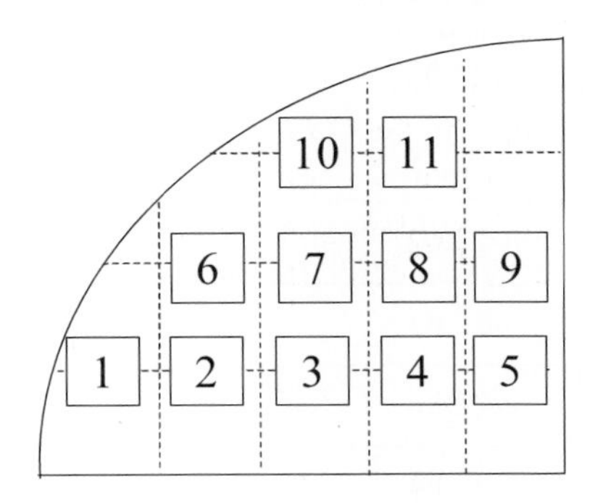

Figure 6-28. Analyzed set of amplifier devices.

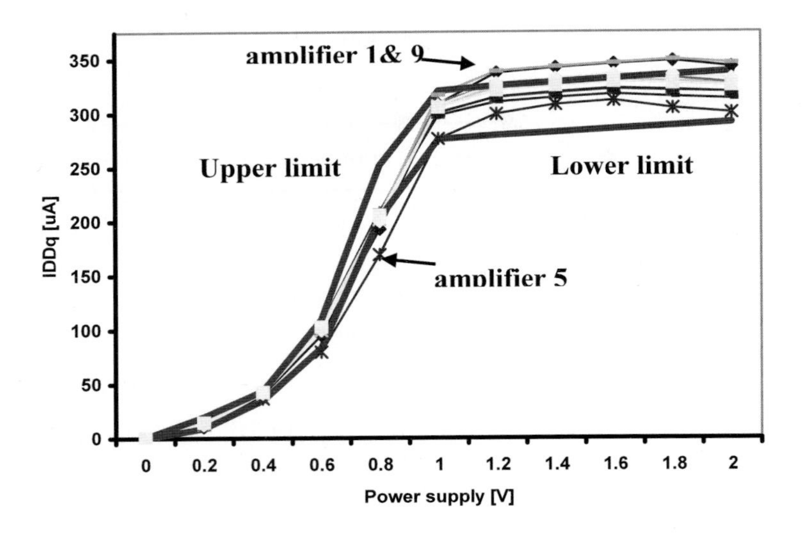

Figure 6-29. Current measurement results.

Table 6-6. Experimental results for voltage gain and quiescent current.

amplifier	Voltage Gain [dB] @2.45GHZ, V_{DD}=1.8V	IDD [uA] @V_{DD}=1.2V, IDD_{ref}=325uA
1	-22.6	337.4
2	-22.97	314.8
4	-22.86	324
5	-25.77	299.3
6	-23.06	315
7	-22.73	326.4
8	-23.02	310.4
9	-22.63	338.8
10	-22.86	327.8
11	-22.84	321.3

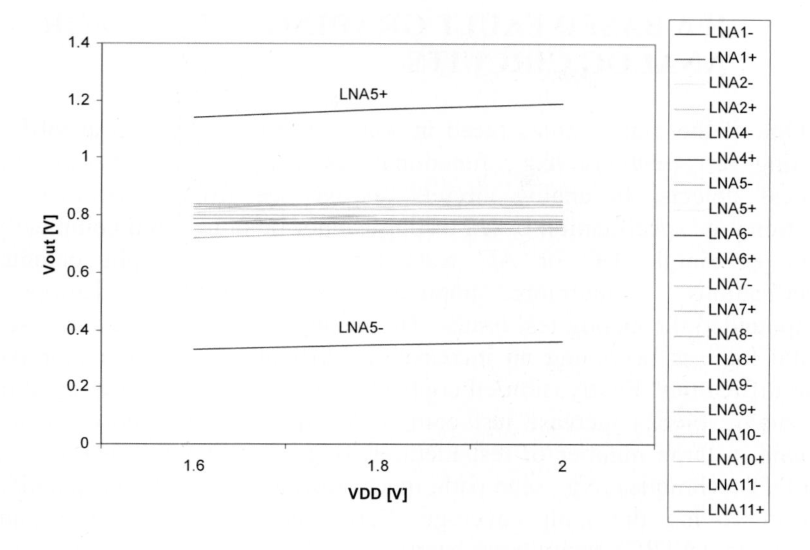

Figure 6-30. Output voltage measurements.

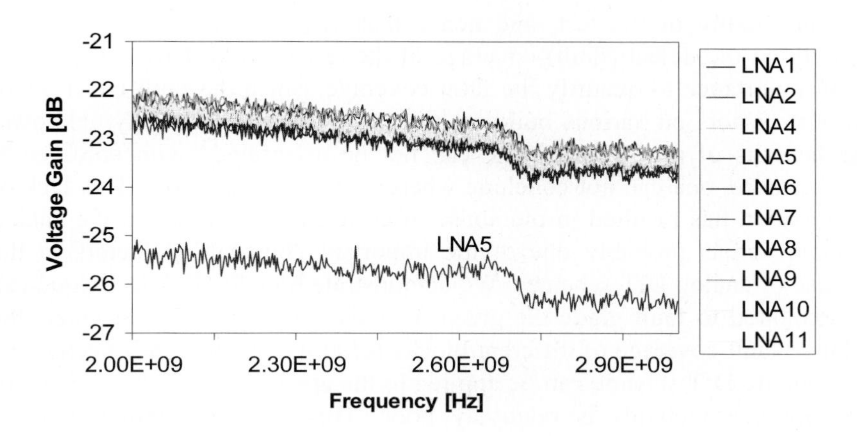

Figure 6-31. Measured voltage gain of the amplifier.

7. IFA BASED FAULT GRADING AND DfT FOR ANALOG CIRCUITS

One of the major issues faced in analog testing is how to quantify the existing test methods (e.g., functional tests) against the manufacturing process defects. In analog circuits, as we see from experiments, the functional or specification based testing cannot be eliminated completely in favor of simple DC or AC tests for circuits with tight parametric specifications. Furthermore, popularity of mixed-signal devices has compounded the analog test issues. The testing of analog blocks in a sea of digital logic is becoming an increasingly difficult task. Two major issues pose difficulties. Firstly, limited controllability and observability conditions for analog blocks increase test complexity and cost. Secondly, in digital domain, a large number of test methods (e.g., functional, structural, I_{DDQ}) and DfT techniques (e.g., scan path, macro test) are available for quantifying and improving the fault coverage. Furthermore, automatic test pattern generation (ATPG) techniques have reduced the test generation cost for digital circuits significantly. Analog testing lacks such tools and techniques. Therefore, analog testing is becoming a bottleneck in testing of mixed-signal ICs in terms of cost and quality.

The quality of the test, and hence that of the tested device, depends heavily on the defect (fault) coverage of the test vectors. Therefore, it is of vital importance to quantify the fault coverage. Since the fault coverage of the test vectors on various building blocks of a design is largely unknown, the benefits of any DfT scheme can not be ascertained with confidence. Furthermore, one can not conclude where DfT is needed most. This lack of information has resulted in the abuse of digital DfT schemes in the analog domain and is probably one of the important contributing factors in the demise of analog DfT schemes. We demonstrate how the IFA technique can be exploited to fault grade the given (conventional) test vectors. Once, the relative fault coverage of different blocks is known by given test vectors, an appropriate DfT scheme can be applied to the areas where fault coverage of existing test methods is relatively poor. This is demonstrated with an example of a flash A/D converter.

7.1 A/D Converter Testing

An A/D converter is normally tested for DC and AC performance. The DC tests typically test for offset voltage and full scale errors. Static differential non-linearity (DNL) and integral non-linearity (INL) measurements are performed by slowly varying the input signal such that the

DC operating point is reached for each measurement. On the other hand, dynamic tests are performed to test for dynamic range, conversion speed, SNR, dynamic DNL, dynamic INL, bit error rate (BER), etc. These dynamic specifications are often tested by performing BER measurement, code density measurement (CDM), beat frequency measurement or SNR or THD measurement.

The code density measurement is an effective way of testing A/D converters. The static DNL and INL of the converter can be computed from this measurement. At the input of an A/D converter a waveform is applied. The amplitude of this waveform is slightly greater than the full scale value of the converter. As the waveform traverses from zero to full amplitude, different output codes appear at the output of the A/D converter. For an accurate measurement at least 8 to 16 codes per level are needed [18,22]. This is achieved by repeating the test for a number of cycles of the waveform. Often a triangular waveform is applied because then every code should have equal density. If a larger or a smaller number of codes is found in the CDM, it shows the presence of poor DNL. A fault is considered detected by CDM if it resulted in more or fewer (pre-specified criterion) occurrences of a given code.

For the SNR, THD and SINAD measurements, a sine wave is applied at the input of a converter and output codes are measured. In order to randomly distribute the quantization error over the measurement, the ratio of signal frequency to the sampling frequency is given by (6.11).

$$\frac{F_{signal}}{F_{sample}} = \frac{M}{N} \tag{6.11}$$

where M and N are mutually prime integers and N is the number of samples taken. Mahoney [17] calls it M/N sampling.

7.2 Description of the Experiment

An 8-bit flash A/D converter [28] is utilized for this experiment. However, the 8-bit flash A/D converter is too complex even for a fault-free simulation at the Spice level. Nevertheless, for the accuracy of the analysis, the circuit level simulation is considered to be an absolute requirement. Therefore, a 3-bit model of the converter at the Spice level was made. This model has only 8 (instead of 256) comparators which could be simulated in reasonable time.

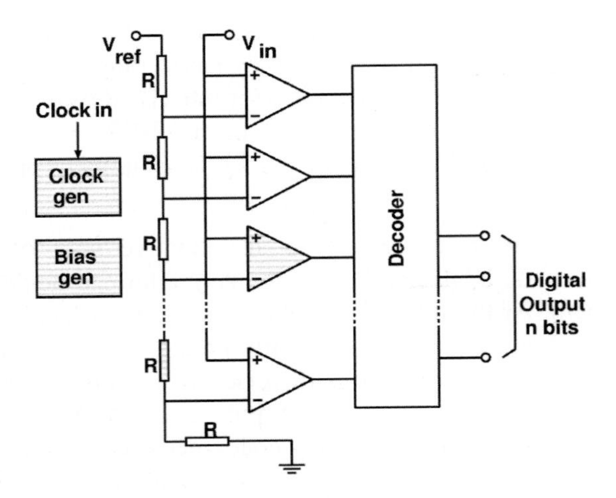

Figure 6-32. Block diagram of a flash A/D converter.

IFA is performed over all basic cells of the converter to find out the fault classes. The 3-bit model has 8 comparators, one clock generator, one biasing generator and one reference ladder. For the fault simulation purposes, a fault is introduced in one of the instances of a basic cell in the model. This is shown in Figure 6-32. For example, faults were introduced one by one in only one out of eight comparators at a time. This was under the single fault assumption in the design. The complete A/D model is simulated for all comparator faults. Once, all faults for the comparator were simulated, faults were introduced in another cell (e.g., clock generator or biasing circuit). In this way, all likely faults in the circuit were simulated.

7.3 Fault Simulation Issues

In analog circuits, fault simulation is a laborious exercise. For each fault, a separate simulation is run. Relatively high degree of human interaction is required for analog fault simulation. Furthermore, there are analog fault simulation issues that should be carefully addressed. The simulation environment is considerably slower from that of a tester. A test that takes fraction of a milli-second on a tester to perform may cost several minutes in simulation environment at Spice level. Furthermore, since fault simulation is to be performed over the complete fault set, the total time for the analysis may become prohibitively large. Due to time constraints, for the CDM, we reduce the average number of codes to 5 and applied a slow ramp so that every 5th clock cycle a new output code is generated. In other words, for 5 clock cycles, the fault-free converter is supposed to have the same output code. Even then, single fault simulation over 3-bit A/D model took 8 CPU (HP 700) minutes with CDM test. We considered a fault detected by CDM if it resulted in more than 7 or less than 3 occurrences of a given code. The SNR, THD and SINAD measurement takes 45-50 CPU minutes for single fault simulation. We selected SINAD instead of SNR as fault detection criterion. SINAD is defined as the signal to noise plus distortion ratio. The fault simulation using BER could not be performed since it would take even more time than the SINAD test. The DNL and INL measurement were carried out using the data of SINAD tests.

Table 6-7. The fault simulation results of the flash A/D converter.

Test / Cell	Code density measurement	SINAD	DNL	INL	Undetected faults
Comparator	112 (157)	19 (45)	25 (45)	21 (45)	17
Clock generator	32 (59)	10 (27)	14 (27)	10 (27)	11
Bias generator	16 (50)	8 (34)	14 (34)	9 (34)	18
Ref. ladder	16 (19)	1 (3)	1 (3)	1 (3)	2

Secondly, we utilized a production test system, MixTest [23], to compute DNL, INL and SINAD of fault simulations. A fault is detected by DNL, INL, or SINAD if the computed value differs from the golden device simulation by a predetermined threshold. The sampling frequency of the converter was 20 MHz. To randomly distribute the quantization errors, a fraction of the sampling frequency {(31/128)x20 MHz} was selected as the input frequency.

Thirdly, the setting up of the thresholds for fault detection in simulation environment must be done carefully. For example, the criterion of 1 LSB for DNL and INL measurements is no longer valid for a 3-bit model of an 8-bit converter. For the original converter, 1 LSB is 2V/256=7.8 mV. For a 3-bit model, 1 LSB amounts to 2V/8= 250 mV that is substantially greater than 7.8 mV. Owing to the constraints of simulation environment, we selected 0.1 LSB or 25 mV as the detection threshold. Though, this may be a little conservative, we assumed that if a fault is detected against a relaxed threshold criterion, it will certainly be detected in the production environment against tighter limits. Similarly, for a 3-bit converter, theoretical SNR should be 19.82 dB and the theoretical SINAD should be somewhat lower than this value. The SINAD for the 3-bit model in fault-free simulation was found to be 18.05 dB. Once again, we took conservative values for the fault detection. A fault is considered detected by SINAD measurement if the SINAD of the converter was less than 17.5 dB.

7.4 Fault Simulation Results

Table 6-7 compiles results of the fault simulation of 3-bit A/D converter model. In the comparator, 157 fault classes were simulated with the CDM test. A set of 112 faults were detected by that test. Owing to large fault simulation time, only those faults that are not detected by CDM (157-112=45) are simulated for SINAD, DNL and INL tests. This is further justified by the fact that CDM test is a simplified version of SINAD, DNL and INL tests. Therefore, if a fault is detected by CDM, it is likely to be detected by these tests. The DNL test was most effective (25/45) in catching rest of the undetected faults in the comparator. INL and SINAD, detected 21/45 and 19/45 faults, respectively, in the comparator. Nearly, 11% of the faults (17/157) in the comparator were not detected by any of these measurements. In the case of the clock generator, 59 fault classes were simulated. The CDM could detect 32 of the simulated faults. The performances of DNL, INL and SINAD, for the remaining undetected faults (27) were relatively poor compared to the comparator. As a result 11 of the 59 clock generator faults remained undetected. The performance of conventional tests was the poorest on the bias generator. A total of 50 fault classes were simulated in the bias generator. The CDM could detect only 16. The performance of DNL was marginally better. It detected 14 out of the remaining 34 undetected faults. On the whole, 36% of the total faults remained undetected. In the reference ladder, 19 fault classes were simulated. The CDM was an effective test and detected 16 fault classes. DNL, INL, and SINAD detected 1 fault class out of the 3 undetected fault classes and 2 fault classes remained undetected.

7.4.1 Analysis

Nearly 20% of the faults in the clock generator and 36% of the faults in the bias generator are not detected by the commonly used A/D specification tests. On the other hand, nearly 90% of the faults in the comparator and resistor ladder network are detected by these tests. The difference in fault coverage is easy to explain. Most of the conventional specification (conversion speed, SNR, DNL, INL, BER, CDM, etc.) tests are targeted toward faithfulness of the data path (i.e., analog input, reference ladder, comparator, decoder, digital output). There is hardly any test that explicitly covers the control path (i.e., clock and the bias generators). These blocks are assumed to be tested implicitly. Poor controllability and observability are other reasons for undetected faults in these blocks. The outputs of these cells are not directly observable. If the faulty bias or clock generator output causes the comparator to behave in an incorrect manner, then the fault is detected by the tests. However, the faults that modify the behavior of the control path only marginally are hard to detect and require testing of the complete dynamic ranges of input amplitude and frequency. Comparators are often designed to withstand the parametric variations of the clock and biasing to optimize the yield. Such a design has a fault masking effect.

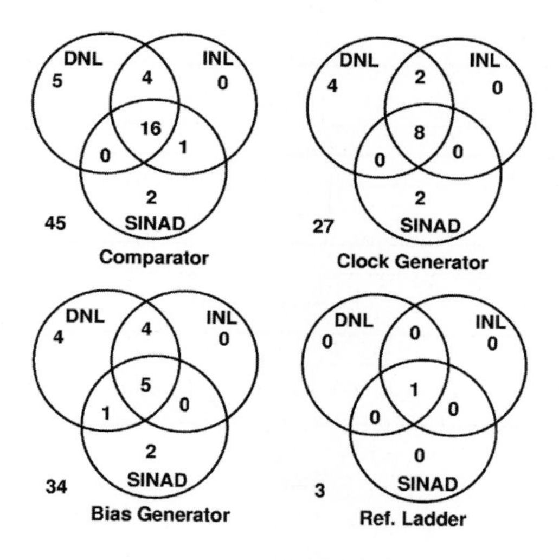

Figure 6-33. Fault detection capabilities of different tests.

Different specification tests differ in fault coverage. The relative fault coverages of different tests are illustrated in Figure 6-33. As explained in previous sub-section, we only considered faults not detected by the CDM test. Though most faults are detected by all tests (DNL, INL, SINAD), DNL is the most effective test. Some faults are only detected by SINAD. However, INL does not detect any fault not detected by the other two tests. Nevertheless, we should keep in mind that INL is not a redundant test. INL is an effective test for detecting parametric variations in the reference ladder. For example, it covers those parametric variations that do not cause appreciable shift in the DNL but affect the whole reference ladder.

7.4.2 DfT Measures

An important question is how the fault coverage can be improved without sacrificing the performance of the converter. Measurement of the quiescent current (I_{DDQ}) may be one solution. Unlike digital circuits, the I_{DDQ} of an analog circuit is not in μA range. Therefore, its detection capability is limited. Alternatively, The A/D converter should be designed such that all high current dissipating paths are either switched off or bypassed for I_{DDQ} testing. Thus, I_{DDQ} test can be made an effective test method in fault detection. However, the design of such a converter requires a non-trivial amount of effort.

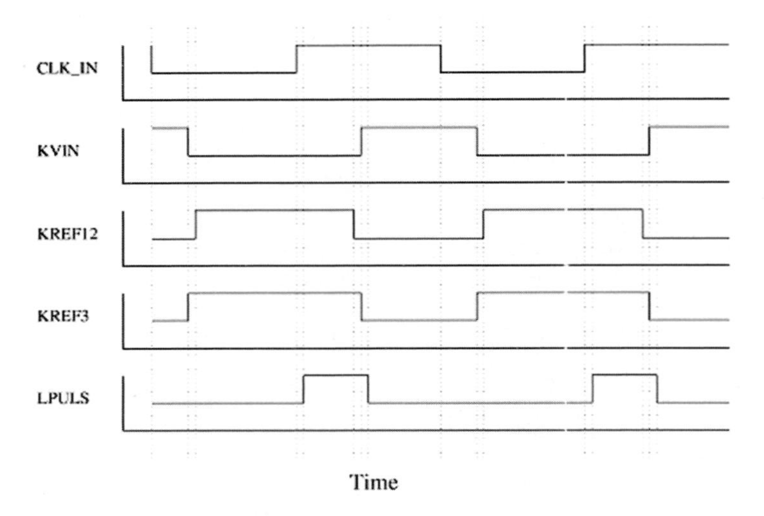

Figure 6-34. Input clock signal and various generated clocks in the analyzed flash A/D converter.

There are innovative voltage DfT techniques that do not cause performance degradation and improve fault coverage. For example, a DfT scheme for the clock driver is explained in Figure 6-34 and Figure 6-35. Figure 6-34 shows 4 clocks generated from the clock generator. These signals are digital in nature but their timing relationship with each other is extremely important for a correct function of the A/D converter. For a DfT solution, we exploit the knowledge of pre-defined timing relationship between different clocks. Typically, a large number of faults degrades the timings of clock signals, if we take a logical AND of these signals (Figure 6-35), we get an output pulse whose position and width are known. Most faults causing timing and/or stuck-at behavior will be detected by the measurement of pulse position and/or width. The number of pulses within the clock cycle and their position from CLK_IN can be the fault detection criteria. Then, more than 95% of the faults influenced output(s) of the clock generator and were detected within 2 clock cycles. This test method detects faults quickly and provides the diagnostic information. There can be a variety of implementations to extract different attributes of periodic signals. It costs approximately 10 logic gates for the particular implementation. The number of gates is a trade-off between the required diagnostics and the cost of implementation. The number of gates can be reduced if the critical signal spacing requirements are known in advance. For example, signals KVIN and KREF12 should be non-overlapping. Hence, only the critical timings are generated by the Boolean operations.

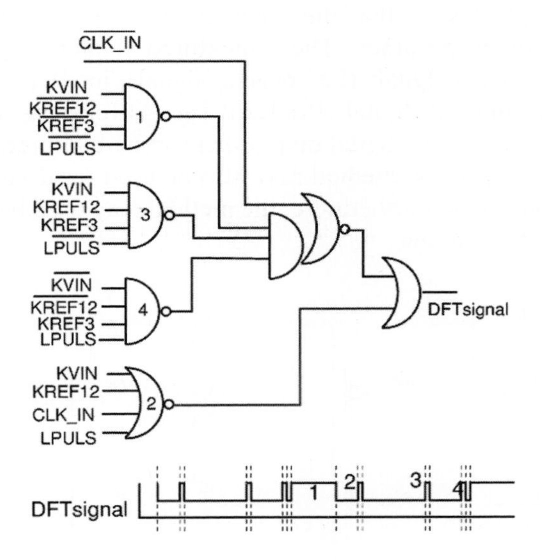

Figure 6-35. A DfT scheme for testing clock generator faults in the analyzed flash A/D.

Similarly, a DfT solution for faults in the bias generator is shown in Figure 6-36. Bias generator provides a set of stable bias signals to the comparator. In the case of the flash A/D converter 4 bias signals are generated. These are stp1 (3.2V), stp2 (3.2V), stp3 (3.4V), and Vbias (2.3V). Each biasing voltage is applied to a p-channel transistor that is individually gated by an n-channel transistor. This scheme allows measurement of quiescent current through individual or multiple paths. Defect-free quiescent currents through the components can be computed. If the presence of a defect in the biasing network influences the voltage level of any of the biasing signal, it is translated into a current that can be measured. Nearly 80% of the bias generator faults are detected by these simple measurements. It is worth mentioning that the popular and expensive (conventional) test method detects only 60% of these faults.

Alternatively, the available infra-structure of the same A/D converter (comparators, etc.) may be utilized to determine the voltage of various internal biasing and clock signals. The same idea can be extended to test external analog blocks preceding the A/D converter. Typically, these blocks are noise shapers, filters, amplifier, etc. The basic idea of this DfT concept is illustrated in Figure 6-37. In the test mode, instead of connecting the normal signal (Vin) to comparators, the desired internal signals of the converter are selectively applied to the comparators. The reference voltage (Vref) is appropriately applied such that the comparators can compare the applied signals with optimum accuracy. These measured signals may include time invariant and/or analog signals (i.e., biasing signals, internal, external nodes of the A/D converter) or digital (Boolean) signals (clocks). The test mode signal is converted into the digital output that can be interpreted against pre-determined values. The test method may also be used for design verification and diagnostic purposes. Furthermore, the method may used be for in-system testing or BIST applications.

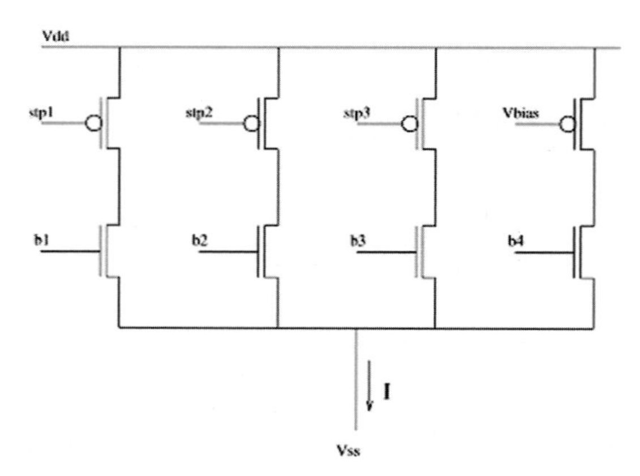

Figure 6-36. A Dft scheme for testing the bias generator faults in the analyzed flash A/D converter.

Figure 6-38 illustrates a schematic of the test controller. It contains a decoder that, depending on the input code (Test), connects a given signal to the comparators. In the figure, signals T1, T2, and T3 form the decoded address. In embedded applications, these signals may be applied through a scan chain. Alternatively, at the cost of adding a simple counter, they may be generated within the converter. Let P1, P2, ... be the periodic signals and let B1, B2, ... be the biasing or time invariant signals for the A/D converter. For testing of an arbitrary analog biasing signal, say B1, the appropriate address value for the test decoder is applied such that B1 is connected to the bank of comparators. Biasing signals are transmitted through transmission gate pairs to minimize signal degradation. Furthermore, biasing signals are DC levels and the effect of transmission gate resistance (R_{on}) in test mode should not be of major significance. Alternatively, signal degradation can be characterized with a signal of known amplitude. The output of the converter gives a digitized value of the bias signal. As mentioned previously, biasing signals are often not tested explicitly and biasing stability with respect to the temperature and other environmental conditions is rarely examined. A biasing circuitry, under the extreme conditions may acquire an oscillating behavior that is very difficult to test. This DfT scheme allows to test the conditions quickly and unambiguously. In a similar fashion other internal analog signals of the A/D converter may be tested. The same method my be utilized to test external analog blocks or analog signals inside them individually. However, when testing a high speed analog signal care should be taken to optimize the test for least signal degradation in the test mode.

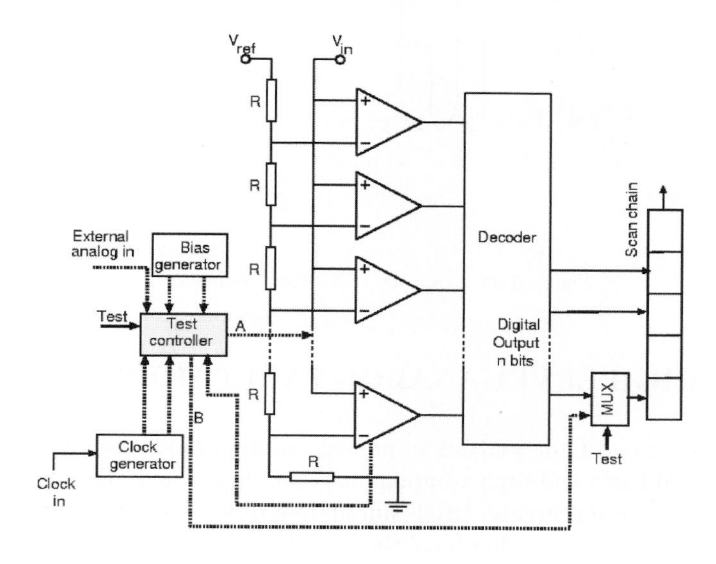

Figure 6-37. A full flash A/D converter block diagram with DfT features.

Testing of digital signals does not require an A/D converter. However, for good functioning of an A/D converter, precise clocking is vital. A way of testing internal clocks is shown in Figure 6-37 and Figure 6-38. Let us assume that clock P1 is to be tested. An appropriate address value for the test decoder, T, is applied. The same test address also enables the output multiplexer such that a part of the digital output contains the clock information that can be shifted out via scan chains. Alternatively, for digital signals, the multiplexer and part of the test controller may be incorporated into the output decoder of the A/D converter such that in the test mode the digital outputs shows various critical digital signals of the comparator. The implementation is simple and does not require further explanation.

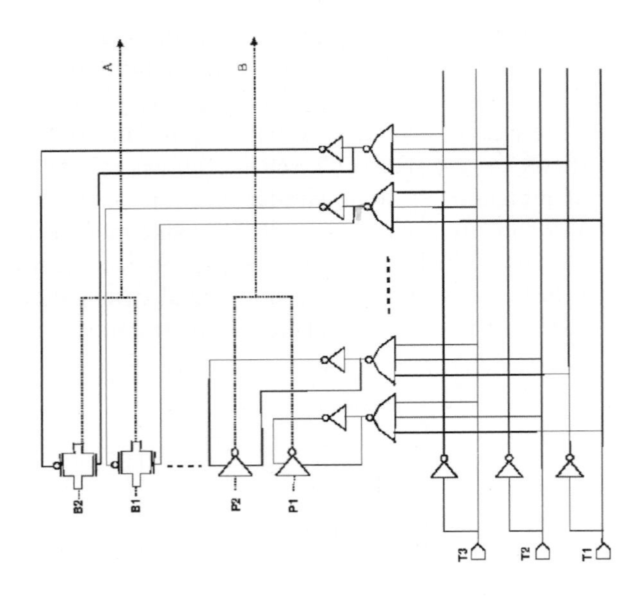

Figure 6-38. Expanded view of test controller.

8. HIGH LEVEL ANALOG FAULT MODELS

Defect-oriented fault analysis of analog building blocks is a tedious task due to lack of tools and high computation time. Spice level fault simulation often requires prohibitively large simulation time. Therefore, there have been attempts to create higher level fault models based upon realistic defects such that the simulation time may be reduced [1,10,16]. Harvey et al. [10] carried out a defect-oriented analysis of a phase locked loop (PLL). Fault free simulation of PLLs takes enormous CPU time. Therefore, higher level fault models were utilized. The PLL was also divided into several macros (Figure 6-39) for which simulation models at behavior-level were developed.

Although the simulation time per macro was reduced significantly as can be seen in Table 6-8, it still took several hours of CPU time to simulate the locking behavior. Circuit-level simulation was not feasible.

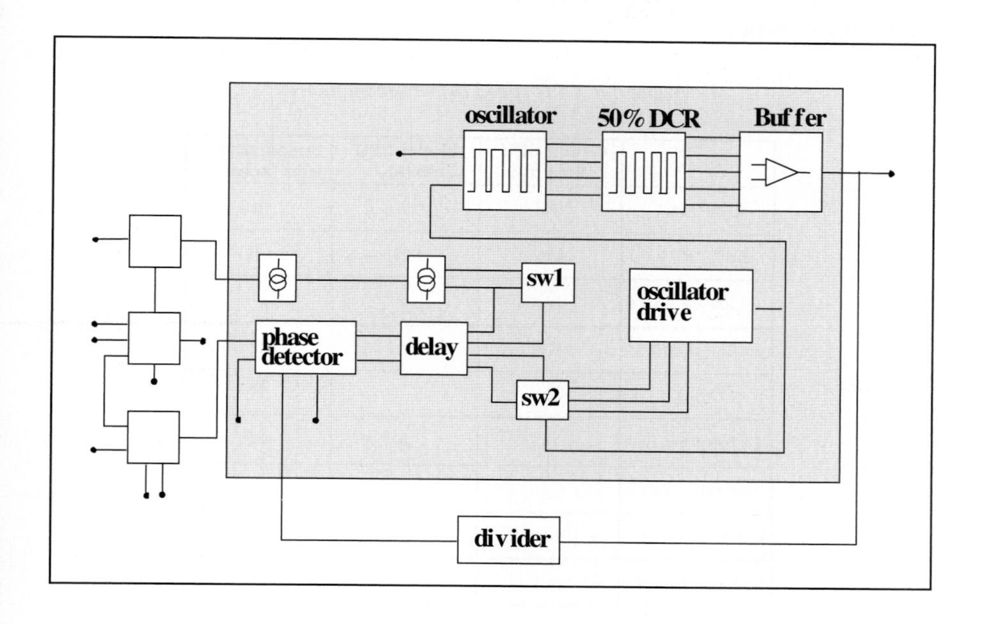

Figure 6-39. Partitioning of the PLL into macros.

For the IFA analysis, faults were inserted into each macro in turn. To ensure correct fault behavior, the macro being analyzed was replaced by its circuit-level description. Most faults caused a hard failure that was already identified in the simulation of the macro being analyzed. Only a few of the faults had to be simulated in a functional way, including all models of the other macros. A complete analysis revealed that the functional test, comprising locking time and capture range measurements, will detect about 93% of the faults. Alternative tests were evaluated with respect to customer quality requirements. The remaining 7% faults can be detected by using power supply voltage levels outside the specified operational range, that extends from 4.5 to 5.5 Volts. Application of a supply voltage of 3 Volts changes circuit sensitivities [6] and thus enables the detection of other faults [10]. Such tests should only be used when the fault-free response can unambiguously be identified.

In a similar manner, an anti side tone module of the PACT IC, used for telephone applications (PCA1070 [2]) was subjected to fault analysis. This module consists of two programmable switched capacitor filters in a feedback loop (see Figure 6-40). Being programmable, the IC can be used in various countries with different statutory requirements, but the necessity of the repetition of functional tests for different settings increases the time required for testing.

Table 6-8. CPU speed improvement factors.

Block	Circuit CPU time (s)	Model CPU time (s)	Speed advantage factor
Oscillator	3082	134	23
50% DCR	803	527	1.5
Buffer	96	4	24
Phase detector	373	14	27
Delay	82	5	16
Current mirror 1	76	3	25
Current mirror 2	106	3	35
Current switch 1	42	3	14
Current switch 2	72	2	36
Oscillator drive	58	6	10

Goal of the analysis of the anti side tone module was to evaluate the effectiveness of the functional tests used in production testing. Again, a macro oriented divide and conquer approach was used. For the analysis of the ZS-filter, other macros were modeled at a level optimized in terms of simulation speed. The assembly of all macros together was used in the fault simulation.

The most complex macro is the ZS-filter, a programmable switched capacitor filter, whose programmed digital setting determines the number of switched capacitors that is activated. It takes a long time to simulate this filter using a circuit simulator due to the combination of the discrete time character of the filter and the functional test signals. Since the simulation is repeated for each fault and for various programmed settings, the required simulation time becomes unacceptable. By replacing the switched capacitors by equivalent models consisting of resistors and current sources, a time continuous equivalent is obtained, which allows for fast simulations. This equivalent model was used to analyze all faults in the decoding logic, operational amplifiers, resistive divider, etc. Faults affecting the clock signals of the switched capacitors and faults inside those capacitors were analyzed separately.

Analysis of 55 most likely faults identified 5 faults that were not detected by functional tests. Two of these could be detected by adding an additional functional test, but the remaining three faults were inherently undetectable. However, further analysis of these five faults revealed that they can be eliminated by minor changes in the routing of the layout.

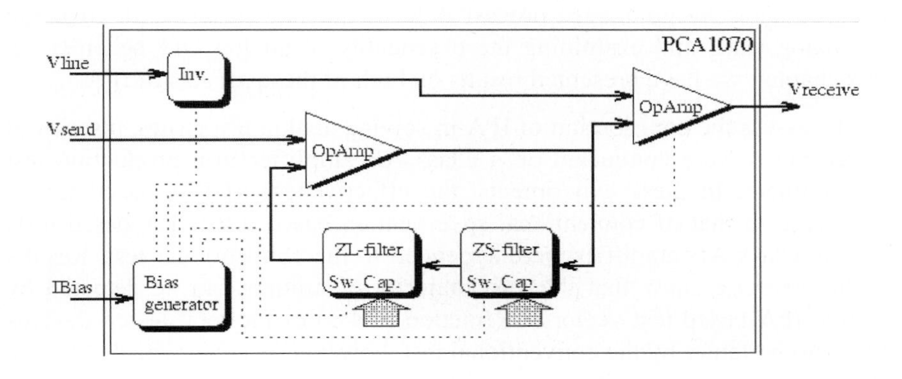

Figure 6-40. The antiside tone module of the PCA 1070.

9. CONCLUSIONS

Analog test issues are more qualitative than quantitative, therefore, digital DfT measures that address quantitative issues are not very successful with analog circuits. Owing to the non-binary nature of circuit operation, analog circuits are influenced by process defects in a different manner than digital circuits. Furthermore, in analog circuits, circuit design and layout techniques are utilized to maximize the performance. Therefore, these circuits show a greater sensitivity to parametric variations compared to their digital counterparts. Many manufacturing process defects that do not influence digital circuit performance, may affect analog circuits significantly. In general, subtle design and manufacturing process sensitivities give rise to many parametric failures. This requires a careful investigation of defects in analog circuits, modeling related aspects and detection strategies. In this chapter, a realistic process defect-oriented simulation-before-test (SBT) testability methodology for analog circuits is proposed. A defect simulator, for sprinkling defects onto the layout, and a circuit simulator, for fault simulation, are the key components of the methodology. The circuits of moderate complexity can be analyzed. However, bigger circuits must be divided for the analysis.

A process defect based analog test approach contrasts with the specification based analog testing. Both approaches have their merits, However, there is a growing consensus that a synergy between the two will result in a better test quality as well test economics. The potential of IFA techniques is assessed on two standpoints: (a) contribution of IFA in testing silicon devices in a production environment and, (b) contribution of IFA in robust analog design against process defects, quantifying the fault coverage of analog tests, and examining the practicality of analog DfT schemes. In this chapter, we have presented results on both of these aspects of IFA.

To assess the contribution of IFA in solving analog test issues, a series of experiments were conducted on a Class AB amplifier in a production test environment. In these experiments, the effectiveness of IFA based test is compared to that of conventional specification based tests. IFA based tests for the Class AB amplifier were appended to the conventional test. Results of the exercise show that the vast majority of failures can be detected by simple IFA based test vectors. A fraction (0.4%) of the total tested devices found to be faulty by the conventional test, however, was not detected by the IFA based test. The subsequent analysis revealed that more than 85% of such escapes were due to un-modeled faults. IFA based tests are simpler compared to the conventional tests and, therefore, can be applied by inexpensive automatic test equipment (ATE) in a production test environment. Both of these aspects result in test cost reduction. The test cost reduction for the Class AB amplifier in the production environment is estimated to be 30%.

These experiments lead to some broad conclusions. Tighter the parametric specifications of an analog circuit, less effective an IFA based test is likely to be. This is because natural process variations with tighter parametric requirements will contribute to a larger number of device failures. These are un-modeled faults/failures, therefore, the effectiveness of IFA based test is lowered. Furthermore, the effectiveness of a process defects based yield loss model diminishes significantly with tightening of the parametric requirements. Hence, the application of IFA based tests to analog circuits with relatively relaxed specifications and higher functional complexity is likely to be more successful. Better control of the manufacturing process (high Cp and Cpk) should also have positive influence on the effectiveness of IFA based test. A limited functional test together with an IFA based test should be another way to avoid such escapes. Thus, strengths of both test methods can be exploited for economic and quality gains.

Results of Vdd ramp study indicate a fault coverage for bridges of 88% using only the power supply ramp test method compared to 75% if only the amplifier's voltage gain is measured, or to 83% if the amplifier's Noise Figure is observed. If a combined functional gain and noise-figure test is carried out then the corresponding fault coverage is about 86%, while the combined structural test of the power supply ramp and the dc-voltage test gives a better coverage of 94%. One can conclude that simple techniques like this one are effective for first test screening.

We have shown that the application of low frequency measurements, namely dc in this case, reduces test complexity without sacrificing fault coverage. These techniques can effectively be used at wafer test to do at least a pre-screening of bad devices. Although this technique was tested on an amplifier its applicability is general enough for types of other circuits. For the present circuit a discrete-step ramp was used, but we can foresee that for other types of circuits like PLLs, the type and slope of the ramp are important. The results shown in this chapter are encouraging and more work is necessary to mature and implement this technique as a standard production testing method.

Assessment of IFA in an analog design environment was carried out using a flash A/D converter. Fault coverage for analog circuits is often not determined. In the absence of fault coverage numbers, effectiveness of any DfT measure cannot be quantified. It is demonstrated that the IFA technique can be exploited to fault grade the flash A/D converter for existing (specification) test practices. The results of the analysis showed that the fault coverage of the specification test was relatively high on the data path. However, it was relatively poor on the control path (clock, bias generators). Almost all specification based tests are targeted towards the faithfulness of the data path and the control path is assumed to be tested implicitly. Separate DfT solutions were proposed for clock and bias generators to improve the fault coverage and to simplify the test.

IFA is has limitations. First and foremost is CPU intensiveness of the method. Application of the analysis requires CAD tools. Circuits of moderate transistor count complexities can only be analyzed by a CAD tool. The circuits with higher complexities must be partitioned or modeled at higher abstraction levels for the analysis.

References

1. B. Atzema, E. Bruls, M. Sachdev and T. Zwemstra, "Computer-Aided Testability Analysis for Analog Circuits," Proceedings of the Workshop on Advances in Analog Circuit Design, April 1995.

2. R. Becker, et al., "PACT - A Programmable Analog CMOS Transmission Circuit for Electronic Telephone Sets," Proceedings of European Solid State Circuits Conference, 1993, pp. 166-169.

3. F.P.M. Beenker, "Testability Concepts for Digital ICs," Ph.D. Thesis, University of Twente, Netherlands, 1994.

4. R.S. Berkowitz, "Conditions for Network-element-value Solvability," IRE Transactions on Circuit Theory, vol. CT-9, pp. 24-29, March 1962.

5. E.M.J.G. Bruls, "Reliability Aspects of Defects Analysis," Proceedings of European Test Conference,1993, pp. 17-26.

6. E. Bruls, "Variable supply voltage testing for analogue CMOS and bipolar circuits," Proceedings of International Test Conference, 1994, pp. 562-571.

7. P. Duhamel and J.C. Rault, "Automatic Test generation Techniques for Analog Circuits and Systems: A Review," IEEE Transactions on Circuits and Systems, vol. CAS-26, no. 7, pp. 411-440, July 1979.

8. N.J. Elias, "The Application of Statistical Simulation to Automated the Analog Test Development," IEEE Transactions on Circuits and Systems, vol. CAS-26, no. 7, pp. 513-517, July 1979.

9. B.R. Epstein, M. Czigler and S.R. Miller, "Fault Detection and Classification in Linear Integrated Circuits: An Application of Discrimination Analysis and Hypothesis Testing," IEEE Transactions on Computer Aided Design of Integrated Circuits and Systems, vol. 12, no. 1, pp. 102-112, January 1993.

10. R.J.A. Harvey, A.M.D. Richardson, E.M.J. Bruls and K. Baker, ``Analogue Fault Simulation Based on Layout Dependent Fault Models,'' Proceedings of International Test Conference, 1994, pp. 641-649.

11. C.F. Hawkins and J.M. Soden, "Electrical Characteristics and Testing Considerations for Gate Oxide Shorts in CMOS ICs", Proceeding of International Test Conference, 1985, pp. 544-555.

12. G.J. Hemink, B.W. Meijer and H.G. Kerkhoff, "TASTE: A Tool for Analog System Testability Evaluation," Proceeding of International Test Conference, 1988, pp. 829-838.

13. W. Hochwald and J.D. Bastian, "A DC Approach for Analog Dictionary Determination," IEEE Transactions on Circuits and Systems, vol. CAS-26, no. 7, pp. 523-529, July 1979.

14. H.H. Huston and C.P. Clarke, "Reliability Defect Detection and Screening During Processing - Theory and Implementation," Proceedings of International Reliability Physics Symposium, 1992, pp. 268-275.

15. A.T. Johnson, Jr., "Efficient Fault Analysis in Linear Analog Circuits," IEEE Transactions on Circuits and Systems, vol. CAS-26, no. 7, pp. 475-484, July 1979.

16. F.C.M. Kuijstermans, M. Sachdev and L. Thijssen, "Defect Oriented Test Methodology for Complex Mixed-Signal Circuits,'' Proceedings of European Design and Test Conference, 1995, pp. 18-23.

17. M. Mahoney, "DSP-Based Testing of Analog and Mixed-Signal Circuits'', Los Alamitos, California: IEEE Computer Society Press, 1987.

18. W. Maly, F.J. Ferguson and J.P. Shen, "Systematic Characterization of Physical Defects for Fault Analysis of MOS IC Cells," Proceeding of International Test Conference, 1984, pp. 390-399.

19. W. Maly, W.R. Moore and A.J. Strojwas, "Yield Loss Mechanisms and Defect Tolerance," SRC-CMU Research Center for Computer Aided Design, Dept. of Electrical and Computer Engineering, Carnegie Mellon University, Pittsburgh, PA 15213.

20. W. Maly, A.J. Strojwas and S.W. Director, "VLSI Yield Prediction and Estimation: A Unified Framework," IEEE Transactions on Computer Aided Design, vol. CAD-5, no.1, pp 114-130, January 1986.

21. W. Maly, "Realistic Fault Modeling for VLSI Testing," 24th ACM/IEEE Design Automation Conference,1987, pp.173-180.

22. A. Meixner and W. Maly, "Fault Modeling for the Testing of Mixed Integrated Circuits," Proceeding of International Test Conference, 1991, pp. 564-572.

23. R. Mehtani, B. Atzema, M. De Jonghe, R. Morren, G. Seuren and T. Zwemstra,``Mix Test: A Mixed-Signal Extension to a Digital Test System,'' Proceedings of International Test Conference, 1993, pp. 945-953.

24. L. Milor and V. Visvanathan, "Detection of Catastrophic Faults in Analog Integrated Circuits," IEEE Transaction on Computer Aided Design of Integrated Circuits and Systems, vol. 8, pp. 114-130, February 1989.

25. L. Milor and A. Sangiovanni-Vincentelli, "Optimal Test Set Design For Analog Circuits", International Conference on Computer Aided Design, 1990, pp. 294-297.

26. N. Navid and A.N. Willson, Jr., "A Theory and an Algorithm for Analog Fault Diagnosis," IEEE Transactions on Circuits and Systems, vol. CAS-26, no. 7, pp. 440-456, July 1979.

27. A. Pahwa and R. Rohrer, "Band Faults: Efficient Approximations to Fault Bands for the Simulation Before Fault Diagnosis of Linear Circuits," IEEE Transactions on Circuits and Systems, vol. CAS-29, no. 2, pp. 81-88, February 1982.

28. M.J.M. Pelgrom and A.C. van Rens, "A 25 Ms/s 8-Bit CMOS ADC for Embedded Applications,'' Proceedings 19th of European Solid State Circuits Conference, 1993, pp. 13-16.

29. R.J. van de Plassche, ``Integrated Analog-to-Digital and Digital-to-Analog Converters," Dordrect: Kluwer Academic Publishers. 1994.

30. R.W. Priester and J.B. Clary, "New Measures of Testability and Test Complexity for Linear Analog Failure Analysis," IEEE Transactions on Circuits and Systems, vol. cas-28, no.11, pp. 1088-1092, November 1981.

31. L. Rapisarda and R.A. Decarlo, "Analog Multifrequency Fault Diagnosis," IEEE Transactions on Circuits and Systems, vol. cas-30, no.4, pp. 223-234, April 1983.

32. R. Rodriguez-Montanes, E.M.J.G. Bruls and J. Figueras, "Bridging Defects Resistance Measurements in CMOS Process," Proceeding of International Test Conference, 1992, pp. 892-899.

33. M. Sachdev, "Catastrophic Defect Oriented Testability Analysis of a Class AB Amplifier," Proceedings of Defect and Fault Tolerance in VLSI Systems, October 1993, pp. 319-326.

34. M. Sachdev, "Defect Oriented Analog Testing: Strengths and Weaknesses,'' Proceedings of 20th European Solid State Circuits Conference, 1994, pp. 224-227.

35. M. Sachdev, "A Defect Oriented Testability Methodology for Analog Circuits," Journal of Electronic Testing: Theory and Applications, vol. 6, no. 3, pp. 265-276, June 1995.

36. M. Sachdev and B. Atzema, "Industrial Relevence of Analog IFA: A Fact or A Fiction," Proceedings of International Test Conference, 1995, pp. 61-70.

37. M. Sachdev, "A DfT Method for Testing Internal and External Signals in A/D Converters", European patent application no. 96202881.7, 1996.

38. R. Saeks, A. Sangiovanni-Vincentelli and V. Vishvanathan, "Diagnosability of Nonlinear Circuits and Systems--Part II: Dynamical case," IEEE Transactions on Circuits and Systems, vol. cas-28, no.11, pp. 1103-1108, November 1981.

39. A.E. Salama, J.A. Starzyk and J.W. Bandler, "A Unified Decomposition Approach for Fault Location in Large Analog Circuits," IEEE Transactions on Circuits and Systems, vol. cas-31, no.7, pp. 609-622, July 1984.

40. J.P. Shen, W. Maly and F.J. Ferguson, "Inductive Fault Analysis of MOS Integrated Circuits," IEEE Design and Test of Computers, vol. 2, no. 6, pp. 13-26, 1985.

41. M. Slamani and B. Kaminska, "Analog Circuit Fault Diagnosis Based on Sensitivity Computation and Functional Testing," IEEE Design & Test of Computers, vol. 9, pp. 30-39, March 1992.

42. J.M. Soden and C.F. Hawkins, "Test Considerations for Gate Oxide Shorts in CMOS ICs," IEEE Design & Test of Computers, vol. 2, pp. 56-64, August 1986.

43. J.M. Soden and C.F. Hawkins, "Electrical Properties and Detection Methods for CMOS IC Defects," Proceedings of European Test Conference, 1989, pp. 159-167.

44. M. Soma, "Fault Modeling and Test Generation for Sample and Hold Circuits," Proceedings of International Symposium on Circuits and Systems, 1991, pp. 2072-2075.

45. M. Soma, "An Experimental Approach to Analog Fault Models," Proceedings of Custom Integrated Circuits Conference, 1991, pp. 13.6.1-13.6.4.

46. M. Soma, "A Design for Test Methodology for Active Analog Filters," Proceedings of International Test Conference, 1990, pp. 183-192.

47. S. Somayayula, E. Sanchez-Sinencio and J. Pineda de Gyvez, "Analog Fault Diagnosis based on Ramping Power Supply Current Signature," *IEEE Trans. On Circuits and Systems-II,* vol. 43, no. 10, pp. 703-712, October 1996.

48. T.M. Souders and G.N. Stenbakken, "A Comprehensive Approach for Modeling and Testing Analog and Mixed Signal Devices," Proceeding of International Test Conference, 1990, pp. 169-176.

49. H. Sriyananda and D.R. Towill, "Fault diagnosis Using Time-Domain Measurements," Proceedings of Radio and Electronic Engineer, vol.9, no. 43, pp. 523-533, September 1973.

50. M. Syrzycki, "Modeling of Spot Defects in MOS Transistors," Proceedings International Test Conference, 1987, pp. 148-157.

51. T.N. Trick, W. Mayeda and A.A. Sakla, "Calculation of Parameter Values from Node Voltage Measurements," IEEE Transactions on Circuits and Systems, vol. CAS-26, no. 7, pp. 466-474, July 1979.

52. P.N. Variyam and A. Chatterjee, "Test Generation for Comprehensive Testing of Linear Analog Circuits using transient response sampling," *Int. Conference on Computer Aided Design*, pp. 382-385, 1997

53. P.N. Variyam and A. Chatterjee, "Specification-Driven Test Generation for Analog Circuits," *IEEE Transactions on Computer-Aided Design of Integrated Circuits and Systems,"* vol. 19, no. 10, pp. 1189-1201, October 2000.

54. V. Vishvanathan and A. Sangiovanni-Vincentelli, "Diagnosability of Nonlinear Circuits and Systems--Part I: The DC case," IEEE Transactions on Circuits and Systems, vol. CAS-28, no.11, pp. 1093-1102, November 1981.

55. K.D. Wagner and T.W. Williams, "Design for Testability of Mixed Signal Integrated Circuits," Proceeding of International Test Conference, 1988, pp. 823-828.

56. A. Walker, W.E. Alexander and P. Lala, "Fault Diagnosis in Analog Circuits Using Elemental Modulation," IEEE Design & Test of Computers, vol. 9, pp. 19-29, March 1992.

57. H. Walker and S.W. Director, "VLASIC: A Catastrophic Fault Yield Simulator for Integrated Circuits," IEEE Transactions on Computer Aided Design of Integrated Circuits and Systems, vol. CAD-5, pp. 541-556, October 1986.

58. R.H.Williams and C.F. Hawkins, "Errors in Testing," Proceeding of International Test Conference, 1990, pp. 1018-1027.

Chapter 7

YIELD ENGINEERING

The advent of high-speed deep submicron circuits with larger die sizes and shorter life cycles lends itself to an increase in fabrication cost. Basically, the economic success of a product may depend on a timely and accurate design for manufacturability (DfM) strategy. Put in other words, an appropriate yield forecast may render significant benefits in both time-to-market and manufacturing cost prediction. Yield forecasting is essential for the development of new products. It effectively shows if a design is feasible of meeting its cost objectives. This aspect is especially crucial if the company operates as a vendor of "finished dice" and not of "finished wafers". Thus, proper yield forecast can give the manufacturer a leading edge over its competitors. This chapter is an introduction to the foundations of functional yield modelling. We will look first at "black-box" models, advance into engineering case studies and conclude with the economical implications of proper yield forecasting.

1. MATHEMATICAL MODELS FOR YIELD PREDICTION

Yield modelling has evolved from simple analytical formulae to complex yield simulations [10, 11] This evolution has gone from empirical formulae, to formulae based on simple statistical analyses, to formulae relying on complex Monte Carlo simulations. Obviously, each improvement copes with more and more complex conditions and it serves to refine the forecasting accuracy of the yield model. Naturally, this accuracy is based on the parameters that the model can handle. It follows then that it is possible to classify the various yield models according to their fidelity, complexity and dimensionality [17]. By fidelity is meant the accuracy in predicting yield compared to actual data; complexity is the accuracy in describing the physical phenomena that cause the yield loss; dimensionality is the number of parameters used in the model. It is obvious that as the dimensionality increases more complex phenomena are modeled resulting in a better fidelity of the model. Generally speaking, the parameters of a yield model are: IC area, defect density, spatial defect distribution on the wafer, defect size distribution, systematic processing errors, IC layout, and naturally the fabrication process parameters.

The choice of model depends on the needs of the problem. For instance, if there is a need to infer about the reasons of yield loss [32] a high dimensional model relying on CAD simulations is desirable. These models rely on CAD tools that allow the user to inspect the IC geometry and to execute algorithms such as dot-throwing or critical-area construction that simulate the possible short and open circuits of the IC [33][14][1][2]. Though models with higher dimensionality have the best fidelity, inspecting a layout of millions and millions of transistors is a very computationally intensive activity. Hence simpler analytical models [28] are primarily used for yield forecasting where there is no need to infer the reasons of yield loss. This does not mean that the model can be inaccurate, it means that the results that are fed back to the user are fewer and usually constrained to knowing the yield figure, defect density and perhaps some parameters describing the spatial defect distribution.

It is erroneous to devise a yield model as a result of some curve fitting between yield and some parameter, let us say, area. A fitting like $Y = a_o + a_1x + a_2 x^2 + a_3x^3 + \ldots$, where x could represent the IC area, will probably work well for the IC for which the fitting was done but it is useless for other ICs because it misses the underlying phenomena causing the yield loss [31][13][4].

We are going to explore most of the yield models available in the literature by means of analyzing the data presented in Table 1. The goal of this study is to examine the advantages and shortcomings of the various yield models and to show why a given model is preferred over others. Table 1 presents the data published by Moore concerning the number of defects per die of a half of a wafer[19]. The original data shows that there are 308 dice per wafer, 618 defects, and 136 defect-free dice. Thus, the yield for this wafer is 136/308, i.e. 0.44. The mean of the distribution, e.g. the number of defects per site, is $\bar{x} = 2.006$ and the standard deviation is σ=2.82.

Table 7-1. Distribution of Defects per die.

D	0	1	2	3	4	5	6	7	8	9	10	11	12	13	14	15	15	17
F	136	50	31	25	12	17	13	7	7	3	2	2	0	1	0	0	1	1

D = Defects per die F = Frequency of occurrence

The first model presented in the literature dates back to 1960 and it was intended to predict the yield of a batch of discrete transistors [34]. The probability of transistor failure was expressed as a ratio of the number of failing transistors to the total number of transistors in the batch. This model is based on the assumption that the probability of failure of each transistor is the same. Let us apply this model to the distribution of Table 7-1. Suppose that there is a pool of defects distributed over a wafer of B chips. We want to find the probability that none of the chips are killed due to a random placement of defects. For our example let k denote the number of defects per site and let P(X= k) be the probability of finding k defects; obviously yield is a particular case when P(X=0). The probability of failure p is in this case given as the inverse of the total number of dice B in the wafer slice, i.e. p=1/B = 1/308. Suppose now that we have N=618 defects, then using the binomial model to predict yield we have

$$P(X=k)=\left(\frac{N!}{k!(N-k)!}\right)(p)^{k}(1-p)^{N-k} \tag{7.1}$$

$$Y=P(X=0)=(1-p)^{N} \tag{7.2}$$

Plugging in numbers we obtain a yield of 0.13. Obviously this is an over pessimistic result and it is quite off from actual data. To make things even worse, let us say now that we have some new products with a die area that is 2, 4, 8 and 9 times the area of the current die and that we want to forecast their yield. To calculate the yield of the data of Table 1 a virtual grid overlaying the wafer map with grids or windows that contain a die multiple, e.g. 2, 4, 8, 9 is analyzed. Then yield is computed as the ratio of the number of defect-free windows to the total number of windows for each window size. Let us attempt to predict yield based on (7.1) assuming that we have the same number of defects N=618.

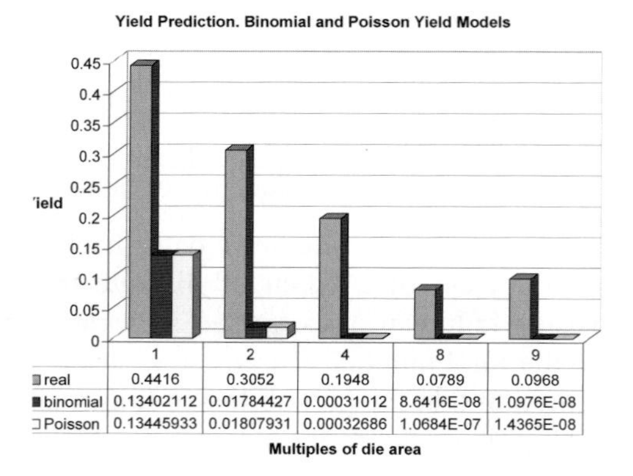

Figure 7-1. Yield prediction for binomial and Poisson models.

The results are plotted in Figure 7-1 where one can see that it is obvious that the binomial model is simply unsuitable. From a statistical standpoint a quick inspection of the distribution shown in Table 1 reveals that its mean is less than its variance. It follows then that the use of the binomial distribution is simply wrong because for a binomial distribution the mean is less or equal than the variance. Also, we are assuming that the number of defects is fixed. This situation seldom occurs in practice and in fact N behaves as another random variable with its own distribution P(N=i). It is possible to use compound Poisson statistics to include the effect of this new variable as follows[29]

$$P(X=k)=\sum_{i=k} P(N=i)\left(\frac{i!}{k!(i-k)!}\right)p^k(1-p)^{i-k} \tag{7.3}$$

Using a Poisson distribution to model the number of defects we have

$$P(N=i)=\frac{e^{-\overline{N}}\,\overline{N}^i}{i!} \tag{7.4}$$

where $\overline{N}$ is the average number of defects per wafer. Substituting (7.4) into (7.3) and after some algebraic manipulation we have

$$P(X=k)=\frac{e^{-p\overline{N}}p^{k}}{k!} \tag{7.5}$$

$$Y=P(X=0)=e^{-p\overline{N}} \tag{7.6}$$

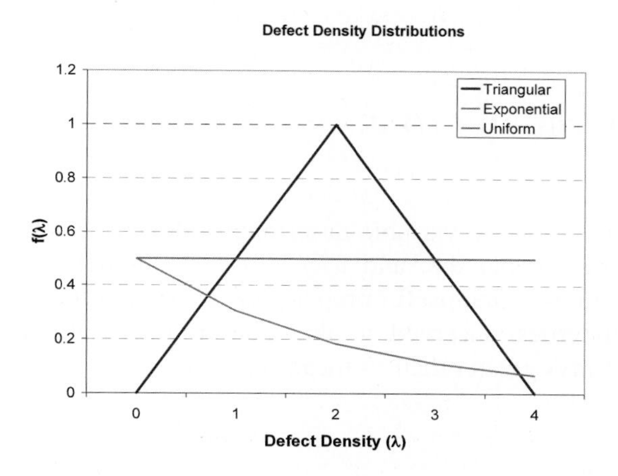

Figure 7-2. Defect Density Distributions.

Noting that $p\overline{N} = \overline{N}/B$ is the average defect density per site we have that after numerical evaluation of (7.5) we obtain a yield of 0.13 as well. A quick inspection of the Poisson distribution properties reveals that its mean equals its variance. For the particular distribution of Table 1 Poisson statistics seem simply unsuitable. However, the simplicity of this formula has appealed many foundries because often enough many of the distributions that are observed follow the Poisson distribution. Let us now project the yield of larger area chips using (7.5). This result is also displayed in Figure 7-1. One notorious aspect is that yield forecast becomes worse for larger chip areas.

The above models fall short from predicting actual yield values because among other things the defect density cannot be correctly modeled by a Poisson distribution. In practice we have that defects are not uniformly distributed; wafer maps show that defects tend to cluster [26][27]. Hence, for cases like this, a Poisson distribution is not suitable because it neglects clustering effects. Moreover, actual data shows that defect densities vary within the wafer, from wafer to wafer, and even from lot to lot[8][7][29].

Thus, we need to find a "wider" distribution than Poisson since for actual data we have that the variance is much greater than the mean. This shortcoming was identified by Murphy who proposed the use of compound statistics to account for varying defect densities as follows [20]

$$P(X = k) = \int_0^\infty \frac{\lambda^k e^{-\lambda}}{k!} f(\lambda) d\lambda \tag{7.7}$$

$$Y = P(X = 0) = \int_0^\infty e^{-\lambda} f(\lambda) d\lambda \tag{7.8}$$

where λ is a generic variable describing a defect density, in number of defects per area or per die, and $f(\lambda)$ is used to model the varying defect density distribution. One useful property of (7.6), and that also satisfies the statistical properties observed in the distribution of Table 1, is that its variance is always greater than its mean, i.e.

$$\bar{x} = \lambda_o \tag{7.9}$$

$$V(x) = \lambda_o + V(\lambda_o) \tag{7.10}$$

where λ_o denotes the average defect density of $f(\lambda)$. Defect density variations have a significant impact on yield. It is thus very important to properly model them. Figure 7-2 shows various plots that attempt to describe the distribution of defect densities. Unable to experimentally verify the defect density distribution, Murphy believed that a Gaussian distribution was appropriate. Rather than using the actual Gaussian distribution, he used a triangular distribution to simplify the mathematics behind it. This distribution is

$$f(\lambda) = \begin{matrix} \lambda/\lambda_o & 0 < \lambda \leqslant \lambda_o \\ 2 - \lambda/\lambda_o & \lambda_o \leqslant \lambda \leqslant 2\lambda_o \end{matrix} \tag{7.11}$$

where λ_o is the average defect density. Introducing (7.11) into (7.8) we obtain the following yield formula

$$Y = P(X=0) = \left(\frac{1-e^{-\lambda_o}}{\lambda_o} \right)^2 \tag{7.12}$$

Evaluating (7.12) with $\lambda_o = p\overline{N} = 618/308 = 2.006$ we obtain a yield of 0.18 which is a better result than the one obtained with (7.4) but it is still an underestimation of actual yield numbers. The poor yield forecast is a result of using the Gaussian distribution to model the defect density. Another alternative is to use a uniform defect density distribution defined as

$$f(\lambda) = 1/2\lambda_o \qquad 0 \le \lambda \le 2\lambda_o \tag{7.13}$$

Introducing (7.13) into (7.8) we obtain

$$Y = \frac{1-e^{-2\lambda_o}}{2\lambda_o} \tag{7.14}$$

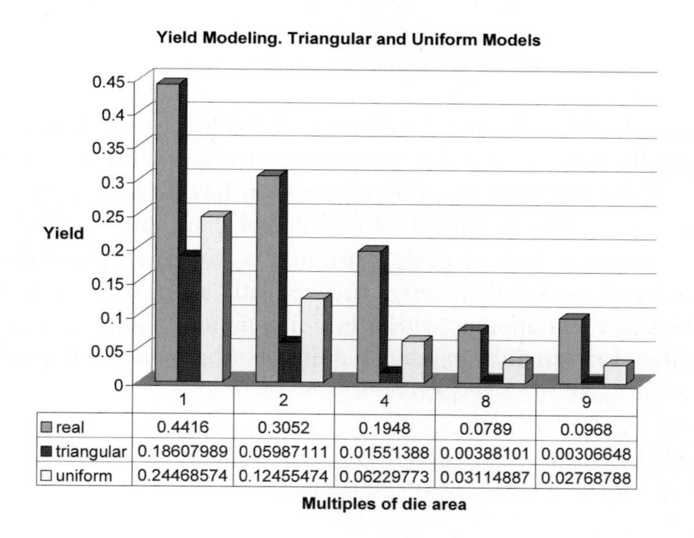

	1	2	4	8	9
real	0.4416	0.3052	0.1948	0.0789	0.0968
triangular	0.18607989	0.05987111	0.01551388	0.00388101	0.00306648
uniform	0.24468574	0.12455474	0.06229773	0.03114887	0.02768788

Figure 7-3. Yield modeling for triangular and uniform defect density models.

Numerically evaluating (7.14) we obtain a corresponding yield of 0.25 which is a forecast much closer to actual data. And so, different defect distributions will render differing yield values. The use of a delta distribution $f(\lambda)=\delta(\lambda-\lambda_o)$ will return the same yield formula of (7.6). Let us use now

(7.12) and (7.14) to predict the yield of larger area chips. These results are shown in Figure 7-3. The observed trend is that these models show also pessimistic yield forecasts.

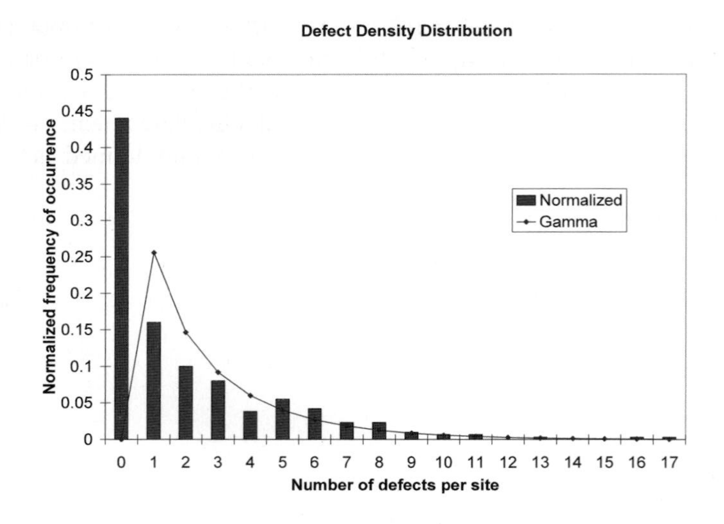

Figure 7-4. Fitting of Gamma distribution.

In general, if we plot the distribution of Table 1 we will observe that it differs greatly from a Gaussian, triangular, or a normal distribution, Figure 7-4. Essentially we have that this distribution has a rather long tail, e.g. a strong variance. This is typical of real defect distributions. Back in 1967, Seeds conjectured that typical distributions consisted basically of two distributions of defect densities, i.e. one with a large population of low defect densities and another with a low population of high defect densities [23]. He therefore used the exponential distribution to model this effect

$$f(\lambda) = \frac{e^{-\lambda/\lambda_o}}{\lambda_o} \tag{7.15}$$

which after introduction in (7.7) renders the following yield formula

$$Y = P(X=0) = \frac{1}{1+\lambda_o} \tag{7.16}$$

Numerically evaluating this result returns a yield of 0.33. Figure 7-5 shows a yield projection using Seed's model for various die areas. A closer to actual data result is obtained now thanks to the proper modeling of the defect density distribution. This was also clearly observed by Okabe and Stapper who used a Gamma distribution for modeling the distribution of defect densities [21][28]; this is shown also in Figure 7-5. The Gamma distribution is as follows:

$$f(\lambda)=\frac{\lambda^{\alpha-1}e^{-\frac{\lambda}{\beta}}}{\Gamma(\alpha)\beta^{\alpha}} \quad (7.17)$$

where α and β are free parameters. The mean and variance of (7.17) are given as

$$E(\lambda)=\lambda_o=\alpha\beta \quad (7.18)$$

$$V(\lambda)=\alpha\beta^2 \quad (7.19)$$

It is possible to show that

$$\alpha=\frac{E^2(\lambda)}{V(\lambda)} \quad (7.20)$$

which is the inverse of the normalized coefficient of variation. This parameter alone describes the spread of the Gamma distribution. Why is this relevant? Because the "length" of a distribution like the one of Table 1 can easily be described with this single parameter. In other words, we are somehow including now the effects of defect clustering in our modeling of defect density distributions. Introducing (7.17) into (7.7) and evaluating the integral results in the negative binomial distribution.

$$P(X=k)=\frac{\Gamma(\alpha+k)\beta^k}{k!\Gamma(\alpha)(1+\beta)^{\alpha+k}} \quad (7.21)$$

Noticing that the expected value of defect densities is $E(\lambda)=\lambda_o=\alpha\beta$, which is more practical to use because it relates to the average defect density observed from real data, we can manipulate (7.21) to compute yield as follows:

$$Y = P(X = 0) = \frac{1}{\left(1 + \frac{\lambda_o}{\alpha}\right)^{\alpha}} \tag{7.22}$$

To numerically evaluate (7.22) we need to be able to compute α from the data of Table 1. Introducing the mean (7.18) and variance (7.19) into (7.9) and (7.10) respectively, we can relate to the mean and variance of the defect density distribution of Table 1 as follows:

$$E(x) = \bar{x} = \lambda_o \tag{7.23}$$

$$V(x) = \lambda_o + V(\lambda_o) = \lambda_o (1 + \frac{\lambda_o}{\alpha}) \tag{7.24}$$

Combining (7.23) and (7.24) and solving for α we have

$$\alpha = \frac{\bar{x}^2}{V(x) - \bar{x}} \tag{7.25}$$

Numerical evaluation of (7.25) gives α = 0.68. Small values of α indicate that the defect distribution has a very long tail, e.g. that there is defect clustering, while the opposite indicates that the placement of defects follows a Poisson distribution. Now that we know α we can numerically evaluate (7.22); this gives a yield of 0.39. This result is closer to actual data yield results. A yield projection for larger chip areas is shown in Figure 7-5. We can see that the yield prediction scales properly with the area of the chip.

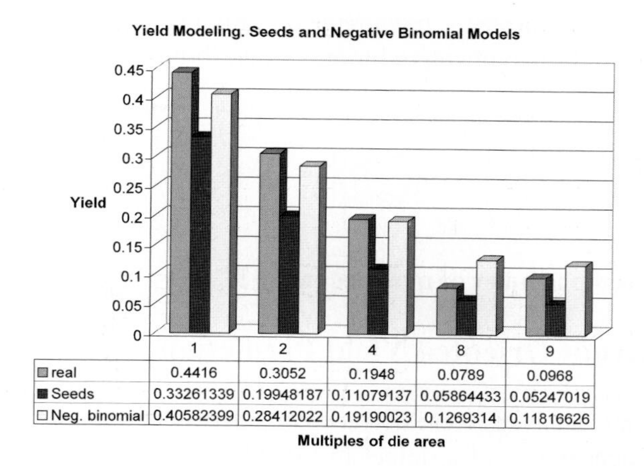

	1	2	4	8	9
real	0.4416	0.3052	0.1948	0.0789	0.0968
Seeds	0.33261339	0.19948187	0.11079137	0.05864433	0.05247019
Neg. binomial	0.40582399	0.28412022	0.19190023	0.1269314	0.11816626

Figure 7-5. Yield of Seeds and Negative Binomial models.

In the previous derivations we have assumed a kind of ideal manufacturing process. In practice there are regions in the wafer that suffer of "systematic" errors like threshold voltage shifts, or shrinking of critical dimensions, etc. Yield models like the ones described before will be inappropriate to deal with this new problem. To account for these systematic errors a gross yield factor Y_o denoting the probability that the chip is free of systematic defects has been used as follows [22]:

$$Y = Y_o \left(1 + \frac{\lambda_o}{\alpha}\right)^{-\alpha} \tag{7.26}$$

There are various ways of estimating the parameters of (7.26). The most common one is with the help of the so-called windowing method described at the beginning of this Section. Yield is calculated accordingly for each sub die multiple such that the values of α, and Y_o can be calculated by curve fitting.

The model that was just derived is known as the large-area defect clustering model. This means that we have dealt with the die as a unit and that sub areas within the die are correlated to the appearance of new defects. For large area chips this model is becoming rather impractical since the occurrence of large-area defects is unlikely. Large area chips imply that defect clusters are smaller than the die area and that these defects will affect the various IC modules in different ways. Therefore, it is possible to compute the yield of each module noting that each one has its own average defect density and clustering parameter α. This principle is used in what is known as the small-area negative binomial distribution [10][9] shown next

$$Y = Y_o \prod_{i=1}^{M} \left(1 + \frac{\lambda_{oi}}{\alpha_i}\right)^{-\alpha_i} \tag{7.27}$$

where M is the number of modules in the IC.

1.1 Layout Oriented Yield Prediction

The previous analyses were derived considering that the IC is a black box and that the occurrence of a defect in this black box is simply catastrophic. As we deepen in our analyses we have to realize that there is more than one kind of defect type and that not all of the area of the IC is susceptible to fail. Recall that spot defects can be classified into two types, the ones that bridge patterns and the ones that break patterns. The electrical consequence of these defects are a "short circuit" and an "open circuit," respectively. It follows then that a defect is catastrophic if it creates an electrical fault[35]. Obviously it suffices to have one catastrophic defect to kill the whole IC. This is the reason why the black-box model is useful.

This is a very important result because now we are in position of having a more accurate yield model which considers the geometrical properties of the IC, i.e. we have a high dimensional yield model. Thus, rather than considering the entire die, we consider now only the fraction that is susceptible to fail. We can do this by modifying the average defect density in (7.26) as follows

$$Y = Y_o \left(1 + \frac{\theta \lambda_o}{\alpha}\right)^{-\alpha} \tag{7.28}$$

Expanding for every module i in the IC, and then for every layer j and then for every defect type k we have

$$Y = Y_o \prod_{i=1}^{M} \prod_{j=1}^{L} \prod_{k=1}^{K} \left(1 + \frac{\theta_{i,j,k} \lambda_{oi,j,k}}{\alpha_{i,j,k}}\right)^{-\alpha_{i,j,k}} \tag{7.29}$$

2. YIELD ENGINEERING

In a typical semiconductor fab environment yield ramp-up of a new process centers around appropriately tuning process steps based on observations collected from enough statistical data coming from Process Control Monitors (PCM) and Process Engineering Monitors (PEM) as well as from specialized yield ramp up test chips. PCMs are used primarily to tune the process recipe, PEMs to adjust understand defectivity data, and the test chips to observe the process tuning on a large scale integration.

It is not unusual to employ test data to find out design issues or manufacturability problems early in the yield ramp up phase. Due to "pollution" of the process, often, one has to identify which wafers and dies are relevant and that can be used to do the forecasting. Recall that yield loss is caused by random defects, process shifts, design sensitivity, and also by measurement related issues such as bad contacts, noise in measurements, etc. In this Section we illustrate several aspects of yield engineering practices based on a test chip for yield ramp-up purposes in CMOS process. This test chip contains several modules, they are: *a)* 1 ROM (krom), *b)* SRAMs (sram), *c)* random dsp-like modules (rdmr), and *d)* modules with a subset of the standard cell library (rdmc). The presented data comes from 52 wafer batches, for a total of 439 wafers and 75000 test chips. To make sure that the data is useful and thus valid, several filters were applied. These filters are as follows: to eliminate process "pollution" only wafers that exhibited a yield higher than 50% (excluding SRAMS) were used; only re-tested wafers were used to eliminate measurement noise; only dies that passed contact, I_{DDQ}, and power up tests were used; only wafers that contain more than 100 good dies were used. Figure 7-6 shows the fab yield of several modules as a function of time; in this case the time line spans over two years. One can see that over time the yield varies and that it is mostly well behaved for most of the modules except for the SRAM. During the yield ramp up phase many process changes or adjustments are made. These experiments are done in so called split-lot matrices where for instance various process steps are swept to render transistor characteristics that go from slow to fast; as well as tuning specific process steps to improve, say, the via yield, the critical dimension yield, etc. One can also see from Figure 7-6 that changes may adversely affect some modules while others are positively tuned. Observe also that towards the end of the ramp-up phase the SRAM yield is controlled as well to peak above 90%.

Figure 7-6 puts in perspective a typical example of daily life yield engineering, e.g. monitoring yield, preventing yield deeps, improving process steps, making sure that the equipment is appropriately maintained, etc.

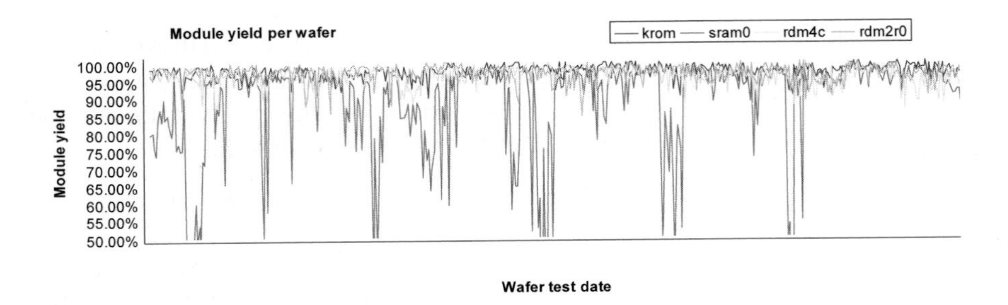

Figure 7-6. Module yield as a function of time.

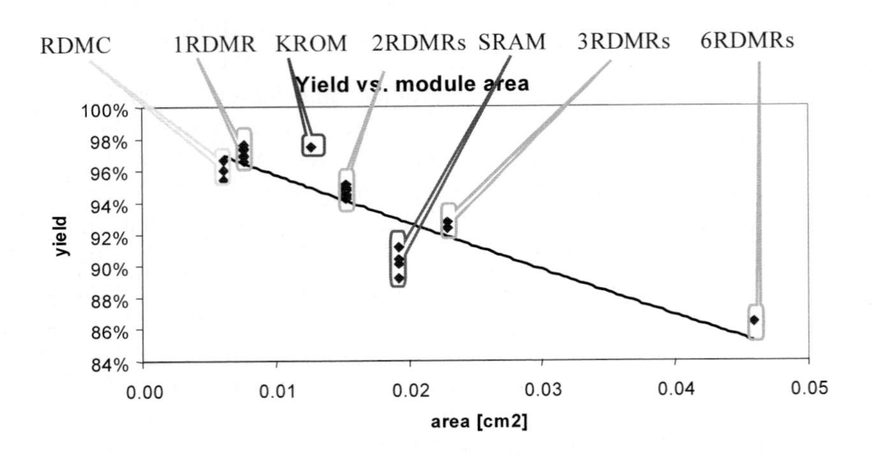

Figure 7-7. Yield vs. module area.

Let us focus now on the specific yield of the modules. For a given "early" time snapshot during the ramp-up process consider the yield per module shown in Figure 7-7. Each dot in the clouds corresponds to an instance of the module in the test chip. Except for the *krom* one can observe the general trend that yield decreases when area increases as discussed at the beginning of this chapter. Also interesting, is that for multiple module instances the yield is not always constant. This mismatch arises not only from defectivity but also from aspects such as location of the module in the test chip, e.g. if it is close to dense module, orientation of the module, die placement in the wafer reticle, etc. From the various modules, the SRAM is the one that has the wider spread; this can be attributed to the fact the module is quite dense and also to fact that the SRAMs are placed at different locations and with different orientations in the test chip.

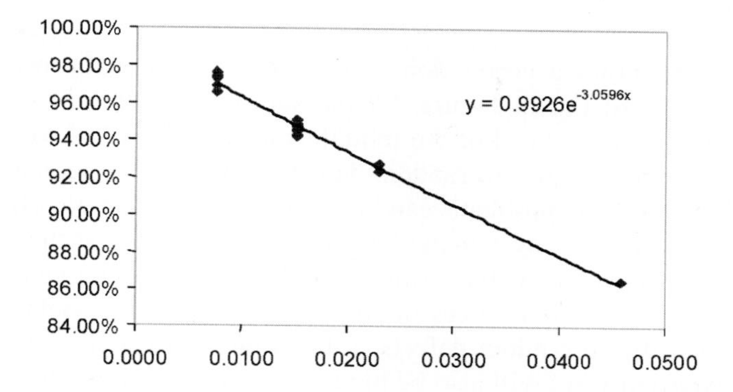

Figure 7-8. Fitting yield measurements to Poisson distribution.

For the measured data a Poisson distribution is a sufficient fit of goodness for yield forecasting. By making use of modules with similar layout characteristics, e.g. logic only without memory, one can quickly extract an estimation of the population's average defect density, D_o. A simple exponential fit to the Poisson curve $Y = Me^{-AD}$ can be carried out to find out the average defect density; in the previous formula M is a so called "fit offset" This factor encloses yield variations due to systematic problems. For this example, $D_o = 3$ defects/cm^2 and $M = 0.993$ as shown in Figure 7-8.

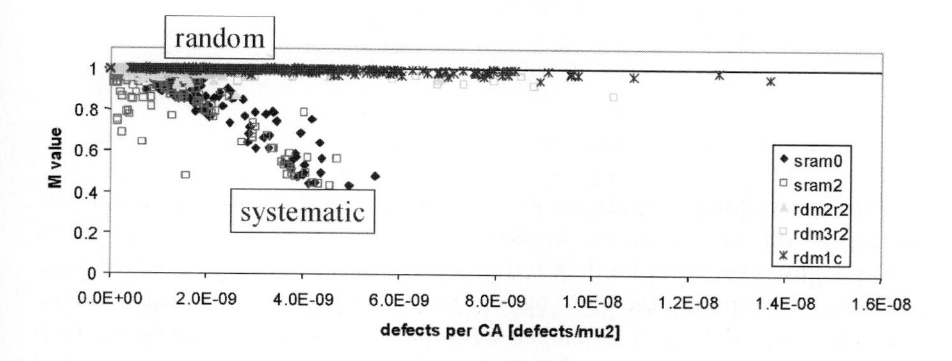

Figure 7-9. Comparing random vs systematic yield losses.

Low values of M are an indication that yield losses are primarily due to systematic problems, while high values are related to random defects. Systematic problems are of particular interest for process engineers because this is a clear indication that some process step is out of control. Much of the work in yield engineering relates to the search and solutions of this kind of problems.

Figure 7-9 shows a correlation of random vs. systematic problems by quantifying M and D_o. In Figure 7-9 the sum of the critical area over all metal layers is used instead of the module's area. This is just to accentuate even more the dependence of random defects on the layout geometry. When M has a low value the problems can be attributed to systematic effects. This stems from the fact the systematic losses affect many dies at the same time or the same dies over a batch of wafers, and thus the resulting yield is low. It should be clear that high values of M reflect the fact that yield losses are primarily because of random defects. It turns out that if defect clustering is high the expected yield will also be high. However, when the distribution of defects is more uniform yield decreases; this is the case observed for *rdm3r2* in Figure 7-9 for relatively large defect densities. In this case more accurate yield models need to be employed such as the negative binomial model discussed in the previous section.

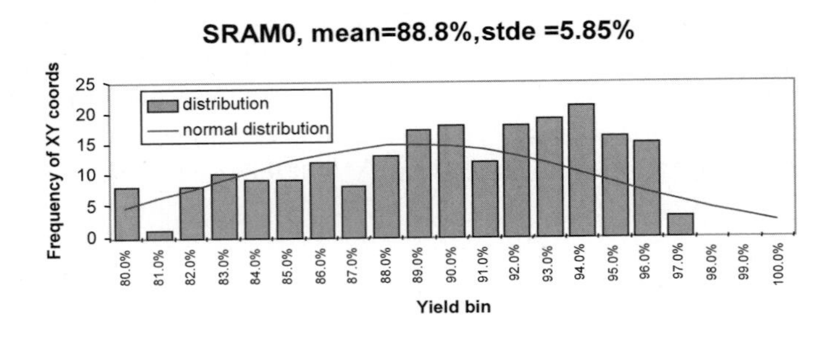

Figure 7-10. Example of manufacturing problems of the type excessive fluctuations.

Wafer stacking operations are also often used to detect systematic problems, for instance, a die location that is always failing, or a low-yield wafer region can easily be detected using this method. Wafer yield stacking consists of computing the yield per die location along all the stacked wafers. Several composite wafer maps can be defined as follows: *Functional map* shows all the die locations that had 100% yield over the set of stacked wafers. The *zero map* is the inverse of the functional map, it basically shows all the locations that had no yield. A *history map* is an uncorrelated wafer map that shows numerically the site yield.

Based on the previous basic maps it is possible to do some yield diagnosis. The following is an example. If the yield is always low and some specific dice always fail then the reason for yield loss is systematic, otherwise it is random. If the yield of the *history* map is low and the yield of the *zero* map is high then the yield loss is *systematic.* If the yield is zero on some of the wafers then *gross manufacturing errors* can be assumed. On the other hand, if the yield is not zero but it is low on all of the wafers then the errors are due to *local* or *global* reasons. If the yield of the *functional* map of some of the wafers is zero then the reason is a *gross manufacturing error.*

Global disturbances can be classified as *design-process mis-centering* and as *excessive fluctuations.* The former is manifested when the majority of the dice have a performance outside an acceptability region. The latter one is manifested when the performance has a very wide spread, which also results in a low number of dice located inside of the acceptability region. In general if the *frequency distribution* of good and bad dies shows a spread of the number of dice along the different site yields then the disturbances are of the kind of *excessive fluctuations.*

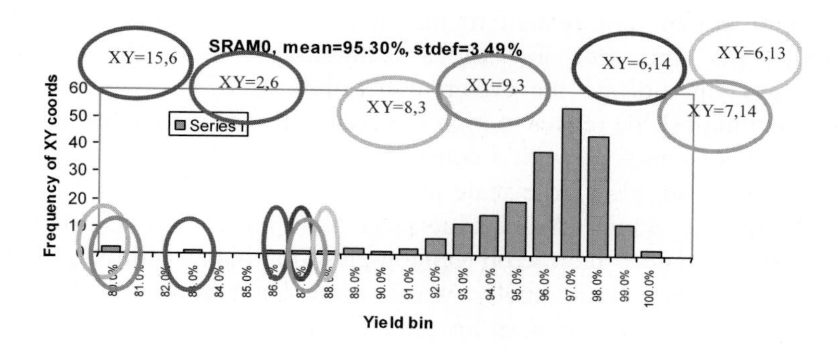

Figure 7-11. Incidence of repeating defects.

Figure 7-10 shows an example for one of the SRAMs of the test chip. Basically, when no data filtering is applied, we have that the site yield is simply smeared out along various yield values when instead it should have been around 97%. This is an example of excessive fluctuations. In the process of yield improvement there are some local problems that manifest themselves as *repeating defects.* Figure 7-11 shows how certain wafer coordinates consistently render a lower than expected site yield.

3. ECONOMICS AND YIELD FORECASTING

The advent of high-speed deep submicron circuits with larger die sizes and shorter life cycles lends itself to an increase in fabrication cost. Basically, the economic success of a product may depend on a timely and accurate design for manufacturability (DfM) strategy. Put in other words, an appropriate yield forecast may render significant benefits in both time-to-market and manufacturing costs. Yield forecasting is essential for the development of new products. It effectively shows if a design is feasible of meeting its cost objectives.

Maly [5] has proposed a simple model to relate yield, area, technological process and economical profitability. The pictorial representation is shown in Figure 7-11a. Here we have that processing complexity is related to the complexity of the semiconductor process, e.g. whether it is CMOS, BiCMOS, or whether it uses four or nine metal layers. A huge capital investment is required to set up such foundries, consequently the impact on chip cost is exponential to write off costs. Yield and area have an exponential relationship as we saw it before. Area and processing complexity are in turn related to the number of transistors that can be integrated per unit area in various technology nodes. The choice of technology will indicate the design integration style. For example, large scale integration (LSI) relates to high integration in a moderate technology – for instance a technology with a constrained number of interconnect layers. On the other hand, very large scale integration (VLSI) relates to submicron technologies, e.g. 65nm CMOS; ultra large scale integration (ULSI) relates to very advanced deep submicron technologies for which a full fledge silicon foundry is needed, and finally wafer scale integration (WSI) relates primarily to large area chips in an ultra clean fab. Figure 7-12b gives an example of how the economical feasibility of a chip can be investigated. A chip with a given transistor count integration is analysed in the example. The integration complexity points to an expected yield and to a minimum processing requirements. These two pointers in turn indicate the tentative cost of such a chip. Profitability can be deduced when the processing and engineering costs are below an acceptable commercial value.

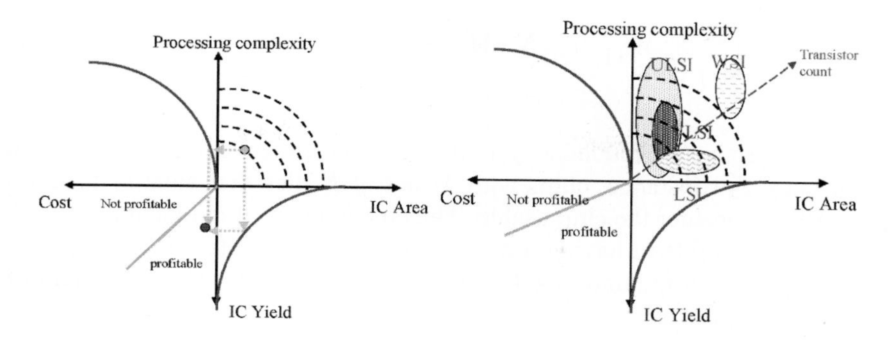

Figure 7-12. Manufacturability analysis of an IC. (a) Setup, (b) example for a VLSI chip.

In the following we will address the economical feasibility of an array processor.

3.1.1 Yield of Partially-Good Chips

Let us consider the case of having a chip architecture with N identical tiles. Suppose that M of these tiles function properly and that $(N\text{-}M)$ tiles are defective. Furthermore, let the probability of finding a fault free tile be denoted by

$$p=Y_o\ exp(-AD) \tag{7.30}$$

where A is the tile area, D is the average number of defects per cm^2, and Y_o is the yield due to global defects. For simplicity and assuming that we are dealing with a mature process we can consider that $Y_o = 1$. This yield model, the Poisson model, underestimates the actual yield. Thus, the results hereby obtained can be considered as a lower bound for the economics study. Let us further assume that every defect is a killer, e.g. if the defect lands on the chip, the chip is regarded as faulty. Let us further assume that random defects do not cluster, or that the cluster is smaller than the total tile area. In a real-life scenario defects are correlated to each other and show clustering. The probability of finding M fault free tiles is obtained as p^M. It also follows that the probability of finding a faulty tile is $(1-p)$ and that the probability of finding $(N\text{-}M)$ faulty tiles is $(1\text{-}p)^{N-M}$. The number of ways in which $(N\text{-}M)$ faulty tiles can occur is given by the binomial coefficient C(N,M). The probability (or yield), Y_{MN}, of finding exactly M fault free tiles out of N tiles can be expressed as

$$Y_{MN} = \binom{N}{M} p^{M} (1-p)^{N-M} \tag{7.31}$$

The objective of redundancy is the replacement of defective tiles with good ones. For instance, chips with M out of N tiles that have to function properly to consider the chip usable. Thus the number of redundant tiles is $R = N - M$. The odds of having a number M, $M+1$, $M+2$, etc. of good tiles on such chips is is a mutually exclusive probabilistic event. Then the yield of such chip is given as

$$Y = \sum_{i=0}^{R} Y_{M+i,N} \tag{7.32}$$

3.1.2 Yield Forecast for an Array of Processors

The study we will present corresponds to an array of homogenous processors, henceforth tiles. There exist two types of tiles in this processor: 1) Compute Tile and 2) Periphery Tile. In this chip, fault tolerance can be achieved at three hierarchical levels: 1) tile to tile, 2) number of IPs that integrate a tile, and 3) within an IP. Tables 2 and 3 show the estimated area per module of the compute and periphery tiles.

Table 7-2. Area estimates of compute tile.

Module	Number	Area/module (mm^2)	Total Area (mm^2)
Module A	5 (min 3)	10.1 (RAM:4mm^2)	50.5
Module B	3	1.1	3.3
DRAM	2x2Mb	7	28
Other	1	6	6

The architecture we are investigating is that of three computing tiles and one periphery tile. The tile-to-tile fault tolerance obeys two rules to consider

Table 7-3. Area estimates of Periphery Tile (mm²).

Processor	
Module A	10.50
Module C	5.00
Module B	10.80
Connectivity	
Interfaces	2.71
Video	
Decoders and processors	14.13
Infrastructure	
DMA and miscellaneous	0.3
Embedded DRAM (4 MB)	15.23

that the chip is a working chip: 1) Periphery Tile must be functional, 2) At least M out of N computing tiles must be functional. Thus, the chip yield can be computed by slightly modifying (2) as follows:

$$Y_{array} = Y_{periphery}\left(\binom{M}{N} Y_{compute}^{M}\left(1 - Y_{compute}\right)^{N-M}\right) \quad (7.33)$$

where $Y_{periphery}$ and $Y_{compute}$ is the yield of one periphery and one compute tile, respectively. Let us now examine $Y_{compute}$. To begin with, redundancy is allowed in the number of A modules of the compute tile. About 40% of the area of one module A is used for RAM. This area can be discarded from the yield computation since it is assumed that the RAM is repairable with a yield, Y_{RAM}, close to 100%. In a similar way, the DRAM area can as well be discarded since it is assumed that it is a repairable memory with yield, Y_{DRAM}, close to 100%. The remaining area is the "killer" area, e.g. the tile becomes faulty if a defect lands in this area. This area is thus 9.9mm^2. One can compute the yield of the compute tile as follows:

$$\begin{aligned} Y_{compute} &= Y_{killer} Y_{A-module} \\ &= \left(\exp(-(A_{B-module} + A_{other} + (1 - Y_{DRAM})A_{DRAM})D)\right) \\ &\quad \left(\binom{M_A}{N_A} Y_A^{M_A}(1 - Y_A)^{N_A - M_A}\right) \end{aligned} \quad (7.34)$$

where D is the average defect density, A is used to designate the area of the corresponding module, M_A is the minimum number of A modules out of a total of N_A modules, and Y_A is used to designate the yield of one module A. The yield of module A is simply computed as

$$Y_A = \exp(-(1 - Y_{RAM})A_{RAM} + A_{logic})D) \quad (7.35)$$

where it is accounted for the fact that the RAM can have a high yield. Let us now turn our attention to the periphery tile. This tile has nine *B* modules, 3 for each compute tile. Redundancy can be afforded here as well. The remaining area is essential for the functional operation of the tile. From Table 7-3 one can consider the area of periphery tile excluding the area of module B and the DRAM's as the killer area. The yield of the periphery tile can be computed in a similar fashion.

$$\begin{aligned} Y_{periphery} &= Y_{killer} Y_{\text{B-module}} \\ &= \left(\exp(-A_{other} D)\right) \\ &\left(\binom{M_B}{N_B} Y_B^{M_B} (1 - Y_B)^{N_B - M_B} \right) \end{aligned} \qquad (7.36)$$

3.1.3 Manufacturing Scenarios

A yield forecast will be performed under two assumptions. In one case we assume that the process is stable and mature and thus the corresponding defect density is low, e.g. D= 0.5def/cm^2. The second case assumes that the process is stable but not yet mature with an average defect density of D=1def/cm^2. In the absence of redundancy the raw yield of the array chip, calculated from (36) and using only the area estimates, results in 30% yield for D=0.5def/cm^2 and 8% yield for D=1def/cm^2. Further it is assumed that the chips are fabricated on a *d=8inch* wafer with a manufacturing price of *g=\$1000/wafer*. Initial mask costs are not taken into account. The manufacturing price of a chip is calculated as

$$price = \left(\frac{\pi (d/2)^2}{A_{tobiko}} \right) Y_{array}\, g \qquad (7.37)$$

Full hardware and at least M *out of 3 compute tiles*

In this case study we examine the probability of obtaining partially good chips. It is assumed that to have a working tile all of its modules (full hardware) must be functional. The results are shown in Table 7-4.

Full hardware and at least 3 out of N compute tiles

Under this scenario it is assumed that all the IPs within a tile must be functional. Redundancy is assumed only at the tile level. Table 7-5 shows the forecast of *at least* 3 good compute tiles. This is a good example that shows the fault tolerance benefits of a homogenous array although in practice the target for the array processor is to have 3 compute tiles. The manufacturing price per chip is also shown. The total array processor's area does not grow linearly from one configuration to another because in the periphery tile only the number of *B* modules increases.

Table 7-4. Yield of partially good chips. Probability of obtaining at least M out of 3 tiles.

	***At least M* out of *3* Tiles**		
	1/3	**2/3**	**3/3**
Yield D = 0.5 def/cm^2	0.69	0.59	0.3
Yield D = 1 def/cm^2	0.45	0.29	0.08

Table 7-5. Yield for full working hardware and at least 3 good compute tiles.

	***At least 3* out of *N* Tiles**			
	3/3	**3/6**	**3/9**	**3/12**
Yield D = 0.5 def/cm^2	0.3	0.77	0.8	0.8
Yield D = 1 def/cm^2	0.088	0.5	0.61	0.64
Number of chips	129	71	48	37
Area (cm^2)	2.43	4.422	6.41	8.4
Price (D=0.5) ($$)	26.12	18.20	25.96	33.64
Price (D=1) ($$)	88.05	28.38	33.66	42.12

Redundancy in the number of A*-modules of compute tile*

In this scenario we look at the case of three compute tiles with partially good A-modules. The case when 3, 4, and 5 out of 5 modules is examined, as well as an alternative scenario with 6 modules. Results are displayed in Table 7-6.

Table 7-6. Yield with 3 compute tiles and redundant A-modules.

	***M* out of *N* A modules**						
	3/5	**4/5**	**5/5**	**3/6**	**4/6**	**5/6**	**6/6**
Yield D = 0.5 def/cm^2	0.6	0.57	0.3	0.61	0.61	0.55	0.26
Yield D = 1 def/cm^2	0.36	0.29	0.088	0.37	0.36	0.27	0.06
Price (D=0.5) ($$)	12.75	13.51	26.12	14.26	14.34	15.58	33.84
Price (D=1) ($$)	21.28	25.9	88.05	23.46	24.3	31.89	131.73

Redundancy in the number of B *modules of the periphery tile*

We examine here the case of full and partial functionality in the compute tile and a tolerance of one failing module *B* per compute tile in the periphery tile.

Table 7-7. Yield with 3 compute tiles and partially good A modules, and one failing B module in the periphery tile.

	***M* out of *N* A modules in compute tile**	
	5/5	**3/5**
Yield D = 0.5 def/cm²	0.34	0.70
Yield D = 1 def/cm²	0.11	0.48
Price (D=0.5) ($$)	22.72	11
Price (D=1) ($$)	66.62	16.1

The yield model used in this analysis provides a lower bound on the forecasts. Worth observing are the big yield differences that result from using average defect densities of 0.5 and 1 defect per cm^2. Table 7-8 summarizes the typical results.

Table 7-8. Expected yield for a 3-compute and 1-periphery tile configuration.

	D = 1 def/cm²	**D = 0.5 def/cm²**
Yield (3/3 tiles)	0.08	0.3
Yield (at least 2/3 tiles)	0.29	0.59
Yield (3/5 A, 3/3 tiles)	0.36	0.6
Price (3/3 tiles) [$]	88.05	26.12
Price (3/5 A, 3/3 tiles) [$]	21.28	12.75

4. CONCLUSIONS

This chapter addressed the fundamentals of yield modeling. Rather than following a strict derivation of the yield formulae, an approach "by example" was followed in which the various yield models were compared against each other. Advantages and improvements over time of the yield models were shown. An essential improvement giving origin to the negative binomial was carried out by Okabe and later extensively used by Stapper. The negative binomial model has proven to be better than other models, although its use in the fab requires of more data acquisition and engineering for the correct fitting of its parameters. Yield has a direct impact on chip cost. As way of example the cost of an ideal multiprocessor was analyzed based on various yield scenarios. Another important area for which yield modeling essential corresponds to the actual improvement of process steps in the fab to guarantee the reliability and reproducibility of the process. This procedure, described as yield engineering in this chapter links to the fundamentals of defect and fault modeling shown in Chapter 2.

References

1. A. Allan and J.A. Walton, "Hierarchical critical area extraction with the EYE tools, " Proc. 1995 IEEE Int. Workshop Defect and Fault Tolerance in VLSI Systems, pp. 28-36, Nov. 1995.
2. I. Chen and A.J. Strojwas, "RYE: A realistic yield simulator for structural faults," Proc. International Test Conference 1987, Washington, DC, USA, 1-3 Sept. 1987, pp.31-42.
3. D.L. Critchow, R.H. Dennard, S. Schuster and E.Y. Rocher, "Design of Insulated-Gate Field-Effect Transistors and Large Scale Integrated Circuits for Memory and Logic Applications," IBM Research Report RC 2290, Oct. 1968.
4. J.A. Cunningham, "The use and evaluation of yield models in integrated manufacturing," IEEE Trans. On Semiconductor Manufacturing., vol. 3, pp. 60-71, May 1990.
5. A. V. Ferris-Prabhu, "Role of defect size distribution in yield modeling," IEEE-Transactions on Electron Devices, vol.ED-32, no.9, p.1727-36, Sept. 1985.
6. A.V. Ferris Prabhu, "Modeling the critical areas in yield forecasts," IEEE J. Solid State Circuits, vol. SC-20, no. 4, pp. 874-878, 1985.
7. A.V. Ferris Prabhu et. al. "Radial Yield variations in semiconductor wafers," IEEE Circuits and Devices Magazine, vol.3, no. 2., pp. 42-47, 1987.
8. A. Gupta, W.A. Porter and J.W. Lathrop, "Defect analysis and yield degradation of integrated circuits," IEEE J. Solid State circuits, vol. SC-9, no. 3, pp. 96-103, 1974.
9. I. Koren, Z. Koren, and C.H. Stapper, "A unified negative binomial distribution for yield analysis of defect tolerant circuits," IEEE Trans. on Computers, vol. 42, pp. 724-737, June 1993.
10. I. Koren, Z. Koren, "Defect tolerance in VLSI circuits: Techniques and yield analysis," Proc. IEEE, vol. 86, no. 9, pp. 1817- 1836, Sept. 1998.
11. W. Maly, "Computer Aided Design for VLSI circuit manufacturability," Proc. IEEE, vol. 78, pp. 356-392, Feb. 1990.
12. H.G. Parks and E.A. Burke, "The nature of defect size distributions in semiconductor processes," IEEE/SEMI Int. Sem. Manufct. Science Symp., pp. 131-135, 1989.
13. J.P. de Gyvez and D.K. Pradhan, Eds., IC Manufacturability: The Art of Process and Design Integration," IEEE Press, 1999.
14. J.P. de Gyvez, "IC defect sensitivity for footprint type spot defects," IEEE Trans. On Computer Aided Design, vol. 11, pp. 638-658, May 1992.
15. W. Maly, A.J. Strojwas and S.W. Director, "VLSI yield prediction and estimation: a unified framework," IEEE Trans. on CAD, vol. CAD-5, no. 1, pp. 114-130, 1986.

16. W. Maly and J. Deszczka, "Yield estimation model for VLSI artwork evaluation, " Electron. Lett., vol 19, no. 6, pp. 226-227, 1983.

17. W. Maly, "Yield models-comparative study," Proc. of the International Workshop Defect and Fault Tolerance in VLSI Systems., vol.2, pp.15-32, Oct. 1989.

18. W. Maly, "Modeling of lithography related losses for CAD of VLSI circuits" IEEE Trans. On Computer Aided Design, vol. CAD-4, no. 3, pp. 166-177, 1985.

19. G.E. Moore, "What levels of LSI is best for you," Electronics, vol. 43, pp.126-130, Feb. 16 1970.

20. B.T. Murphy, "Cost Size optima of monolithic integrated circuits," Proc. IEEE, vol. 52, no. 12, pp. 1537-1545, 1964.

21. T. Okabe, M. Nagat, S. Shimada, "Analysis on yield of integrated circuits and a new expression for the yield," Electron. Eng. Jpn., vol. 92, no. 6, pp. 135-141, 1972.

22. O. Paz and T.R. Lawson, "Modification of Poisson statistics: Modeling defects induced by diffusion," IEEE J. Solid-State Circuits, vol. SSC-12, pp. 540-546, Oct. 1977.

23. R.B. Seeds, "Yield economic and logistic models for complex digital arrays," IEEE Int. Conv. Rec., pt. 6, pp. 61-66, April 1967.

24. Z. Stamenkovic and N. Stojadinovic, "New defect size distribution function for estimation of chip critical area in integrated circuit yield models," Electron. Lett., vol. 28, no. 6, pp. 528-530, Mar. 1992.

25. C.H. Stapper and R.J. Rosner, "Integrated circuit yield management and yield analysis: development and implementation," IEEE Trans. On Semiconductor Manufacturing, vol. 8, no. 2, May 1995.

26. C.H. Stapper, "Yield Model for fault clusters within integrated circuits," IBM J. Res. Dev. Vol. 28, no. 5, pp. 636-640, 1984.

27. C.H. Stapper, "On Yield, fault distributions and clustering of particles," IBM J. Res. Dev., vol. 30, no. 3, pp. 326-338, 1986.

28. C.H. Stapper, F.M. Armstrong and K. Saji, "Integrated circuit yield statistics," Proc. IEEE, vol. 71, pp. 453-470, Apr. 1983.

29. C.H. Stapper, "The effects of wafer to wafer defect density variations on integrated circuit defect and fault distirbutions, " IBM J. Res. Dev., vol. 29, no. 1, pp. 87-97, 1985.

30. C.H. Stapper, "Modeling of integrated circuit sensitivities," IBM J. Res. Dev., vol. 27, no. 6, pp. 549-557, 1983.

31. C.H. Stapper, "Fact and fiction in yield modeling," Microelectronics Journal, vol. 20, nos. 1-2, pp. 129-151, 1989.

32. C.H. Stapper, "LSI yield modeling and process monitoring," IBM J. Res. Develop., vol. 20, pp. 228-234, May 1976.

33. D.M.H. Walker, Yield Simulation for Integrated Circuits, Ph.D. Thesis, Carnegie Mellon University, CMU-CS-86143, pp. 138-139, July 1986.

34. T.J. Wallmark, "Design Considerations for integrated Electron Devices," Proc. IRE, vol. 48, no. 3, pp. 293-300, 1960.

35. T. Yanagawa, "Yield degradation of integrated circuits due to spot defects," IEEE Trans. On Electron. Devices, vol. ED-19, no. 2, pp. 190-197, 1972.

Chapter 8

CONCLUSION

Concluding this book we summarize the accomplishments. We began with emphasizing the relevance of testing in general and structural testing in particular. The main contributions of defect-oriented testing are summarized and at the same time its limitations are also highlighted. Furthermore, future trends and research directions are recommended.

1. TEST AND YIELD ENGINEERING COMPLEXITY IN NANO-METRIC TECHNOLOGIES

Imperfections in the manufacturing process necessitate testing of the manufactured ICs. The fundamental objective of testing is to distinguish between good and faulty ICs. This objective can be achieved in several ways. Earlier, when ICs were relatively less complex, this objective was achieved by functional testing. However, as the complexity of the fabricated ICs increased, it was soon discovered that the application of the functional test is rather expensive in test resources and is inefficient in catching the manufacturing process imperfections. The exponential increase in the cost of functional testing led to tests that are not functional in nature, but are aimed to detect the possible faulty conditions in ICs. The circuit is analyzed for faulty conditions and tests are generated to test for such conditions. Like any other analysis, this fault analysis also requires a model (or abstraction) to represent likely faults in ICs with an acceptable level of accuracy. This type of fault model based testing is known as structural testing. The name structural test comes from two counts. First, the testing is carried out to validate the structural composition of the processed design rather than its function and, second, the test methodology has a structural basis, i.e., the fault model for test generation.

Structural testing gained popularity in the 70s and the 80s when LSI complexity forced researchers to pay attention to test cost reduction. Structural DfT methodologies like scan path and level sensitive scan design (LSSD) emerged for digital circuits. These DfT methods became popular because their application could change distributed sequential logic into a big unified shift-register for testing purposes. As a result, the overall test complexity is reduced. Owing to these techniques, the test generation and the fault grading for complex digital circuits became possible.

At the same time significant research effort was directed on fault models. It was discovered that classical, logic level SA fault model in particular, and voltage-based test methods in general are inadequate to meet rising quality and reliability expectations on integrated circuits. Many commonly occurring defects like gate oxide defects often are not detected by logic tests. Therefore, such escaped defects are quality and reliability hazards. This increased quality awareness brought in new test techniques like quiescent current measurements or I_{DDQ} test as it is popularly known, in the test flow for digital CMOS ICs. Arguably I_{DDQ} testing was an effective test method for catching manufacturing process defects in 90s. However, in the last several years, increased integration, higher transistor leakage current owing to transistor scaling has eroded much of the defect detection capabilities of I_{DDQ} testing. Recent work on ΔI_{DDQ}, and transient current testing hold some promise.

Test complexity in nanometer technologies can broadly be segregated into (i) test complexity associated with large number ($>10^8$) of transistors, and (ii) test complexity associated with parametric issues of large number of transistors. The former is associated with sheer number of transistors that must be controlled and their behavior must be observed. The latter is related to ensure specifications. In this broad context and in our humble opinion, the center of test gravity is shifting towards the latter. There are several reasons for this assertion. Realization of many manufacturing steps is much more difficult in nano-metric technologies compared to previous generations. Therefore, subtle variations in process may cause unacceptable defect levels or parametric variations which must be tested. Secondly, specifications have become aggressive with ever-shrinking margins. Therefore, role of the test has become increasing crucial in ensuring quality expectations.

A System on a Chip, as advanced as it could be, but with low yield is simply a showstopper! The advent of Systems on a Chip (SoC) with larger die sizes and shorter life cycles questions yield and design engineering strategies. The SoC's commercial success depends on a timely and accurate design for manufacturability (DfM) strategy given the increase in fabrication costs and manufacturability risks of nanometer technologies. Yield prediction is essential for the development of new SoCs as it effectively shows if a design is likely to meet its cost and engineering objectives. Appropriate system-yield forecasts provide significant benefits for both time-to-market and system-level engineering decisions. The latter addresses primarily the SoC's requirements in terms of redundancy and fault tolerance. Nowadays accurate design-yield forecasts are only possible when the layout is available. However, given the complexity of modern SoC's, yield forecasts should be made available at an early phase of the design and not when the design is ready. Early-design-yield forecast enables the evaluation of tradeoffs between manufacturability and the choices of interconnect routing, power and speed. Thus, the formulation of appropriate early-yield models that blend the statistics of the fabrication line with design parameters is an open challenge.

Although functional yield remains the main focus of attention, nanometer technologies exhibit a higher dependence on parametric and systematic yield losses. Transistor orientation, stepper field dependence, and intra-die electrical parameter differences such as random threshold voltage and via resistance variations pose serious problems for designs based on low-voltage low-power premises, e.g. clock skews, excessive leakage current, out of spec critical-path delays, stability of flip flops and memory cells, etc. Likewise, the impact of spot defects is not longer constrained to hard faults. Spot defects can affect the wire impedance and intra wire capacitance giving origin to signal integrity problems such as cross-talk and timing closure of high-speed designs. Rigorous investigations involving "new-defect" to fault mappings and efficient DfM strategies are required for future sub-nanometer designs/technologies. An open DfM challenge is the implementation of built-in *"adaptive process regulators"* that help the SoC deal with the impact of process variability during its normal operation. Examples of process regulators are closed-loop control systems for minimization of threshold voltage mismatch, leakage current, clock skew, etc. The failure of nanometer technologies to continue with constant process tolerances gives origin to significant challenges for design and test technologies that can deal with process and litho variability. As the variation of fundamental parameters such as channel length, threshold voltage, thin oxide thickness and interconnect dimensions goes well beyond acceptable limits, new circuit topologies, logic and layout optimizations and test methodologies are needed. In addition to process variability, future interconnect optimizations must account for signal and supply noise, thermal gradients, EMI and substrate coupling. Research efforts in this focus area delve into the incorporation of real-time adaptive schemes for the minimization of process variability effects and improvement of timing and signal integrity closure.

2. ROLE OF DEFECT-ORIENTED TESTING

Defect-oriented testing can play a role in reducing both types of test complexity alluded in previous section. The test complexity associated with the larger number of transistors can be rationalized. Faults, and hence tests, may evolve considering the layout and likely defects. It is pertinent to mention again the work of Zachariah and Chakravarty [10] and Krishnaswamy et al [3] at Intel Corporation (see Chapter 4, Section 7). Zachariah and Chakravarty developed a methodology of extracting bridging faults for million transistor circuits. Krishnaswamy et al used Zachariah and Chakravarty's tool to analyze a complete Pentium 4 layout containing more than 40 million transistors.

In principle, the defect-oriented test method is independent of technology and design style. The method may be used for digital, analog or mixed signal circuits in purely CMOS, BiCMOS or bipolar technologies. In spite of the method's wide applicability, we restricted ourselves to the CMOS environment because of the overwhelming benefits and popularity of the CMOS technology. Furthermore, the level of information about defects that is available for CMOS technology is unmatched in other technologies. Considering these practical factors, researchers have analyzed a wide variety of CMOS circuits through the defect-oriented test methods. These circuits include purely digital circuits, quasi-digital/analog circuits such as DRAM, SRAM; and analog circuits (Class AB amplifier, A/D converter).

2.1 Strengths of Defect-oriented Testing

Defect information can be exploited in many ways during the development of an IC. Defect-oriented testing is receiving substantial attention in the industry. It has been applied to a complex, mixed-signal and large volume single chip TV IC at Philips Semiconductors to improve the yield and quality [5]. Similarly, other companies and institutions are paying attention to defect-oriented testing. The salient advantages of this method are as follows:

- **Shorter and efficient production tests**

The application of defect-oriented method often results in shorter and effective production tests. Tests are directed towards a particular class of defects likely to occur in the production environment and cause yield loss in a particular circuit type. For example, the work of Dekker et al. [2] on testing of SRAMs and that of Wright et al., [9] on CAMs illustrates the potential of the method.

- **Improved and robust design**

The information on what can go wrong in a basic cell can be exploited to improve the design so that the detection of difficult to test defects is simplified. The work of Sachdev on memory address decoder (see Chapter 5, Section 4) is an example where layout of the basic cells was modified to improve the detection of stuck-open faults.

- **Defect based DfT techniques**

The defect information can also be exploited to devise innovative test modes such that defects are quickly and easily tested. For example, scan chain transparency results in testing of defects in scan chains quickly and efficiently. Similarly, an I_{DDQ} based parallel RAM test methodology results in efficient detection of defects. In analog circuits also it is possible to devise DfT techniques to test hard-to-detect defects. Furthermore, one can fault grade analog tests for realistic defects.

2.2 Limitations of Defect-oriented Testing

Every test method has its constraints and limitations. Defect-oriented test method is no exception. Owing to CAD tool limitations, the application of this method is limited to relatively small macros and building blocks. At present, research effort is directed towards relaxing such constraints. Relatively large number of defects should be sprinkled onto the analyzed layout to generate a realistic fault list. As a result a large number of simulations must be performed to ascertain the effectiveness of the given set of tests. For accurate results, these simulations need to be performed at the Spice level requiring large amount of CPU resources. Alternatively, a high level model requires substantially lower effort.

3. FUTURE DIRECTIONS

It is always difficult to predict the future. It is best to look backwards to predict the future trends. Testing has come a long way. In the early days of the semiconductor industry, testing was merely a verification of functionality. Quality and economics issues were not aggressively pursued. In late 70s and 80s, growing semiconductor industry understood the futility of functional testing and started to pay attention to fault modeling, structural testing, DfT techniques, etc. Alternative test techniques, like I_{DDQ}, emerged as "quality" supplements. At the same time, quality and economics became core business issues. With the increasing competition, it was no longer sufficient to be able to produce a product. The focus shifted on how efficiently and economically one can make the product. Testing is recognized as the bottleneck and the last checkpoint for ensuring product quality and reliability. As a result, in the first half of 90s, we witnessed a number of studies reporting benefits of incorporating non-Boolean (I_{DDQ}) tests in the test suite.

What next? ICs are going to be more complex and faster. For example, Intel has reported a microprocessor with one billion transistors [6]. Clock frequencies of microprocessors will exceed 5 GHz in future. Production is being ramped up in 65 nm technology. It is clear that the integration capability has kept up with the Moore's law. In the future this trend is likely to slow down. Despite these advances testing is expected to stay as the bottleneck. At the same time, the nanometer test issues (reduced power supply voltage, reduced noise margin and increased transistor sub-threshold current) will force researchers to look for innovative test solutions.

The application of adaptive techniques to control either or both power supply (V_{DD}) and threshold voltage (V_t) has gained increased attention. This stems from the fact that modern electronics are hampered by the variation of fundamental process and performance parameters such as threshold voltage and power consumption. Design technologies such as AMD's PowerNow!, Transmeta's LongRun, Intel's Enhanced SpeedStep, to mention some instances, are vivid examples of commercial ICs that use power management based on power supply scaling. In addition to these commercial accomplishments, chip demonstrators with V_{DD} and V_t scaling capabilities have also been reported in the literature archival. Other reported uses of V_{DD} and V_t scaling, besides power management in processors, are in testing [4], product binning [8], and yield tuning [1]. Furthermore, a very effective way to minimize leakage power consumption is to turn-off the parts of a chip, which are not active by means of power switches. Power switches bring another dimension to testing, e.g. how should the switch and the core wrapped with switches be tested?

The implementation of circuits and systems in new nanometer technologies requires new ideas to make the system performance predictable. It is clear that the performance of an SoC implemented in a, say, 90nm technology or beyond, may severely be hampered by excessive transistor leakage, by the impact of local and global process variability, and by reduced noise margins. Future chips will have means to adapt online and in real-time design parameters such as power supply and frequency of operation under constrained performance conditions.

Finally, in the analog arena we have that test cost per unit and test equipment capital cost dominate manufacturing test methodology decisions. Typical test costs for analog circuits as a percentage of the total manufacturing cost in are approximately 5% for most products rising to 10% for mixed-signal types, to 15% for devices such as tuners, to projected values of 50% for future RF devices. This rising cost trend is quite alarming. In addition to the cost issue, a paradigm shift into silicon or MCM high-integration is constraining the access to RF ports and thus the applicability of traditional RF test methods is more and more problematic. Moreover, present testing approaches focus on parametric aspects while actual rejects are functional. Robust methods are missing, and as such the need for structural testing, or defect-oriented test strategies are needed [7]. Time to market of future RF-IPs will resemble the digital's. Consequently, it is also expected that under the vision of reusability, future RF-IPs will include DfT and possibly BIST, and that the currently comfortable upper limit on test development time cycles will have to decrease.

References

1. T. Chen and S. Naffziger, "Comparison of Adaptive Body Bias (ABB) and Adaptive Supply Voltage (ASV) for Improving Delay and Leakage Under the Presence of Process Variation", IEEE Transactions on VLSI Systems, October 2003, Vol. 11, No. 5, pp. 888-899.
2. R. Dekker, F. Beenker and L. Thijssen, "Fault Modeling and Test Algorithm Development for Static Random Access Memories," Proceedings of International Test Conference, 1988, pp. 343-352.
3. Krishnaswamy V., Ma A.B., Vishakantaiah P., "A study of bridging defect probabilities on a Pentium (TM) 4 CPU," IEEE Int. Test Conf., 2001, pp. 688-695.
4. T. Miyake, et al. , "Design Methodology of High Performance Microprocessor using Ultra-Low Threshold Voltage CMOS," Proc. of IEEE Custom Integrated Circuits Conference, 2001, pp. 275-278.
5. L. Nederlof, "One Chip TV," Proceedings of International Solid State Circuits Conference, 1996, pp. 26-29.
6. E. Rusu, S. Tam, H. Muljono, D. Ayers, J. Chang, "A Dual-Core Multi-Threaded Xeon Processor with 16MB L3 Cache," Proceedings of IEEE International Solid State Circuits Conference, pp. 102-103, 2006.

7. E. Silva, J. Pineda de Gyvez and G. Gronthoud, "Functional vs. Multi-VDD Testing of RF Circuits," Int. Test Conference, Nov. 2005.

8. J. Tschanz, J. Kao, S. Narendra, R. Nair, D. Antoniadis, A. Chandrakasan, and Vivek De, "Adaptive body bias for reducing impacts of die-to-die and within-die parameter variations on microprocessor frequency and leakage", IEEE Solid-State Circuits Conference, February 2002, Vol.1, pp. 422-478.

9. D. Wright, and M. Sachdev, "Transistor-Level Fault Analysis and Test Algorithm Development for Ternary Dynamic Content Addressable Memories," IEEE International Test Conference, September 2003.

10. S. T. Zacariah and S. Chakarvarty, "A Scalable and Efficient Methodology to Extract Two Node Bridges from Large Industrial Circuits," Proceedings of IEEE International Test Conference, pp. 750-759, November 2000.

Index

M

N

O

P

Q

R

S

T

U

V

W

Y